AF581793

Freshwater Algal Toxins

Freshwater Algal Toxins: Monitoring and Toxicity Profile

Editors

Angeles Jos
Ana M. Cameán

MDPI • Basel • Beijing • Wuhan • Barcelona • Belgrade • Manchester • Tokyo • Cluj • Tianjin

Editors

Angeles Jos
University of Sevilla
Spain

Ana M. Cameán
University of Sevilla
Spain

Editorial Office
MDPI
St. Alban-Anlage 66
4052 Basel, Switzerland

This is a reprint of articles from the Special Issue published online in the open access journal *Toxins* (ISSN 2072-6651) (available at: https://www.mdpi.com/journal/toxins/special_issues/Freshwater_Toxicity).

For citation purposes, cite each article independently as indicated on the article page online and as indicated below:

LastName, A.A.; LastName, B.B.; LastName, C.C. Article Title. *Journal Name* **Year**, *Volume Number*, Page Range.

ISBN 978-3-03943-679-8 (Hbk)
ISBN 978-3-03943-680-4 (PDF)

Contents

About the Editors

Angeles Jos is a Full Professor of Toxicology in the Faculty of Pharmacy at the University of Seville. Born and raised in Seville, she has developed her professional career mainly at the University of Seville with postdoctoral work at the University of Bern (Switzerland). She is a senior scientist in the Toxicology (CTS-358) research group at the Department of Food Science, Toxicology and Legal Medicine. Her research interests are in the field of food safety, particularly in the hazard characterization of different toxicants present in food. Among them, cyanotoxins (mainly microcystins and cylindrospermopsin) have a pivotal role in her line of research, studying their toxic effects using both in vitro and in vivo methods, investigating their toxic mechanisms (genotoxicity, oxidative stress, etc.), and developing analytical methods for their determination in water and food samples.

Ana M. Cameán obtained a Ph.D. in Pharmacy from the University of Seville (US) (1985) and has been a Professor of Toxicology at US since 2005. She has developed her teaching and research career in the Toxicology Section of the Department of Food Science, Toxicology and Legal Medicine at US. She has been the leader of the Toxicology (CTS-358) research group since its creation. Her research interests are in the area of food safety, focusing mainly on the study of cyanotoxins (microcystins, cylindrospermopsin) present in water and food, evaluating their transference through the development and validation of methods for their determination, and studying their toxic effects with in vitro and in vivo models as well as the effects of cooking, histopathological alterations, bioaccessibility, etc. In parallel, evaluating the safety of various natural products in food packaging or as antioxidants in feed is also of interest.

Editorial

Freshwater Algal Toxins: Monitoring and Toxicity Profile

Angeles Jos * and Ana M. Cameán

Area of Toxicology, Faculty of Pharmacy, University of Sevilla, C/Profesor García González 2, 41012 Sevilla, Spain; camean@us.es

* Correspondence: angelesjos@us.es

Received: 18 September 2020; Accepted: 23 September 2020; Published: 13 October 2020

Climate change and human activities are more and more affecting the dynamics of phytoplankton communities. Among them, cyanobacterial abundance has increased disproportionately relative to other phytoplankton, and this trend is likely to continue in the coming decades. This fact has deleterious effects on ecosystem biodiversity but also adversely affects drinking water supplies, livestock watering, crop yields, aquaculture, etc. Thus, the proliferation of cyanobacterial blooms represents an economic issue and, more importantly, human and animal health risks due to the common production of potent toxins, cyanotoxins. Moreover, these risks are increased due to their accumulation potential and their transference to the food chain.

In spite of the worldwide increasing occurrence of cyanotoxins, they are still underestimated in regulations, with Microcystin-LR (MC-LR) as the main one (and in many cases the only one) considered. However, risk management of cyanotoxins is only possible after a thorough risk evaluation, and for that purpose, toxicity and exposure data are required. Thus, occurrence and monitoring information is of key importance, and new data in relation to the conditions that favor cyanobacterial growth and cyanotoxins production are welcome in order to prevent their appearance.

On the other hand, in regard to toxicity, the scientific literature shows already a wide array of adverse effects cyanobacterial toxins can induce. However, there are still health consequences not investigated deeply enough as well as many data gaps in different aspects regarding different targets of cyanobacteria toxicity, from plants to animals and humans. Thus, the aim of this special issue was to gather new studies that could contribute in the risk evaluation process of cyanotoxins. This goal was achieved with a compilation of 11 articles (10 research papers and a review).

Among the articles focused on monitoring issues, Hartnell et al. [1] investigated the cyanobacterial abundance and the MC profiles in two southern British lakes. They could not correlate the elevated MC concentrations found with the number of cyanobacterial cells, but the linear regression analysis performed suggested that temperature and dissolved oxygen could explain the variability of MC across both reservoirs. They concluded that there is a need to develop inclusive, multifactor holistic water management strategies to control cyanobacterial risks in freshwater bodies.

Taranu et al. [2] applied multivariate canonical analyses and regression tree analyses to identify how different congeners (MC-LA, -LR, -RR, and -YR) varied with changes in meteorological and nutrient conditions over time and space. They found that MC-LR was associated with strong winds, warm temperatures, and nutrient-rich conditions, whereas MC-LA, for example, tended to dominate under intermediate winds and nutrient-poor conditions. Thus, the knowledge of the environmental factors leading to the formation of different MC congeners in freshwaters is necessary to assess the duration and the degree of toxin exposure under future global change.

Tamele and Vasconcelos [3] performed a review about the MCs incidence in the drinking water of Mozambique. This report showed that the few studies done in Maputo and Gaza provinces indicated the occurrence of MC-LR, -YR, and -RR at concentrations above the maximum limit recommended by the World Health Organization. Authors evidenced the need to implement an operational monitoring program of MCs in order to reduce or avoid the possible cases of intoxications since the drinking water quality control tests in the country do not include a MCs test.

Other aspects were covered by Perez and Chu [4], LeBlanc et al. [5], and Galetović et al. [6]. Perez and Chu [4] focused their study on zinc metal resistance and stress response in a toxigenic cyanobacterium, *Microcystis aeruginosa* UTEX LB 2385, by monitoring cells with $ZnCl_2$ treatment. Among their results, they found that *M. aeruginosa* UTEX LB 2385 was able to survive $ZnCl_2$ concentrations of up to 0.25 mg/L, with increasing biomass through 15 days. A persistent yield of the cyanotoxin MC-LR (µg/cell) was observed in all $ZnCl_2$ treated cells by 15 days, indicating that this cyanotoxin remains present in the environment even with low cell concentrations.

Leblanc et al. [5] isolated a new MC, [D-Leu^1]MC-LY, and other related ones, from *Microcystis aeruginosa* strain CPCC-464. The compound was characterized by ^{1}H and ^{13}C NMR spectroscopy, liquid chromatography–high resolution tandem mass spectrometry (LC–HRMS/MS), and UV spectroscopy. Moreover, [D-Leu^1]MC-LY showed a potency similar to MC-LR in the protein phosphatase 2A inhibition assay. The authors concluded that [D-Leu^1]-containing MCs may be more common in cyanobacterial blooms than is generally appreciated but are easily overlooked with standard targeted LC–MS/MS screening methods.

Finally, regarding monitoring-related aspects, Galetović et al. [6] reported the absence of cyanotoxins in Llayta, edible Nostocaceae colonies from the Andes Highlands, by using molecular and chemical methods. Thus, they concluded that Llayta could be considered a safe ingredient for human consumption.

Cyanotoxins toxicity has been an important topic in this special issue. The study of Llana-Ruiz-Cabello et al. [7] explored the susceptibility of two green vegetables, spinach and lettuce, to the cyanotoxins MC and cylindrospermopsin (CYN), individually and in mixture. The study revealed growth inhibition of the aerial part in both species when treated with 50 µg/L of MC, CYN, and CYN/MC mixture. MC showed to be more harmful to plant growth than CYN. Additionally, CYN, but not MC, was translocated from the roots to the leaves. CYN and MC affected the levels of minerals, particularly in plant roots.

Foss et al. [8] described a case report in which MC intoxications of canines were diagnosed through interpretation of clinicopathological abnormalities, pathological examination of tissues, microscopy, and analytical MC testing of antemortem/postmortem samples. The described cases represent the first use of urine as an antemortem, non-invasive specimen to diagnose MC toxicosis. Authors concluded that antemortem diagnostic testing to confirm MC intoxication cases is crucial for providing optimal supportive care and mitigating MC exposure.

Leticia Díez-Quijada et al. [9] explored the genotoxic effects of MC-LR and CYN combinations in rats. Their results revealed an increase in micronucleous formation in bone marrow. However, no DNA strand breaks nor oxidative DNA damage were induced, as shown in the comet assays. The histopathological study indicated alterations only in the highest dose group. Therefore, the combined exposure to cyanotoxins may induce genotoxic and histopathological damage in vivo.

Finally, two articles pointed out that MC-LR exacerbated the severity of illnesses such as non-alcoholic fatty liver disease [10] and dextran sulfate sodium (DSS)-induced colitis [11] in animal models. This is an important aspect, as the exposure to MC-LR, even at levels that are below the no observed adverse effect level (NOAEL) established in healthy animals, can worsen pre-existing pathologies.

All these studies have contributed to extend the knowledge on cyanotoxins and complete those published in our previous special issue, "Cyanobacteria and Cyanotoxins: New Advances and Future Challenges" [12]. Moreover, the 20 special issues dealing with this topic published thus far in the journal *Toxins* demonstrate the interest cyanobacteria and cyanotoxins have in the scientific community.

Funding: This research received no external funding.

Acknowledgments: The co-editors are very grateful to all the authors who contributed to this Special Issue Freshwater Algal Toxins: Monitoring and Toxicity Profile. We greatly appreciate the efforts carried out by external expert peer reviewers, whose rigorous evaluations of all the submitted manuscripts contributed to improve the quality of this Special Issue. The co-editors also wish to thank the Ministerio de Ciencia e Innovación of Spain (PID2019-104890RB-I00, MICINN, UE) and the Faculty of Pharmacy of the Universidad de Sevilla. Finally, the co-editors very much appreciate the good organization and the constant support of the MDPI editorial team and staff.

Conflicts of Interest: The authors declare no conflict of interest.

References

1. Hartnell, D.M.; Chapman, I.J.; Taylor, N.G.H.; Esteban, G.F.; Turner, A.D.; Franklin, D.J. Cyanobacterial Abundance and Microcystin Profiles in Two Southern British Lakes: The Importance of Abiotic and Biotic Interactions. *Toxins* **2020**, *12*, 503. [CrossRef] [PubMed]
2. Taranu, Z.E.; Pick, F.R.; Creed, I.F.; Zastepa, A.; Watson, S.B. Meteorological and Nutrient Conditions Influence Microcystin Congeners in Freshwaters. *Toxins* **2019**, *11*, 620. [CrossRef]
3. Tamele, I.J.; Vasconcelos, V. Microcystin Incidence in the Drinking Water of Mozambique: Challenges for Public Health Protection. *Toxins* **2020**, *12*, 368. [CrossRef] [PubMed]
4. Perez, J.L.; Chu, T.-C. Effect of Zinc on Microcystis aeruginosa UTEX LB 2385 and Its Toxin Production. *Toxins* **2020**, *12*, 92. [CrossRef]
5. Leblanc, P.; Merkley, N.; Thomas, K.; Lewis, N.I.; Békri, K.; Renaud, S.L.; Pick, F.R.; McCarron, P.; Miles, C.O.; Quilliam, M. Isolation and Characterization of [D-Leu1]microcystin-LY from Microcystis aeruginosa CPCC-464. *Toxins* **2020**, *12*, 77. [CrossRef]
6. Galetović, A.; Azevedo, J.; Castelo-Branco, R.; Oliveira, F.; Gómez-Silva, B.; Vasconcelos, V. Absence of Cyanotoxins in Llayta, Edible Nostocaceae Colonies from the Andes Highlands. *Toxins* **2020**, *12*, 382. [CrossRef]
7. Llana-Ruiz-Cabello, M.; Jos, A.; Cameán, A.; Oliveira, F.; Felpeto, A.B.; Machado, J.; Azevedo, J.; Pinto, E.; Almeida, A.; Campos, A.; et al. Analysis of the Use of Cylindrospermopsin and/or Microcystin-Contaminated Water in the Growth, Mineral Content, and Contamination of Spinacia oleracea and Lactuca sativa. *Toxins* **2019**, *11*, 624. [CrossRef] [PubMed]
8. Foss, A.J.; Aubel, M.T.; Gallagher, B.; Mettee, N.; Miller, A.; Fogelson, S.B. Diagnosing Microcystin Intoxication of Canines: Clinicopathological Indications, Pathological Characteristics, and Analytical Detection in Postmortem and Antemortem Samples. *Toxins* **2019**, *11*, 456. [CrossRef] [PubMed]
9. Díez-Quijada, L.; Medrano-Padial, C.; Llana-Ruiz-Cabello, M.; Cătunescu, G.M.; Moyano, R.; Risalde, M.A.; Cameán, A.; Jos, A. Cylindrospermopsin-Microcystin-LR Combinations May Induce Genotoxic and Histopathological Damage in Rats. *Toxins* **2020**, *12*, 348. [CrossRef]
10. Lad, A.; Su, R.C.; Breidenbach, J.D.; Stemmer, P.; Carruthers, N.J.; Sanchez, N.K.; Khalaf, F.K.; Zhang, S.; Kleinhenz, A.L.; Dube, P.; et al. Chronic Low Dose Oral Exposure to Microcystin-LR Exacerbates Hepatic Injury in a Murine Model of Non-Alcoholic Fatty Liver Disease. *Toxins* **2019**, *11*, 486. [CrossRef] [PubMed]
11. Su, R.C.; Blomquist, T.M.; Kleinhenz, A.L.; Khalaf, F.K.; Dube, P.; Lad, A.; Breidenbach, J.D.; Mohammed, C.J.; Zhang, S.; Baum, C.E.; et al. Exposure to the Harmful Algal Bloom (HAB) Toxin Microcystin-LR (MC-LR) Prolongs and Increases Severity of Dextran Sulfate Sodium (DSS)-Induced Colitis. *Toxins* **2019**, *11*, 371. [CrossRef]
12. Cameán, A.M.; Jos, A. *Cyanobacteria and Cyanotoxins: New Advances and Future Challenges*, 1st ed.; MDPI: Basel, Switzerland, 2020; pp. 1–248.

Article

Cyanobacterial Abundance and Microcystin Profiles in Two Southern British Lakes: The Importance of Abiotic and Biotic Interactions

David M. Hartnell [1,2,*], Ian J. Chapman [2,3], Nick G. H. Taylor [1], Genoveva F. Esteban [2], Andrew D. Turner [1] and Daniel J. Franklin [2]

1 The Centre for Environment, Fisheries and Aquaculture Science (Cefas), The Nothe, Barrack Road, Weymouth, Dorset DT4 8UB, UK; nick.taylor@cefas.co.uk (N.G.H.T.); andrew.turner@cefas.co.uk (A.D.T.)

2 Centre for Ecology, Environment and Sustainability, Faculty of Science & Technology, Bournemouth University, Fern Barrow, Poole, Dorset BH12 5BB, UK; ijchapman@outlook.com (I.J.C.); gesteban@bournemouth.ac.uk (G.F.E.); dfranklin@bournemouth.ac.uk (D.J.F.)

3 New South Wales Shellfish Program, NSW Food Authority, Taree 2430, Australia

* Correspondence: david.hartnell@cefas.co.uk; Tel.: +44-1305-206600

Received: 5 June 2020; Accepted: 30 July 2020; Published: 5 August 2020

Abstract: Freshwater cyanobacteria blooms represent a risk to ecological and human health through induction of anoxia and release of potent toxins; both conditions require water management to mitigate risks. Many cyanobacteria taxa may produce microcystins, a group of toxic cyclic heptapeptides. Understanding the relationships between the abiotic drivers of microcystins and their occurrence would assist in the implementation of targeted, cost-effective solutions to maintain safe drinking and recreational waters. Cyanobacteria and microcystins were measured by flow cytometry and liquid chromatography coupled to tandem mass spectrometry in two interconnected reservoirs varying in age and management regimes, in southern Britain over a 12-month period. Microcystins were detected in both reservoirs, with significantly higher concentrations in the southern lake (maximum concentration >7 μg L^{-1}). Elevated microcystin concentrations were not positively correlated with numbers of cyanobacterial cells, but multiple linear regression analysis suggested temperature and dissolved oxygen explained a significant amount of the variability in microcystin across both reservoirs. The presence of a managed fishery in one lake was associated with decreased microcystin levels, suggestive of top down control on cyanobacterial populations. This study supports the need to develop inclusive, multifactor holistic water management strategies to control cyanobacterial risks in freshwater bodies.

Keywords: flow cytometry; liquid chromatography coupled to tandem mass spectrometry (LC-MS/MS); cyanotoxins; risk assessment; management strategies; modelling

Key Contribution: In two similar lakes, microcystin levels and cyanobacterial communities were significantly different; one interesting observation was the introduced omnivorous fish, which appeared to reduce toxin levels in one lake.

1. Introduction

Cyanobacteria blooms are a global problem in freshwater ecosystems [1–3]. A range of factors have been reported to influence the abundance and likelihood of bloom formation in freshwater systems, notably increased temperature, and nutrient enrichment [4–7]. A proportion estimated as between 40–70%, of cyanobacteria blooms are reported to occur concomitantly with elevated levels of cyanobacterial toxins (microcystins) [8–10]. Microcystins are known to be responsible for toxic events globally, most frequently reported are wild animal, livestock, and pet deaths with numerous accounts in the literature, from both more and less economically developed nations [8,11–14].

In most countries, control plans for public health risks associated with exposure to cyanobacterial toxins are based on assessments of cyanobacterial cell presence and density in the event of bloom formation. In the United Kingdom, assessments and management recommendations are made by national agencies (the Environment Agency (EA) in England and the Scottish Environmental Protection Agency (SEPA) in Scotland). In both administrations, samples are collected reactively from the water column in response to visual bloom occurrence and cyanobacterial are identified to genus level and cells are counted microscopically to determine the cell density in terms of number of cells per millilitre of water. Samples containing >20,000 cells mL^{-1} trigger actions such as preventative closures or restrictions on usage. The presence of cyanobacterial scums on the water surface automatically indicates the need for responsive action, as scum formation is known to increase the likelihood of adverse health effects by factors of up to 1000 [15] and, in the UK, would typically result in measures to prevent exposure of humans and animals [16]. Systematic, or risk-based routine monitoring of water bodies for cyanobacteria is not undertaken in the UK; consequently the incidence, intensity and seasonality of cyanobacterial blooms is not well known [10,17]. Furthermore, whilst the presence of elevated cyanobacterial cells enables identification of potential risks, toxin production during blooms formation is not certain [8,15]. Turner et al. [10] found that only 18% of samples containing cyanobacterial cells exceeding action state thresholds contained microcystins above the WHO medium health criterion of 20 μg L^{-1} in freshwater bodies in England. Therefore, management actions driven by elevated cyanobacterial cell counts may be unnecessary when blooms are formed from non-toxic species and may have unnecessary detrimental economic impacts. Figure 1 shows the occurrence and magnitude data combined with that of Turner et al. [10]; all data were collected in 2016.

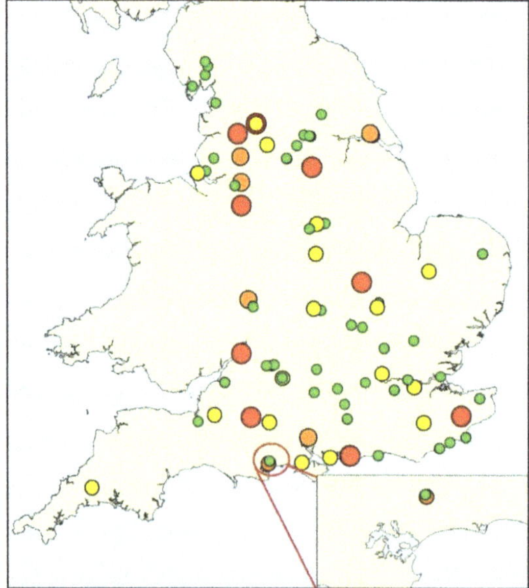

Figure 1. Occurrence and magnitude of total microcystins recorded from England and Wales in 2016. (red: >100 μg/L; orange: 20–100 μg/L; yellow: 2–20 μg/L; green: <2 μg/L). Insert, microcystin data and location of the study site, Longham Lakes, Bournemouth, Dorset, UK. (adapted from Turner et al. [10]).

Few studies have examined the prevalence and levels of microcystin toxins and variants globally. In the European Multi-Lake Survey, toxin profile data from 26 European countries from lakes with a history of eutrophication were analysed, together with environmental parameters. The authors reported direct and indirect effect of temperature on toxin concentrations and profiles, concluding that whilst few geographical patterns could be discerned, increasing lake temperatures could drive changes in the distribution of cyanobacterial toxins, possibly selecting for a few toxic species [18]. In a study on the array of microcystins during cyanobacterial blooms in Lake Victoria, Tanzania, East Africa, Miles et al. [19] reported a distinctive, complex toxin profile signature during bloom events which has also been confirmed in Ugandan and Kenyan regions of Lake Victoria [19,20]. In a systematic study to assess microcystins in freshwater lakes in England, Turner et al. [10] revealed complex toxin profiles with occurrence of toxin clusters unrelated to cyanobacterial species and no correlation with environmental parameters. These data are suggestive of complex ecosystems, with levels and signatures of microcystin and variants potentially influenced by geographical range but with the impact of environmental factors unclear.

It has been reported that light intensity, temperature, nutrients, and hydrodynamics influence the occurrence and density of cyanobacterial blooms [8,21]. Several studies have attempted to model cyanobacterial concentrations using meteorological, hydrological, and environmental parameters [21–25]. In most studies, the predictive ability of models with respect to risk management has been limited not least because the relationship between the presence and increase in cyanobacterial cells is not always correlated with an increase in the occurrence of toxins [10,26,27]. Notwithstanding this, Carvalho et al. [24] demonstrated that statistical models applied to phytoplankton data from 134 lakes in the UK could be used to describe lakes that may be susceptible to cyanobacterial blooms events. It is evident that understanding the key environmental drivers that favour cyanobacterial abundance and potentially toxic events would facilitate proactive rather than reactive monitoring and management strategies to reduce the public and animal health risks.

In this study, two freshwater reservoirs were routinely monitored by light microscopy, flow cytometry, and liquid chromatography coupled to tandem mass spectrometry over a 12-month period. Measurements of cyanobacterial cells and a range of biological and chemical factors were examined to explore the potential of providing a predictive tool for water management.

2. Results

Water measurements and samples were collected and analysed from the May 16, 2016 until May 31, 2017. As stratification was not observed, the data from each depth were combined to produce an average for each measurement at the time of sampling, except for turbidity (NTU), where the lake bottom measurement was disregarded due to sediment disturbance from the horizontal sampler.

2.1. Study Site

Longham Lakes consists of two freshwater reservoirs, used as a nature reserve and recreational fishery within the borough boundaries of Bournemouth (Figures 1 and 2). The two lakes located at national grid reference SZ 06237 98079 are man-made, fed by the River Stour and provide an auxiliary water supply to the Bournemouth-Poole conurbation. The northern lake was completed in 2003, has a perimeter of 1400 m, and an area of 97,000 m^2. The southern lake is connected to the northern lake; it was completed in 2010, has a perimeter of 2050 m, and an area of 250,000 m^2. The maximum depth for both lakes is approximately 14 m and they both have an average depth of 2.9 m. Longham Lakes is managed by Bournemouth Water, which is part of South West Water. Lake water chemistry and phytoplankton are constantly monitored, and weekly water samples are taken.

2.2. Chemical and Biological Parameters

Table 1 shows data collected over the 12-month study period, mean, medium, maxima and minima for total microcystins, *Microcystis* cells, phycocyanin fluorescence, temperature, turbidity, dissolved oxygen, pH, chlorophyll *a*, *b* and total carotenoids are given for the two lakes. A null hypothesis that no differences between biological and chemical measurements were observable between the two lakes was tested at the $p = 0.05$ significance level using a series of Student's t-tests. No significant differences between temperature, pH, or turbidity were observed between the two lakes over the study period ($p > 0.05$); however, significant and highly significant differences between the two lakes across the sampling period were observed for dissolved oxygen ($p < 0.001$), chlorophyll *a* and *b* levels ($p < 0.01$, $p < 0.001$), and carotenoids ($p < 0.001$) with dissolved oxygen demonstrating the most difference between lakes.

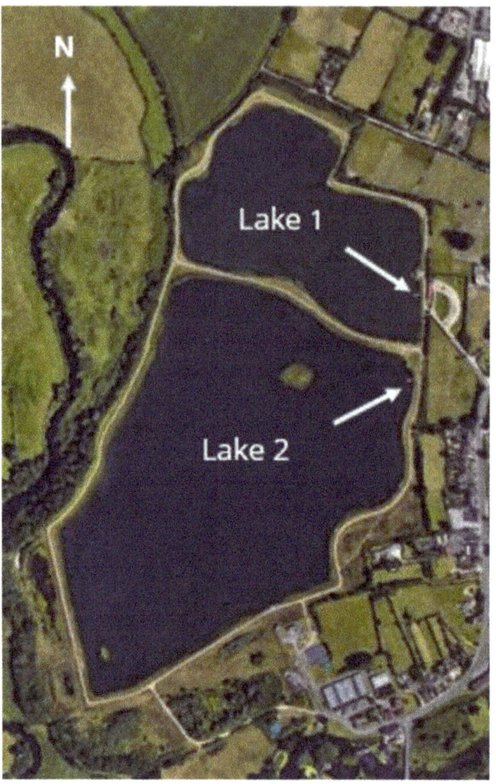

Figure 2. Aerial view of Longham Lakes with sampling point marked by arrows in Lake 1 (northern) and Lake 2 (southern).

Table 1. Biological and chemical measurements from Longham Lakes 1 and 2, between 16 May 2016 and 31 May 2017.

Parameter	Lake 1 (Northern)				Lake 2 (Southern)				Student
	Low	Mean	Median	High	Low	Mean	Median	High	*t*-Test
Total microcystins (μg L^{-1})	nd [1]	0.497	nd [1]	1.922	nd [1]	1.524	nd [1]	7.089	$p < 0.01$
Microcystis cells (cells mL^{-1})	251	6874	2826	51,384	258	1403	1012	12,204	$p < 0.001$
Phycocyanin (Cells mL^{-1})	109	1425	836	7649	20	1924	705	10,290	$p > 0.05$
Temperature (°C)	5.57	14.96	16.51	21.64	5.81	15.13	15.88	21.51	$p > 0.05$
Turbidity (NTU)	−0.40	2.25	1.50	8.90	−1.60	1.51	0.70	8.20	$p > 0.05$
Dissolved Oxygen (mg L^{-1})	6.14	12.19	12.07	24.16	9.44	12.74	12.99	18.98	$p < 0.001$
pH	7.52	8.44	8.47	9.29	8.06	8.52	8.54	8.97	$p > 0.05$
Chlorophyll *a* (mg mL^{-1})	0.44	3.821	2.398	15.373	0.042	1.315	0.969	4.056	$p < 0.01$
Chlorophyll *b* (mg mL^{-1})	0.41	2.296	2.206	6.752	nd	1.294	1.143	4.367	$p < 0.001$
Total Carotenoids (mg mL^{-1})	nd	1.200	0.676	6.295	nd	0.260	0.135	1.467	$p < 0.001$

[1] Limit of detection (LOD) for MC-LR = 0.0013 ± 0.0011 ng mL^{-1} [10].

2.3. Identification and Enumeration of Phytoplankton by Light Microscopy

A wide range of phytoplankton genera were identified in both lakes between August 2016 and May 2017, a number of chlorophytes and diatoms were only identified to the class level. *Microcystis* cells were recorded in both lakes, maximum >7000 (lake 1) & >8000 cells mL^{-1} (lake 2); other cyanobacteria included *Anabaena*, maximum >34,000 (lake 1) & >21,000 cells mL^{-1} (lake 2), *Aphanizomenon*, maximum >2500 (lake 1) & >20,000 cells mL^{-1} (lake 2), and *Oscillatoria*, maximum >6000 cells mL^{-1} (lake 1 & 2). The non-cyanobacteria identified were *Asterionella, Euglena, Pediastrum, Scenedesmus, Tabellaria,* and *Volvox* (Figure 3). No correlation was found between cyanobacteria identified and counted by either light microscopy or flow cytometry with microcystins detected (data not shown).

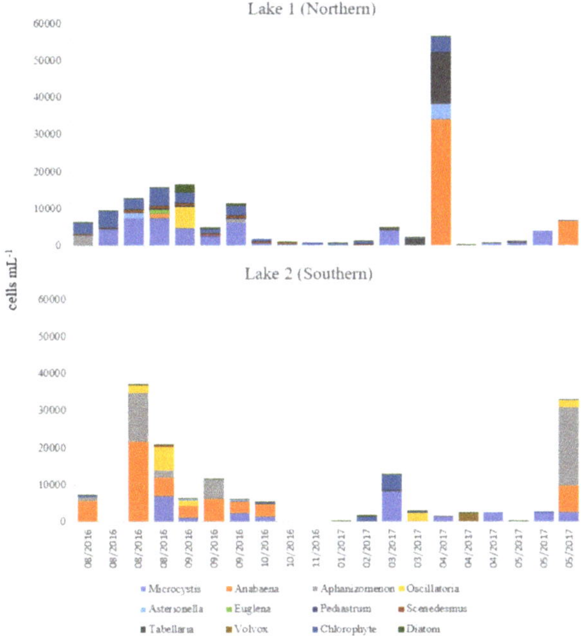

Figure 3. Stacked bar chart showing the date, number, and taxa of phytoplankton identified in Longham Lakes 1 & 2 by light microscope.

2.4. Comparison of Counts of Microcystis Cells by Flow Cytometry and Microscopic Method

In both lakes, counts of *Microcystis* cells by flow cytometry were consistently higher and no zero counts were registered as compared to counts by light microscope (Figure 4). A strong correlation between the two methods was observed in lake 1 when tested with a Pearson product moment correlation ($PC = 0.763$, $p = 0.001$), but correlation was not observed between the two methods in lake 2 ($PC = -0.048$, $p = 0.864$).

2.5. Determination of Microcystis Cells and Microcystin Concentrations

Figure 5 shows the *Microcystis* cells and total microcystins measured over the study period at both lakes. *Microcystis* cells were detected in both lakes by flow cytometry throughout the sampling period, increasing in July/August in lake 1 and in August in lake 2. An order of magnitude more *Microcystis* cells were detected in lake 1 than lake 2. Mean *Microcystis* cells in lake 1 were 6874 mL^{-1} with a median of 2826 and range of 251 (23 May2016) to 51,384 mL^{-1} (14 July2016). In Lake 2, mean *Microcystis* cells were 1403 mL^{-1} with a median of 1012 and range of 258 (19 December 2016) to 12,204 mL^{-1} (07 March 2017) (Table 1).

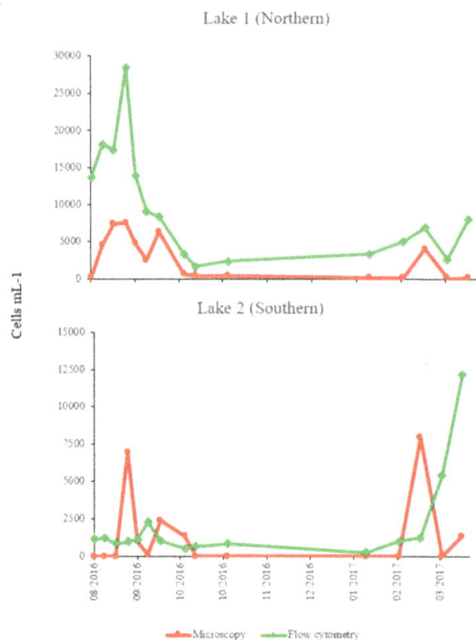

Figure 4. Comparison of counts of *Microcystis* cells in both lakes at Longham, as counted by flow cytometry and microscope methods over the study period.

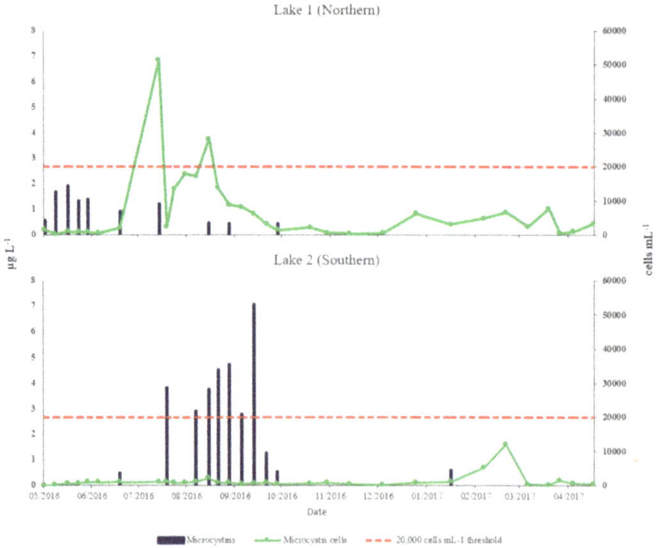

Figure 5. Seasonal variation recorded at Longham Lake (1 & 2) of *Microcystis* cells (cells mL^{-1}) by flow cytometry (right-hand axis) and total microcystins quantified by liquid chromatography coupled to tandem mass spectrometry (µg L^{-1}) (left-hand axis). Red line indicates UK cyanobacterial cell density action threshold [16].

Microcystin variants were detected in both lakes but were consistently lower in lake 1 than lake 2. Total microcystin variants and quantities detected are shown in Figure 5. In total, 7 microcystin variants were detected in Lake 1. These comprise of MC-LR, MC-LA, MC-LY, MC-LF, MC-LW, MC-YR, and Asp^3 MC-LR/[Dha^7] MC-LR. The microcystin variant detected at the highest concentrations at lake 1 was MC-LF in June and July. Six microcystin variants were detected in Lake 2 (MC-LR, MC-RR, MC-LA, MC-LY, MC-YR and Asp^3 MC-LR/[Dha^7] MC-LR) (Figure 6). In Lake 2, MC-YR was detected at highest concentrations during August and September; similar, but slightly lower levels of variant MC-LR was detectable between August and October (Figure 6). Mean total microcystins were 0.5 μg L^{-1} (<LOD to 1.9 μg L^{-1}) in lake 1 and 1.5 μg L^{-1} (<LOD to 7.1 μg L^{-1}) in lake 2. Maximum levels in Lake 1 were detected between 23 May 2016 and 14 July 2016; microcystins were rarely detected between 3 August 2016 and 28 November 2016. Maximum levels in Lake 2 were detected between 3 August 2016 and 28 September 2016; microcystins were rarely detected between 23 May 2016 & 17 July 2016 and between 3 November 2016 and 31 May 2017.

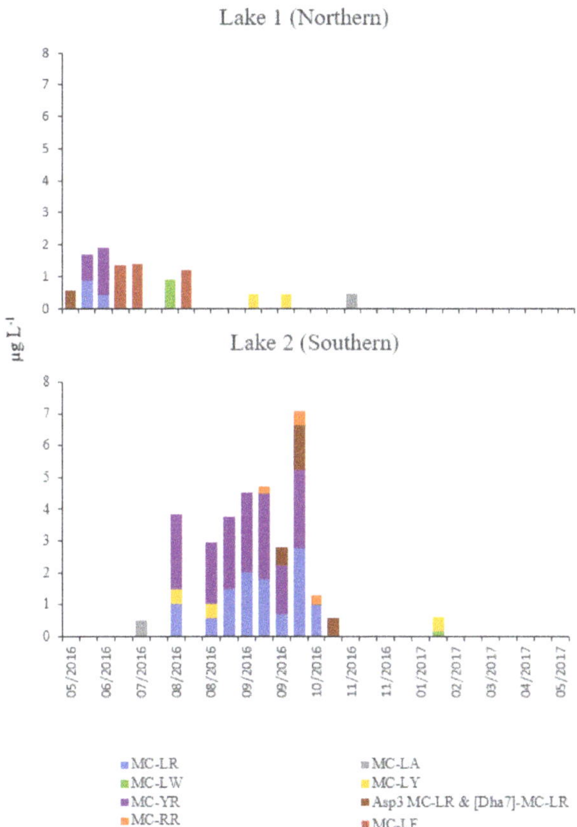

Figure 6. Microcystin variants qualified and quantified by liquid chromatography coupled to tandem mass spectrometry (μg L^{-1}) from water samples collected at Longham Lakes (1 & 2).

A null hypothesis that no differences between *Microcystis* cells, total microcystins, and phycocyanin (cyanobacteria cells) measurements were observable between the two lakes was tested using a series of Student's *t*-tests. Total microcystins and *Microcystis* cells were significantly and highly significantly different between the two lakes respectively ($p < 0.01$, $p < 0.001$) (Table 1).

2.6. The Ability of Chemical and Biological Parameters to Predict Presence of Microcystins

Regression analysis suggested that lake, temperature, and dissolved oxygen explained a significant amount of the variability observed in the microcystin values across the two lakes and study period (adjusted R-squared: 0.251, F-statistic: 5.68 on 3 and 39 DF, p-value 0.003). Table 1, model 1 predicted microcystin values in to be 2.18 times higher in lake 2 than in lake 1, and that microcystin levels increase with temperature and decrease with increases in dissolved oxygen. No statistically significant interaction terms were observed between any of these variables. Examination of diagnostic plots showed the model fit to be poor, with a high number of zero values observed for microcystin being highly influential on the model fit. No standard data transformations (e.g., log, square root, quadratic) improved the model fit, so a further model (Table 2, model 2) was fit to the subset of the data for which microcystin was detected (i.e., microcystin > 0). The stepwise model building process resulted in the same variables being selected for inclusion in model 2 as for model 1; however, a larger amount of the variability in the observed microcystin values (adjusted R-squared: 0.566, F-statistic: 9.275 on 3 and 16 DF, p-value < 0.001) was now explained and examination of the diagnostic plots associated with model 2 showed improved fit. As expected, given the influence of the zero microcystin values in model 1, the effect size associated with the explanatory variables is considerably different in model 2, increasing by around 1.5 times in all cases. Subsequent logistic regression analysis was not able to identify any variables that showed a statistically significant association with the presence/absence of microcystin (i.e., the occurrence of microcystin above the limits of detection).

Table 2. Multiple linear regression predictors for microcystin levels in both lakes. Model 1 includes observations where microcystin was not detected (i.e., microcystin = 0), model 2 shows results for data relating to positive microcystin observations only (i.e., microcystin > 0).

Parameter	Estimate	Std. Error	t-Value	Pr(>|t|)
Model 1: Zero microcystin values included (n = 43)				
Intercept	1.446	1.295	1.116	0.271
Lake	2.183	0.583	3.748	0.001
Dissolved O_2	−0.418	0.143	2.920	0.006
Temperature	0.173	0.062	2.780	0.008
Model 2: Zero microcystin values removed (n = 20)				
Intercept	1.408	2.358	0.597	0.559
Lake	3.523	0.698	5.051	0.000
Dissolved O_2	−0.596	0.194	3.076	0.007
Temperature	0.295	0.119	2.470	0.025

3. Discussion

In this study, relationships between biological and chemical parameters, cyanobacterial taxa with specific reference to *Microcystis* spp., using microscopy and flow cytometry were examined in two lowland lakes in southern Britain. Total microcystins, microcystin variants and toxin profiles were also determined using ultra-high-pressure liquid chromatography coupled to mass spectrometry. *Microcystis* spp. and microcystins were detected and quantified in both lakes. The objective of the study was to attempt to identify drivers of microcystin elevation with a view to better informing water risk management strategies to protect ecological and human health.

Levels of microcystin variants were not correlated with numbers of *Microcystis* cells and were significantly higher in lake 2 than in lake 1. Microcystin levels did not exceed the WHO medium health threshold of 20 μg L^{-1}. In 2017, the WHO [15] published chronic health threshold levels of 1 μg L^{-1} MC-LR for lifelong drinking water consumption; 16% of samples from lake 2 exceeded the MC-LR chronic threshold value. Quantifiable levels of toxins were detected in 48% of the samples; identical toxin detection frequencies were observed, but total levels and profiles differed significantly between the lakes. Toxin concentrations ranged from not detected to 7.1 μg L^{-1} and concentrations were similar to those reported by several authors for waterbodies in the absence of scums in the Lower Great Lakes [28], England, and Wales [10] and selected European water bodies [18].

In lake 1, total microcystins were approximately an order of magnitude lower than in lake 2, despite proportionally higher isolation frequencies and levels of *Microcystis* cells (lake 1, maximum density 51,384 cells mL^{-1}). *Microcystis* cells did not exceed 8000 cells mL^{-1} in lake 2 and constituted a relatively minor fraction of the total estimates of cyanobacteria. It is probable that *Microcystis* cells at both lakes were non-toxin producing strains and that toxins detected in this study were produced by species other than *Microcystis*. It is well documented that the *Microcystis* blooms vary in their toxin profiles [10,15,29]. The ability for microcystin production in *Microcystis* spp. and other cyanobacterial species is genetically determined [30,31]. Several studies have reported that strains isolated from geographically and temporally distinct *Microcystis* spp. populations are clonal and therefore likely to be either microcystin producers or non-producers [5,32]. The hypothesis that species other than *Microcystis* spp. were responsible for toxin production in the study lakes is supported by the presence of other cyanobacteria, e.g., *Anabaena* spp., *Aphanizomenon* spp., and *Oscillatoria* spp. at elevated levels particularly in lake 2. At lake 2, on the 3 occasions where the UK threshold action limits were exceeded, less than 8.5% of the estimated cyanobacterial populations comprised *Microcystis* cells. In a review of cyanobacterial bloom taxa in the UK, Howard et al. [33] recorded that the dominant species in addition to *Microcystis* spp., were *Oscillatoria, Planktothrix, Anabaena, Pseudanabaena,* and *Gomphosphaeria*. The dominance of these species amongst phytoplankton communities in samples from natural lakes and reservoirs has been subsequently confirmed in England and Wales [10,34] and Scotland [17]. *Anabaena* spp., *Aphanizomenon* spp., and *Oscillatoria* spp., are well known as toxin-producing species and common members of phytoplanktonic communities [29].

Maximum total microcystin level recorded was in lake 2 (7.1 $\mu g\ L^{-1}$), with mean levels in lake 1 of 0.497 $\mu g\ L^{-1}$ and 1.524 $\mu g\ L^{-1}$ in lake 2. Total microcystins and *Microcystis* cells were significantly and highly significantly different between the two lakes, respectively (total microcystins higher in lake 2; *Microcystis* spp. cells higher in lake 1). Of the multiple microcystin variants described, MC-LR has been most extensively studied and is reported to be between 3 and 10 times more toxic than other microcystin congeners [28]. Analysis of the microcystin toxin profiles between the two study lakes indicated differences in variants, both in terms of variants, proportions, and levels. In lake 1, where relatively low levels of microcystins were determined, MC-LR was detected at low levels ($<1\ \mu g\ L^{-1}$) and the dominant variant was the more hydrophobic MC-LF. This finding is in accordance with Turner et al. [10], who reported MC-LF as the highest mean proportion of profiles from *Aphanizomenon* sp. and *Oscillatoria* sp. in analyses of freshwater bodies in England and Wales. This adds some support to the premise that *Microcystis* cells present in lake 1 were non-toxin producers. In lake 2, MC-YR was the dominant congener closely followed by MC-LR, with MC-RR and Asp^3-MC-LR appearing towards the end of the period of toxin prevalence. Microcystin profiles determined in samples taken from weekly monitoring during August and September were similar in relative proportions, indicating potentially that clonal or semi-clonal populations were responsible for toxin production. Similar data have been generated from studies in Greece [35], throughout Europe via the European Multi Lake Survey [18] and Finland [36], the latter associated with toxins produced by species of *Anabaena*. Whilst data on the relative proportions of microcystin variants as measured by LC-MS/MS are relatively sparse, levels and variants of microcystins presented here are consistent with the recorded literature and can help to unravel the structure and function of cyanobacterial populations.

In recent years, many authors have studied the drivers for elevated cyanotoxin levels and/or occurrence of cyanobacterial blooms in natural lakes, reservoirs [18,37–39], and aquaculture systems [40]. In a large-scale study assessing the continental scale distribution of cyanotoxins across Europe, Mantzouki et al. [18] demonstrated that temperature (rather than nutrient availability or euphotic depth) was generally responsible for distributional characteristics of cyanotoxins. Likewise, concordant observations with respect to temperature were recorded by Elliot [39] in an assessment of climate change on pelagic freshwater cyanobacteria. The author demonstrated increased relative cyanobacteria abundance concurrently with increased water temperature, together with decreased flushing rates and increased nutrient loading. Similarly, Sinden and Sinang [40] identified temperature, in combination with elevated water pH, as key environmental factors influencing proliferation of cyanobacteria and toxicity in Malaysian aquaculture ponds. In the present study, the presence of correlations between a range of biological and chemical parameters were tested against presence of microcystins. Perhaps not surprisingly given the low levels of microcystins detected in lake 1, no correlations were observed. For lake 2, where moderate levels of microcystins were present during the summer and autumn, water temperatures did not correlate; however, a decrease in dissolved oxygen was closely associated with presence and levels of microcystins. A decrease in dissolved oxygen concurrent with decomposition of cyanobacterial blooms has been reported previously [6,41] and thus has the potential to indicate onset of toxicity derived from lysed cells within a rapidly blooming population. Perturbations in dissolved oxygen are frequently used as an indicator of water quality and eutrophication [42]. Conversely, with respect to the correlations of microcystin production, in a comprehensive review of biological and chemical factors, Dai et al. [42] postulated that light intensity and temperature were the most important physical factors with nitrogen and phosphorus as the critical chemical drivers of harmful algal blooms and microcystins. The authors also noted the complex interactions with biotic factors, suggesting that predator-prey relationships in phytoplanktonic communities may promote microcystin production and release [42].

To explain the strength of the association between the measured parameters and their potential future ability to inform predictive models for microcystin events, stepwise multiple linear regression analysis was applied to all chemical and biological parameters. Across both lakes, microcystins increased with temperature and decreased with dissolved oxygen. Using a similar statistical approach as a precursor to the development of empirical predictive models for cyanobacterial, Beaulieu et al. [43] showed that total nitrogen and water temperature provided the best model and explained 25% of cyanobacterial biomass. Using these explanatory variables, the authors developed competing path models, which showed that both nitrogen and temperature were indirectly (and directly) linked to cyanobacteria by interactions with total algal biomass. Model outputs predicted an average doubling of cyanobacterial biomass with a 3.3 °C rise in water temperature. In contrast, Carvalho et al. [24] showed that significant explanatory variables were dissolved organic carbon and pH, and that furthermore nutrient concentrations were not a primary explanatory variable. In this study the relatively high levels of variability explained by combined biological and chemical parameters via regression analysis indicated promise and areas for future investigation.

Although in this study, no single variable could be considered predictive of microcystin production nor could all the variability between the lakes be explained by combinations of measurements, the major identifiable difference between the two water bodies was the presence of a managed fishery at lake 1. Omnivorous fishes; common carp (*Cyprinus carpio*), bream (*Abramis brama*), tench (*Tinca tinca*) and pike (*Esox lucius*), and substantial natural populations of roach (*Rutilus rutilus*), and rudd (*Scardinius erythrophthalmus*) were present in lake 1, whereas no managed fishery or introductions of fish species operated at lake 2. Larval, fry, and fingerlings of stocked species favour zooplankton, with certain high nutrient species (rotifers etc.) making up a substantial component of adult fish's food supply. Larval stages and young age classes of *R. rutilus* and *S. erythrophthalmus*, numerous during summer months are almost exclusively zooplanktoniverous, with a feeding preference for rotifers [44]. Rotifers and other small zooplankton, such as cyclopoid copepods and cladocerans, are selective grazers that can coexist with bloom forming cyanobacteria and are reported to periodically exhibit top-down control [45]. Interestingly, several authors have suggested that *Microcystis* spp. represent a less attractive foodstuff for zooplankton due inter alia to colony formation [45–47], which together with a reduction in grazing pressure on other cyanobacterial taxa may have created a competitive advantage for the non-toxic *Microcystis*-like cell populations, such as unknown species of picoplankton in lake 1. *R. rutilus* abundance has previously been implicated in ichthyo-eutrophication of reservoirs in the Czech Republic [44]; this has been attributed to a range of factors but may also include reduction in grazing rates by zooplankton, which in turn create advantageous conditions for blooms [48]. Furthermore, eutrophication can change microbial loops and therefore may inhibit antagonistic microorganisms (viruses, bacteria, microalgae, microfungal, and amoeboid taxa) present within *Microcystis* colonies in the operation of bottom-up controls [45].

Environmental drivers for toxin production from cyanobacterial taxa are complex and intricacies of phytoplankton communities, cryptic ecological interactions, and the presence of non-toxin producing cyanobacterial strains make this a challenging area of risk management. This study demonstrated that no single variable could be used to predict microcystin levels but supported the use of multiple measurements in the development of more holistic predictive models. In this study, multiple measurements were dissolved oxygen, turbidity, phycocyanin, temperature pH, chlorophyll *a, b,* and total carotenoids. The potential for enhanced fish stocks to exert top down control on toxic cyanobacterial populations was an interesting observation and is indicative of wider multifaceted trophic relationships influencing cyanobacterial population dynamics and toxin production. Whilst challenging at an environmentally relevant scale [21], improved predictive ability and modelling will provide more efficient, proactive management of water bodies impacted by toxic cyanobacteria and in turn have positive ecological, public, and animal health benefits.

4. Materials and Methods

4.1. Sample Collection and Water Parameter Measurements

Water samples were collected with chemical and biological measurements between May 16 2016 and May 31 2017, from both northern and southern lakes at points A & B (Figure 2). The lakes were sampled weekly in the spring, summer, and autumn, with sampling frequency dropping to every three weeks in the winter. Measurements of dissolved oxygen, turbidity, phycocyanin, temperature, pH, and salinity, were made by multiparameter probe (6600 V2, YSI, Xylem Analytics, Singapore) from the surface and at 1 m intervals to the bottom. Water samples for flow cytometer and toxin analysis were collected from the same depths using a 2.2 L horizontal sampler. Water samples for microscope analysis were collected from the surface of each lake only. Water samples were in put in opaque bottles and stored in a cool box for return to the laboratory.

4.2. Cell Discrimination by Flow Cytometry

Water samples were analysed for using a flow cytometer (C6, BD Accuri, San Jose, CA, USA), samples were aliquoted (≈2 mL) into a 5 mL sample tube, and homogenised by vortex. A 5-min custom fluidic setting was selected of 25 µm core and 100 µL/min, with a threshold of 80,000 au on forward scatter (FSC) signal. Unicellular *Microcystis* cells were resolved by size and three auto-florescence channels. Following Chapman [49], side scatter (SSC) signal was used as an indicator cell size and verified by calibration beads (PPS-6K, Spherotech, Lake Forest, IL, USA), gated between 75,000 to 700,000 au. Yellow auto-florescence (FL2) was used as an indicator of phycoerythrin and carotenoids, gated at 40 to 2000 au. Red auto-florescence (FL3) was used as an indicator of chlorophyll, gated between 180,000 to 2,200,000 au. Finally, far-red auto-florescence was used as an indicator of phycocyanin, gated between 11,000 to 500,000 au. The method was optomised using 6 *Microcystis*, 2 non-*Microcystis* cyanobacteria, and 2 eukaryotic algal reference strains.

4.3. Toxin Analysis by Liquid Chromatography Coupled to Tandem Mass Spectrometry

In triplicate, 200 mL of water sample was filtered (CFC, Whatman, Maidstone, UK), filter papers were wrapped individually in aluminium foil and preserved at −80 °C. On analysis, filter papers were subjected to three cycles of freeze-thawing before submersion in 10 mL of 80% aqueous methanol. Samples were left in the dark at 4–6 °C for 24 h, before ~0.5 mL was aliquoted into a LCMS certified vial. Toxin analysis was carried by ultra-high-performance liquid chromatography (UHPLC) (Acquity, Waters, Manchester, UK) coupled to a tandem quadruple mass spectrometer

(Xevo TQ, Waters, Manchester, UK). All instrument solvents and chemicals were of LC-MS-grade (Fisher Optima, Thermo Fisher, Manchester, UK). Reference toxins used for the detection method included the microcystin analogues MC-RR, MC-LA, MC-LY, MC-LF, MC-LW, MC-YR, MC-WR, MC-HilR, MC-HtyR, MC-LR & Asp^3-MC-LR (Enzo Life Sciences, Exeter, UK) and [Dha^7]-MC-LR and matrix reference material of blue-green algae (RM-BGA, Lot 201301) containing a range of microcystins (Institute of Biotoxin Metrology, National Research Council Canada). Analysis of microcystins was conducted following the method by Turner et al. [50].

4.4. Identification and Enumeration of Phytoplankton by Light Microscopy

Surface water samples were aliquoted into a 15 mL centrifuge tube; the full tube was sealed then inverted and the lid struck on the bench several times to burst any gas flotation vesicles within cells. Centrifuge tubes were then placed upright and stored at 4–6 °C in the dark for two days for the phytoplankton to settle out. The top 14 mL was carefully removed by pipette to not disturb the sedimented phytoplankton; the remaining 1 mL was vortexed and transferred to a Sedgwick-rafter counting chamber. The phytoplankton in a minimum of 10 of the 1000 grid squares were identified and enumerated by light microscope (BX51, Olympus, Tokyo, Japan) at 40× and 100× magnification.

4.5. Multiple Linear Regression Model

Relationships between the presence and level of microcystin and potential predictor variables were explored visually using plot functions and the strength of these relationships assessed using multiple linear regression models after making appropriate transformations data (if required) to ensure its distribution met the test assumptions. A logistic regression model (GLM assuming a binomial distribution and log link function) was also applied to determine whether any of the environmental variables measured were able to reliably predict the presence/absence of microcystin rather than its levels. For both the linear and logistic regression analysis, univariable analysis was first performed and then a multivariable model was built using a forward stepwise approach in which variables and combinations that led to statistically significantly reductions ($p \leq 0.05$) in the Akaike information criteria (AIC) were retained in the model. The presence of two-way interaction effects was explored for all variable combinations but were only retained if their inclusion resulted in a significant reduction in AIC. All analysis and data visualisations were conducted in R version 4.0.2 [51].

Author Contributions: D.M.H., I.J.C., G.F.E., A.D.T., and D.J.F. designed and conceived the study, D.M.H., and I.J.C. performed the field and laboratory work, N.G.H.T. provided formal analysis G.F.E., D.J.F., and A.D.T. provided supervision, D.M.H. prepared the original draft, D.M.H., I.J.C., G.F.E., A.D.T., and D.J.F. reviewed the paper, and D.M.H. and N.G.H.T. wrote the revised drafts. All authors have read and agreed to the published version of the manuscript.

Funding: This research was funded by Bournemouth University post graduate studentships. Analytical chemical methods were funded by Cefas Seedcorn (Contract DP305).

Acknowledgments: The authors would like to thank Lewis Coates for preparing Figure 1, and Bournemouth Water, especially Ian Hayward the bailiff at Longham Lakes. D.M.H. would like to thank Rachel Hartnell for reviewing the original draft of this manuscript.

Conflicts of Interest: The authors declare no conflict of interest.

References

1. Codd, G.A. Cyanobacterial toxins, the perception of water quality, and the prioritisation of eutrophication control. *Ecol. Eng.* **2000**, *16*, 51–60. [CrossRef]
2. Paerl, H.W.; Fulton, R.S.; Moisander, P.H.; Dyble, J. Harmful freshwater algal blooms, with an emphasis on cyanobacteria. *Sci. World J.* **2001**, *1*, 76–113. [CrossRef] [PubMed]
3. Paerl, H.W.; Paul, V.J. Climate change: Links to global expansion of harmful cyanobacteria. *Water Res.* **2012**, *46*, 1349–1363.

4. Robarts, R.D.; Zohary, T. Temperature effects on photosynthetic capacity, respiration and growth rates of bloom-forming cyanobacteria. *New Zealand J. Mar. Fresh.* **1987**, *12*, 391–399. [CrossRef]
5. Paerl, H.W.; Huisman, J. Blooms like it hot. *Science* **2008**, *320*, 57–58. [CrossRef] [PubMed]
6. O'Neil, J.M.; Davis, T.W.; Burford, M.A.; Gobler, C.J. The rise of harmful cyanobacteria blooms: The potential roles of eutrophication and climate change. *Harmful Algae* **2012**, *14*, 313–334. [CrossRef]
7. Rigosi, A.; Carey, C.C.; Ibelings, B.W.; Brookes, J.D. The interaction between climate warming and eutrophication to promote cyanobacteria is dependent on trophic state and varies among taxa. *Limnol. Oceanogr.* **2014**, *59*, 99–114. [CrossRef]
8. Lawton, L.A.; Codd, G.A. Cyanobacterial (blue-green algal) toxins and their significance in UK and European waters. *Water Environ. J.* **1991**, *5*, 460–465. [CrossRef]
9. Chorus, I. Cyanotoxin occurrence in freshwaters—A summary of survey results from different countries. In *Cyanotoxins—Occurrence, Causes, Consequences*; Chorus, I., Ed.; Springer: Berlin, Germany, 2001; pp. 75–78.
10. Turner, A.D.; Dhanji-Rapkova, M.; O'Neill, A.; Coates, L.; Lewis, A.; Lewis, K. Analysis of Microcystins in Cyanobacterial Blooms from Freshwater Bodies in England. *Toxins* **2018**, *10*, 39. [CrossRef]
11. Kuiper-Goodman, T.; Falconer, I.O.M.; Fitzgerald, J. Human health aspects. In *Toxic Cyanobacteria in Water: A Guide to Their Public Health Consequences, Monitoring and Management*; Chorus, I., Bartram, J., Eds.; E & F N Spon: London, UK, 1999; pp. 125–160.
12. Van Ginkel, C.E. *Toxic Algal Incident in the Grootdraai Dam*; Institute for Water Quality Studies, Department of Water Affairs and Forestry: Pretoria, South Africa, 2011.
13. Alonso-Andicoberry, C.; García-Viliada, L.; Lopez-Rodas, V.; Costas, E. Catastrophic mortality of flamingos in a Spanish national park caused by cyanobacteria. *Vet. Rec.* **2002**, *151*, 706–707.
14. Qin, B.; Zhu, G.; Gao, G.; Zhang, Y.; Li, W.; Paerl, H.W.; Carmichael, W.W. A drinking water crisis in Lake Taihu, China: Linkage to climatic variability and lake management. *J. Environ. Manage* **2010**, *45*, 105–112. [CrossRef]
15. World Health Organization (WHO). *Guidelines for Drinking-Water Quality: Fourth Edition Incorporating the First Addendum*; WHO: Geneva, Switzerland, 2017.
16. Water UK. *Water UK Technical Briefing Note: Blue Green Algal Toxins in Drinking Water*; Water UK Position Paper; Water UK: London, UK, 2006.
17. Krokowski, J.T.; Lang, P.; Bell, A.; Broad, N.; Clayton, J.; Milne, I.; Nicolson, M.; Ross, A.; Ross, N. A review of the incidence of cyanobacteria (blue-green algae) in surface eaters in Scotland including potential effects of climate change, with a list of the common species and new records from the Scottish Environmental Protection Agency. *Glasg. Nat.* **2012**, *25*, 99–104.
18. Mantzouki, E.; Lürling, M.; Fastner, J.; de Senerpont Domis, L. Temperature effects explain continental scale distribution of cyanobacterial toxins. *Toxins* **2018**, *10*, 156. [CrossRef]
19. Miles, C.O.; Sandvik, M.; Nonga, H.E.; Rundberget, T.; Wilkins, A.L.; Rise, F.; Ballot, A. Thiol derivatization for LC-MS identification of microcystins in complex matrices. *J. Environ. Sci. Technol.* **2012**, *46*, 8937–8944. [CrossRef]
20. Okello, W.; Ostermaier, V.; Portmann, C.; Gademann, K.; Kurmayer, R. Spatial isolation favours the divergence in microcystin net production by Microcystis in Ugandan freshwater lakes. *Wat. Res.* **2010**, *44*, 2803–2814. [CrossRef]
21. Oliver, R.L.; Hamilton, D.P.; Brookes, J.D.; Ganf, G.G. Physiology, blooms and prediction of planktonic cyanobacteria. In *Ecology of Cyanobacteria II*; Whitton, B.A., Ed.; Springer: Dordrecht, Germany, 2012; pp. 155–194.
22. Downing, J.A.; Watson, S.B.; McCauley, E. Predicting cyanobacteria dominance in lakes. *Can. J. Fish. Aquat. Sci.* **2001**, *58*, 1905–1908. [CrossRef]
23. Howard, A.; Easthope, M.P. Application of a model to predict cyanobacterial growth patterns in response to climatic change at Farmoor Reservoir, Oxfordshire, UK. *Sci. Total Environ.* **2002**, *282*, 459–469. [CrossRef]
24. Carvalho, L.; Miller, C.A.; Scott, E.M.; Codd, G.A.; Davies, P.S.; Tyler, A.N. Cyanobacterial blooms: Statistical models describing risk factors for national-scale lake assessment and lake management. *Sci. Total Environ.* **2011**, *409*, 5353–5358. [CrossRef]
25. Rigosi, A.; Hanson, P.; Hamilton, D.P.; Hipsey, M.; Rusak, J.A.; Bois, J.; Sparber, K.; Chorus, I.; Watkinson, A.J.; Qin, B.; et al. Determining the probability of cyanobacterial blooms: The application of Bayesian networks in multiple lake systems. *Ecol. Appl.* **2015**, *25*, 186–199. [CrossRef] [PubMed]

26. Sivonen, K.; Jones, G. Cyanobacterial toxins. In *Toxic Cyanobacteria in Water: A Guide to Public Health Significance, Monitoring and Management*; Chorus, I., Bartram, J., Eds.; E & F N Spon: London, UK, 1999; pp. 41–111.
27. Mekebri, A.; Blondina, G.J.; Crane, D.B. Method validation of microcystins in water and tissue by enhanced liquid chromatography tandem mass spectrometry. *J. Chromatogr. A* **2009**, *1216*, 3147–3155. [CrossRef] [PubMed]
28. Dyble, J.; Fahnenstiel, G.L.; Litaker, R.W.; Millie, D.F.; Tester, P.A. Microcystin concentrations and genetic diversity of *Microcystis* in the lower Great Lakes. *Environ. Toxicol.* **2008**, *23*, 507–516. [CrossRef] [PubMed]
29. Sivonen, K. Cyanobacterial toxins. In *Encyclopedia of Microbiology*, 3rd ed.; Schaechter, M., Ed.; Academic Press: London, UK, 2009; pp. 290–307.
30. Janse, I.; Kardinaal, W.E.A.; Meima, M.; Fastner, J.; Visser, P.M.; Zwart, G. Toxic and nontoxic Microcystis colonies in natural populations can be differentiated on the basis of rRNA gene internal transcribed spacer diversity. *Appl. Environ. Microbiol.* **2004**, *70*, 3979–3987. [CrossRef] [PubMed]
31. Meyer, K.A.; Davis, T.W.; Watson, S.B.; Denef, V.J.; Berry, M.A.; Dick, G.J. Genome sequences of lower Great Lakes Microcystis sp. reveal strain-specific genes that are present and expressed in western Lake Erie blooms. *PLoS ONE* **2017**, *12*, e0183859. [CrossRef] [PubMed]
32. Tillett, D.; Parker, D.L.; Neilan, B.A. Detection of toxigenicity by a probe for the microcystin synthetase A gene (mcyA) of the cyanobacterial genus Microcystis: Comparison of toxicities with 16S rRNA and phycocyanin operon (phycocyanin intergenic spacer) phylogenies. *Appl. Environ. Microbiol.* **2001**, *67*, 2810–2818. [CrossRef]
33. Howard, A.; McDonald, A.T.; Kneale, P.E.; Whitehead, P.G. Cyanobacterial (blue-green algal) blooms in the UK: A review of the current situation and potential management options. *Prog. Physic. Geograp.* **1996**, *20*, 53–61. [CrossRef]
34. Krokowski, J.T.; Jamieson, J. A decade of monitoring and management of freshwater algae, in particular cyanobacteria in England and Wales. *Freshwat. Forum* **2002**, *18*, 3–12.
35. Zervou, S.K.; Christophoridis, C.; Kaloudis, T.; Triantis, T.M.; Hiskia, A. New SPE-LC-MS/MS method for simultaneous determination of multi-class cyanobacterial and algal toxins. *J. Hazard. Mater.* **2017**, *323*, 56–66. [CrossRef]
36. Halinen, K.; Jokela, J.; Fewer, D.P.; Wahlsten, M.; Sivonen, K. Direct evidence for production of microcystins by *Anabaena* strains from the Baltic Sea. *Appl. Environ. Microbiol.* **2007**, *73*, 6543–6550. [CrossRef]
37. Mowe, M.A.; Porojan, C.; Abbas, F.; Mitrovic, S.M.; Lim, R.P.; Furey, A.; Yeo, D.C. Rising temperatures may increase growth rates and microcystin production in tropical *Microcystis* species. *Harmful Algae* **2015**, *50*, 88–98. [CrossRef]
38. Ferguson, C.A.; Carvalho, L.; Scott, E.M.; Bowman, A.W.; Kirika, A. Assessing ecological responses to environmental change using statistical models. *J. Appl. Ecol.* **2008**, *45*, 193–203. [CrossRef]
39. Elliott, J.A. Is the future blue-green? A review of the current model predictions of how climate change could affect pelagic freshwater cyanobacteria. *Wat. Res.* **2012**, *46*, 1364–1371. [CrossRef] [PubMed]
40. Sinden, A.; Sinang, S.C. Cyanobacteria in aquaculture systems: Linking the occurrence, abundance and toxicity with rising temperatures. *Int. J. Environ. Sci. Technol.* **2016**, *13*, 2855–2862. [CrossRef]
41. Paerl, H.W.; Otten, T.G. Blooms bite the hand that feeds them. *Science* **2013**, *342*, 433–434. [CrossRef]
42. Dai, R.; Wang, P.; Jia, P.; Zhang, Y.; Chu, X.; Wang, Y. A review on factors affecting microcystins production by algae in aquatic environments. *World J. Microbiol. Biotechnol.* **2016**, *32*, 51. [CrossRef] [PubMed]
43. Beaulieu, M.; Pick, F.; Gregory-Eaves, I. Nutrients and water temperature are significant predictors of cyanobacterial biomass in a 1147 lakes data set. *Limnol. Oceanogr.* **2013**, *58*, 1736–1746. [CrossRef]
44. Zapletal, T.; Mares, J.; Jurajda, P.; Vseticková, L. The food of roach, Rutilus rutilus (Actinopterygii: Cypriniformes: Cyprinidae), in a biomanipulated water supply reservoir. *Acta Ichthyol. Piscat.* **2014**, *44*, 15. [CrossRef]
45. Van Wichelen, J.; Vanormelingen, P.; Codd, G.A.; Vyverman, W. The common bloom-forming cyanobacterium Microcystis is prone to a wide array of microbial antagonists. *Harmful Algae* **2016**, *55*, 97–111. [CrossRef]
46. Rohrlack, T.; Dittmann, E.; Börner, T.; Christoffersen, K. Effects of cell-bound microcystins on survival and feeding of *Daphnia* spp. *Appl. Environ. Microbiol.* **2001**, *67*, 3523–3529. [CrossRef]
47. Lotocka, M. Toxic effect of cyanobacterial blooms on the grazing activity of *Daphnia magna* Straus. *Oceanologia* **2001**, *43*, 441–453.

48. Zurawell, R.W.; Chen, H.; Burke, J.M.; Prepas, E.E. Hepatotoxic cyanobacteria: A review of the biological importance of microcystins in freshwater environments. *J. Toxicol. Environ. Health Part B* **2005**, *8*, 1–37. [CrossRef]
49. Chapman, I.J. Developing Pioneering New Tools to Detect and Control the Toxic Alga Microcystis in Lakes and Reservoirs. Ph.D. Thesis, Bournemouth University, Poole, UK, 2016.
50. Turner, A.D.; Waack, J.; Lewis, A.; Edwards, C.; Lawton, L. Development and single-laboratory validation of a UHPLC-MS/MS method for quantitation of microcystins and nodularin in natural water, cyanobacteria, shellfish and algal supplement tablet powders. *J. Chromatogr. B* **2018**, *1074*, 111–123. [CrossRef] [PubMed]
51. R Core Team. *R: A Language and Environment for Statistical Computing*; R Foundation for Statistical Computing: Vienna, Austria, 2020; Available online: https://www.R-project.org/ (accessed on 14 July 2020).

Article

Meteorological and Nutrient Conditions Influence Microcystin Congeners in Freshwaters

Zofia E. Taranu [1,2,*], Frances R. Pick [1], Irena F. Creed [3], Arthur Zastepa [4] and Sue B. Watson [5]

1 Department of Biology, University of Ottawa, Ottawa K1N 6N5, ON, Canada; Frances.Pick@uottawa.ca
2 Aquatic Contaminants Research Division, Environment and Climate Change Canada, Montreal H2Y 2E7, QC, Canada
3 School of Environment and Sustainability, University of Saskatchewan, Saskatoon S7N 5C8, SK, Canada; icreed@uwo.ca
4 Canada Centre for Inland Waters, Environment and Climate Change Canada, Burlington, ON L7S 1A1, Canada; arthur.zastepa@canada.ca
5 Department of Biology, University of Waterloo, Waterloo, ON N2L 3G1, Canada; jkswatson@shaw.ca
* Correspondence: zofia.taranu@gmail.com

Received: 1 October 2019; Accepted: 23 October 2019; Published: 26 October 2019

Abstract: Cyanobacterial blooms increasingly impair inland waters, with the potential for a concurrent increase in cyanotoxins that have been linked to animal and human mortalities. Microcystins (MCs) are among the most commonly detected cyanotoxins, but little is known about the distribution of different MC congeners despite large differences in their biomagnification, persistence, and toxicity. Using raw-water intake data from sites around the Great Lakes basin, we applied multivariate canonical analyses and regression tree analyses to identify how different congeners (MC-LA, -LR, -RR, and -YR) varied with changes in meteorological and nutrient conditions over time (10 years) and space (longitude range: 77°2′60 to 94°29′23 W). We found that MC-LR was associated with strong winds, warm temperatures, and nutrient-rich conditions, whereas the equally toxic yet less commonly studied MC-LA tended to dominate under intermediate winds, wetter, and nutrient-poor conditions. A global synthesis of lake data in the peer-reviewed literature showed that the composition of MC congeners differs among regions, with MC-LA more commonly reported in North America than Europe. Global patterns of MC congeners tended to vary with lake nutrient conditions and lake morphometry. Ultimately, knowledge of the environmental factors leading to the formation of different MC congeners in freshwaters is necessary to assess the duration and degree of toxin exposure under future global change.

Keywords: Cyanotoxins; microcystin congeners; MC-LA; nutrients; climate; Great Lakes; raw water intake; multivariate statistics; long-term monitoring

Key Contribution: We showed that, regionally, microcystin congener composition varied systematically with environmental change (primarily in response to changes in wind speed, with secondary effects of nutrients and temperature). In turn, this regional-scale heterogeneity helped explain global-scale patterns and the reported dichotomy in congener dominance between Europe and North America.

1. Introduction

The incidence of severe cyanobacterial blooms is increasing worldwide [1–5], along with the risk of exposure to cyanobacterial toxins [6–8]. Cyanotoxins are found on all continents [9], though among the suite of cyanotoxins that occur in freshwater ecosystems, microcystins (MCs) are the most commonly reported [9] and the most diverse, with over 240 different structural variants (congeners) identified to date [10]. Despite considerable research on MCs, little is known about the global or multi-region occurrence of the different congeners, and how their distribution relates to environmental conditions (though see key work by [11,12]). This inability to predict the spatial (and temporal) variability in MC congeners is a major concern given that numerous wildlife and livestock fatalities have been linked to exposure to these cyanotoxins [13,14]. Humans are also at risk. People living in close proximity to lakes with frequent MC-producing cyanobacterial blooms or exposed to contaminated drinking or recreational water have been reported to experience various health problems such as muscle pain and gastrointestinal, skin and ear irritations [15]. There is also mounting evidence of chronic health problems associated with MC exposure, including a higher incidence of non-alcoholic liver cancer [16–19]. More recently, MCs have been shown to transfer up the pelagic food chain to fish [20], with additional long-term implications for human health [21,22].

Although new MC congeners are continually being identified [10], only a small number are monitored on a regular basis, largely due to the lack of standards and the need to use advanced LC/MS/MS techniques to identify multiple congeners. More routinely, total MC concentrations are reported as MC-LR equivalents based on enzyme-linked immunosorbent assays (ELISA) or protein phosphatase inhibition bioassays (PPI). Among MC congeners, MC-LR is considered the most common and widely distributed [23,24], however, the use of ELISA kits by many agencies may create a bias as the method has a lower reactivity to MC congeners other than MC-LR. This bias would be most extreme in regions where MC-LR is not the most common congener or not among the MC congeners detected (e.g., [25,26]), leading to an underestimation of total MC concentrations. For instance, MC-RR and -YR have been increasingly reported globally (e.g., [27]), and there are reports of MC-LA dominance in Canada (e.g., in Canada [28,29]) and in the US (e.g., Midwest [24]), suggesting that these variants are more common than previously thought [30]. From a toxicity point of view, the occurrence of MC-LA dominating blooms in some North American lakes can have important consequences on animal health. MC-LA has been shown to be as toxic as MC-LR [31,32] and as persistent if not more so than MC-LR. For example, [33] found that MC-LA persisted throughout the recreational season (9.5 weeks) in a small temperate Canadian lake, long after the disappearance of surface cyanobacterial blooms (visible for five weeks), whereas MC-LR was found to decline much more rapidly (two to four weeks) in other lakes. MC-LA may also penetrate lipid membranes more readily (more hydrophobic) than MC-LR or -RR, increasing its likelihood of bioaccumulation, and it has been directly tied to wildlife mortalities [14]. MC-LA is also less readily removed by carbon filtration than MC-LR, -RR, and -YR [32], posing additional challenges for water treatment facilities.

Compared to the large number of studies based on total microcystins, there are fewer studies on the dynamics of specific MC congeners, and these studies suggest there may be a wide range in congener composition among lakes [24,26]. Indeed, while MC-LA appears to be an important player in North American lakes, it seems less common in European lakes [7]. Though this difference may be due to differences in standards available for MC congener analysis, recent work suggests that regional differences in environmental conditions may explain some of the spatial variability in cyanotoxin composition. For instance, environmental factors such as water temperature, light regime, and thermal stratification were shown to be significant drivers of the cyanotoxin composition observed across a broad-scale synthesis of data from 137 European lakes [12]. These spatial patterns in cyanotoxin composition may admittedly reflect the dominance of certain cyanobacterial strains or species, which are themselves driven by environmental gradients (e.g., [28]). However, patterns in species composition may not necessarily determine patterns in MC congener composition. MCs are produced by a number of cyanobacteria (e.g., *Dolichospermum*, *Microcystis*, and *Planktothrix*), each of which can produce several MC congeners simultaneously [26,34–36]. In addition, there is a wide variation in congener composition among strains of the same species. Furthermore, given that guidelines associated with drinking water consumption are based on MC concentrations, and not species composition [12,37], and that MC congeners can vary in toxicity by one to two orders of magnitude [32,38], the questions of which MC congener dominates and why are important to resolve directly. Agencies would also benefit from robust predictions of impending toxic blooms and knowledge of which cyanotoxins are likely to occur under which routinely monitored environmental conditions [39]. Knowledge of the persistence of MC congeners in natural systems would also be important to reduce the discrepancy between lake closures and the period of potential toxin exposure [33,40]. Overall, a shift in focus from the analysis of one (MC-LR) to many (MC congener composition) will likely have important implications for the protection and management of freshwater resources.

In this study, we modelled changes in the relative concentration of different MC congeners (MC-LA, -LR, -RR, and -YR) in response to routinely monitored environmental factors using a raw water intake dataset collected by a single agency (the Ontario Ministry of the Environment, Conservation and Parks (OMECP)) using the same analytical methods. This entailed both spatial (19 intake sites situated across 12 main water bodies) and temporal (seven months sampled over 10 years) analyses to identify the importance of changes in meteorological conditions and lake nutrient concentrations on the variation of different MC congeners across the Great Lakes region of North America. We examined the potential role of weather-related variables including air temperature, precipitation, wind speed and wind direction as these have previously been shown to enhance cyanobacterial dominance and bloom formation [41–43]. We also examined the potential role of major nutrients (phosphorus (P) and nitrogen (N)) as they are strong predictors of cyanobacterial biomass, cyanobacterial dominance, and cyanotoxins in freshwaters [44,45]. Lastly, we considered characteristics of each intake site (depth and distance from shore) as well as dreissenid mussel control measures (raw water pre-chlorination) which may affect MCs [46,47]. To place this regional analysis within a global context, we conducted a synthesis of the peer-reviewed literature to test for systematic patterns of MC congener dominance across multiple regions.

2. Results

2.1. Regional Analysis of the Great Lakes Intake Sites

The time period for which analytical data were available differed among MC congeners. Most were only quantified from 2013 onwards, MC-LR, -RR, and -YR were quantified from 2004 onwards, and MC-LA from 2006 onwards. We thus restricted our statistical analyses to years when the four most dominant congeners (MC-LA, -LR, -RR, and -YR) were measured (i.e., 2006–2016). We also restricted our analysis to ice-free months (April to November). This provided us with a regional raw water intake database spanning a 10-year period, collected during the open water season from 12 main water bodies in Ontario, Canada (Laurentian Great Lakes, Lake of the Woods, and other Ontario lakes and rivers, Figure 1). For the years and months examined, intake sites were sampled, on average (mean and median), four times per month (range: one to nine days per month), for a total of 2002 unique days-months-years-sites covering the years 2006–2016 and 19 intake sites (as some water bodies had multiple intake sites).

Despite differences in the time of sampling across the main water bodies (e.g., complete sampling in Lake Ontario, intermittent in Detroit River and Lake Erie, and late-onset of sampling in Lake St. Clair (Figure 2)), we observed important spatial patterns in MC congener composition within the Great Lakes and the surrounding region. When averaging the concentrations of each congener across all sampling dates, we noted the variability in congener dominance among water sources (Figure S1). Some water bodies (e.g., Lake St. Clair) had greater MC-LA concentrations, while others (e.g., Lake Ontario) had greater MC-LR or MC-RR concentrations across time points. In general, MC-LA was higher at intake sites located along the Detroit River, Lake Erie, and Lake St. Clair, and was identified by the indicator value index (*indval*, Dufrêne and Legendre 1997) as having high fidelity and specificity to these sites (Figure 1), whereas MC-LR was an indicator of sites along Lake Ontario's Bay of Quinte. These dominance patterns were also consistent from year to year within a given water body (Figure S2), notably so among the most routinely monitored sites with highest MC concentrations (i.e., intakes on Lake Ontario's Bay of Quinte (Bayside, Belleville, Deseronto, and Picton), Lake Erie (Elgin, Essex, Pelee, and Union), Lake St. Clair (Lakeshore and Stoney Point) and the Detroit River).

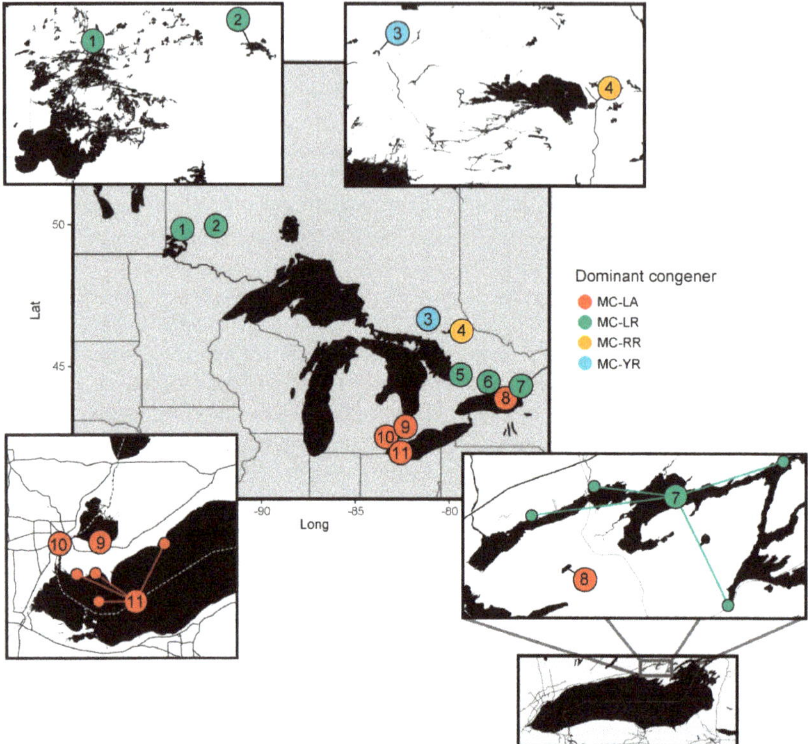

Figure 1. Map of main water bodies with raw water intake samples analyzed by the OMECP. Points are colour coded by dominant congener as determined by an indicator analysis (*indval* function from the {labdsv} package in R), which showed that Robin Lake (**8**), Lake St. Clair (**9**), Detroit River (**10**) and Lake Erie (**11**) were dominated by MC-LA (red), Lake of the Woods (**1**), Lake Wabigoon (**2**), Lake Couchiching (**5**), the Otonabee River (**6**), and Lake Ontario's Bay of Quinte (**7**) were dominated by MC-LR (green), Ramsey Lake (**3**) was dominated by MC-YR (blue), and Lake Nipissing (**4**) by MC-RR (orange). Note: There are four intake sites located along Lake Erie (Elgin, Essex, Union, and Pelee island), four along the Bay of Quinte (Bayside, Belleville, Deseronto, and Picton) and two along Lake St. Clair (Lakeshore and Stoney Point). All other main water bodies only have one intake site.

When examining all unique sampling dates (i.e., not averaged across years within a water body, nor across observations within a given year and water body) in the four most frequently monitored water bodies (Bay of Quinte, Lake Erie, Lake St. Clair and the Detroit River), we noted that the variance in MC-LA and -LR (the two dominant congeners), greatly increased after the year 2012 in the Bay of Quinte, with similar patterns in Lake Erie, Lake St. Clair and the Detroit River (Figure S3), though the sampling was more sporadic for the latter three prior to 2012. For the years monitored, the concentration of MC-LA likewise increased over time in Lakes Erie, St. Clair and the Detroit River (Figure S3). Consequently, the relative concentration of each MC congener changed over time (Figure 2), whereby MC-LA concentrations approached and even surpassed those of MC-LR and -RR after 2012.

Environmental factors likewise varied over space and time in the four most frequently sampled water bodies of the Great Lakes basin. On the dates sampled, the Bay of Quinte region tended to experience warmer conditions than the other sites, whereas the Detroit River and Lake St. Clair region tended to experience higher wind and wetter conditions (Table 1). Meteorological conditions also varied over the years monitored and across stations (Figure S4). Lower wind speed, decreased precipitation, and higher maximum temperatures were recorded after 2012 at the meteorological station near the Detroit River and Lake St. Clair. Similar drops in wind speed and precipitation were recorded in 2015 and 2016 at the weather stations near Lake Erie and the Bay of Quinte (Figure S4). Across sites and time points, nutrient concentrations were typically within the oligo-mesotrophic range, with minimum, median, and maximum TP concentrations of 5, 21 and 675 μg/L, respectively. TP concentrations tended to be higher along the Bay of Quinte, whereas TN was highest at the Lake St. Clair intake sites (Table 2). There was a tendency for TP to decrease over time across the four most frequently monitored water bodies, whereas TN only decreased in the Bay of Quinte, Lake Erie and Lake St. Clair sites (Figure S3). No significant trend in TN was detected in the Detroit River. The depth of the intake sites, their distance from shore, and the use of pre-chlorination to control for dreissenid mussels (pre-chlorination in 59% of sites) also differed among intake sites and main water bodies (Figure S5, Table 3).

Figure 2. Relationships between dominant congeners from the Laurentian Great Lakes basin raw water intake data. Log-transformed concentrations of (**a**) MC-RR and (**b**) MC-LA vs. MC-LR.

As mentioned previously (and shown in Figure S2), most water bodies were only intermittently monitored. To thus provide a more robust investigation of the distribution pattern of MC congeners in time and space, we restricted further statistical analyses to the four water bodies most frequently sampled by the OMECP (Bay of Quinte, Lakes Erie and St. Clair, and the Detroit River). In terms of relationships among MC congener composition and environmental change, a multivariate canonical ordination (redundancy analysis, RDA) showed that the relative abundance of each congener varied with meteorological conditions, nutrient concentrations, chlorine treatment (Yes/No), and the distance of the intake sites from shore (Figure 3). The relative abundance of MC-LR and -RR increased as mean monthly maximum temperatures and TP increased, and as the direction of the wind with the highest speed changed from southwesterly winds (200°) to northern winds (360°). In contrast, the relative abundance of MC-LA increased as maximum wind speeds decreased. Furthermore, intakes closer to shore tended to have higher MC-LR and -RR, and chlorinated sites tended to have higher MC-LA (Figure 3). The multivariate linear model examined with this RDA accounted for a small portion of the total variance in MC congener composition (R^2-adj = 0.09). We also identified a clustering among observations from the same water body (Figure 3) and including water body as a co-variate in the RDA accounted for an additional 7% of the variance in MC composition (figure not shown). A limitation of the RDA, however, is that observations with missing environmental data were omitted prior to analysis and that the relationships were based on linear regressions.

Table 1. Summary statistics of the climate data (monthly averages from April to November, for the 2006–2016) from weather stations near the four main water bodies most frequently monitored by the Ontario Ministry of the Environment, Conservation and Parks (OMECP).

Environmental Factor	Percentile	Stn near L Ontario (N = 456)	Stn near L St. Clair and Detroit R (N = 236 & 118)	Stn near L Erie (N = 476)
MAT	min	−7.5	−9.3	−13.7
	5th	−2.6	−4.4	−9.2
	35th	8.8	6.7	2.3
	50th	14.3	13.0	10.3
	75th	23.5	20.9	21.4
	95th	27.4	24.8	25.5
	max	28.8	26.5	27.5
Precipitation	min	0.8	0.8	0.6
	5th	1.3	1.9	1.4
	35th	48.1	54.1	5.5
	50th	70.5	73.8	42.6
	75th	98.0	110.2	73.2
	95th	125.3	166.3	122.7
	max	195.6	207.0	150.0
Wind direction	min	2.0	1.0	1.0
	5th	4.0	7.2	2.8
	35th	24.0	21.0	19.0
	50th	25.0	23.0	20.5
	75th	28.0	26.0	22.8
	95th	31.0	29.0	34.3
	max	35.0	34.0	36.0
Wind speed	min	32.3	36.1	36.2
	5th	33.7	37.3	39.3
	35th	52.5	52.0	45.1
	50th	57.0	60.0	51.0
	75th	69.0	74.0	65.0
	95th	91.0	95.3	78.5
	max	100.0	111.0	100.0

MAT = maximum air temperature, and N = total sample size (only one station used for each water body, to the exception of Lake St. Clair and Detroit River where the same station was used due to proximity of both water bodies). Due to some NAs, N = 384 observations for wind direction in Lake Erie, whereas all other climate variables have N = 476 observations in Lake Erie.

The use of a multivariate regression tree (MRT) analysis helped identify significant non-linear responses to environmental factors, as well as their interactions, which together accounted for an additional 25% of the variance in MC congener composition (R^2 = 0.34, Figure 4). Among the variables tested, wind speed was the most important explanatory variable in predicting congener composition (Figure 4, tree node 1), but temperature, precipitation, and nutrients had significant secondary effects. When average monthly winds were very stable (maximum wind speed <36 km/hr on average), MC-LR dominated, followed by MC-RR (Figure 4, node 5). When the wind was relatively stable (maximum wind speed between 36 and 51 km/hr) coupled with low average monthly precipitation (<1.6 mm), MC-LR still dominated, followed by MC-RR (Figure 4, node 5). In contrast, when intermediate winds (36 and 51 km/hr) were coupled with precipitation above 1.6 mm, MC-LA was the main MC encountered, with more MC-LA than other congeners when temperatures were cooler (<13 °C). Under wet but warmer conditions (≥13 °C), MC-LA, -LR, and -RR were generally more abundant when TP was at least in the mesotrophic range (TP > 8.5 μg/L) (Figure 4, node 8), but MC-LA was greatest when TN was below 169 µg/L (Figure S6). Under high wind speeds (maximum wind speed > 51 km/hr), combined with warm air temperatures (MAT > 14 °C on average) and as the direction of wind with the highest speed changed from southwest to more northern winds, MC-LR once again dominated, especially under nutrient-rich conditions (TP > 26 µg/L), followed by MC-RR (Figure 4, node 4).

Table 2. Summary statistics of nutrient data (months: April–November, years: 2006–2016) from the four main water bodies most frequently monitored by OMECP.

Water Source	Percentile	L Ontario (*N* = 771 *n sites* = 4)	L St. Clair (*N* = 128 *n sites* = 2)	Detroit R (*N* = 30 *n site* = 1)	L Erie (*N* = 195 *n sites* = 4)
TN	min	25.0	79.0	112.0	104.0
	5th	180.2	209.0	139.6	170.6
	35th	460.0	390.0	320.2	330.0
	50th	510.0	490.0	345.0	416.5
	75th	610.0	888.0	413.0	576.0
	95th	869.8	1982.0	558.8	730.9
	max	2896.0	4020.0	1340.0	1244.0
TP	min	6.0	6.0	6.0	5.0
	5th	12.0	10.0	7.0	7.0
	35th	22.0	15.0	11.0	12.0
	50th	27.0	18.0	12.0	14.0
	75th	39.0	25.0	15.5	19.0
	95th	63.0	36.0	28.5	42.2
	max	675.0	66.0	179.0	139.0

TN = total nitrogen (µg/L), TP = total phosphorus (µg/L), *N* = total sample size, and *n sites* = number of raw water intake sites within each main water body.

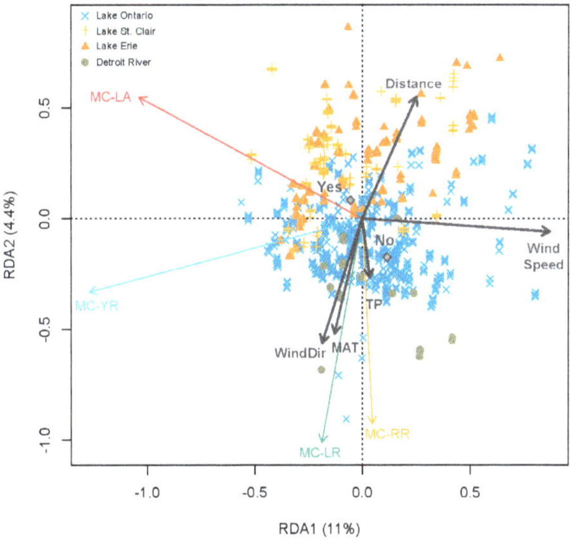

Figure 3. Redundancy analysis of the relationships between the relative abundance of microcystin congeners (i.e., MCL-A, -LR, -RR, and -YR) and environmental factors (i.e., meteorological conditions, nutrient concentrations, intake location, and chlorine treatment) from the most frequently sampled water bodies of the Laurentian Great Lakes basin raw water intake data. Quantitative environmental factors were scaled and centered to reduce variance prior to analysis. The centroids of the qualitative variable (categorical variable for chlorine treatment, Yes/No) are shown by the grey diamonds.

Table 3. Summary of morphometric and nutrient (total phosphorus (TP)) variables for the Laurentian Great Lakes basin (raw water intake sites analyzed by OMECP) and global synthesis of the literature. *N* represents the number of main water bodies.

	Great Lakes Basin Dataset			Global Dataset					
	All Intake Sites			N & S America			Africa, Asia, Europe & Oceana		
	Depth at Intake (m)	Distance from Shore (m)	TP (μg/L)	Surface Area (ha)	Maximum Depth (m)	TP (μg/L)	Surface Area (ha)	Maximum Depth (m)	TP (μg/L)
Range	0–12	0–1300	5–675	4.38–72,500	1–44	2–2047	0.7–305,000	1–175	2–14,420
Median	5	340	20	127	5.4	41.5	521	7.4	69
Mean	5.1	320	27	1633	7.7	150	11065	17	700.7
N	17	16	17	65	69	60	108	110	95

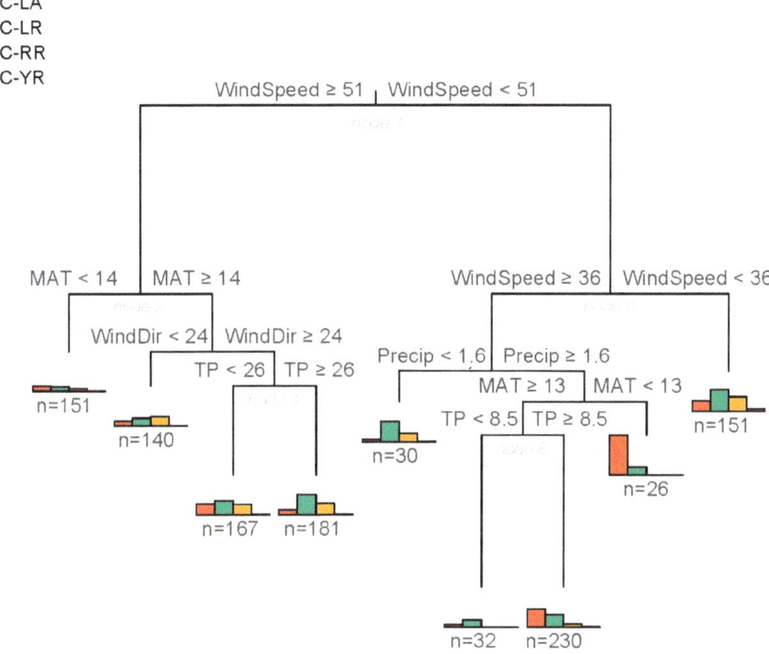

Figure 4. Multivariate regression tree (MRT) of congener relative abundance from the most frequently sampled water bodies of the Laurentian Great Lakes basin, constrained by meteorological conditions (MAT = maximum temperature, WindSpeed = maximum wind speed, WindDir = direction of wind with the highest speed, Precip = precipitation) and nutrient concentrations (TP = total phosphorus).

To tease apart potential effects of geographical location from those of the landscape-scale environmental gradients, we used univariate mixed-effect regression trees with either the concentration of MC-LA or -LR, the two most common congeners detected in the region, as the response variables (the analysis provided similar results when using the relative abundance of MC congeners instead of absolute concentrations, and although not shown MC-RR behaved as MC-LR). The univariate trees showed that MC-LR concentrations were highest under warm, high nutrient conditions (Figure 5a) and that MC-LR concentrations were higher on average in the Bay of Quinte (Figure 5b). In contrast, MC-LA concentrations were higher on average in Lake St. Clair, and primarily related to wind speed (Figure 5c,d). The concentrations of both MC congeners, where greatest from 2013 onwards (Figure 5).

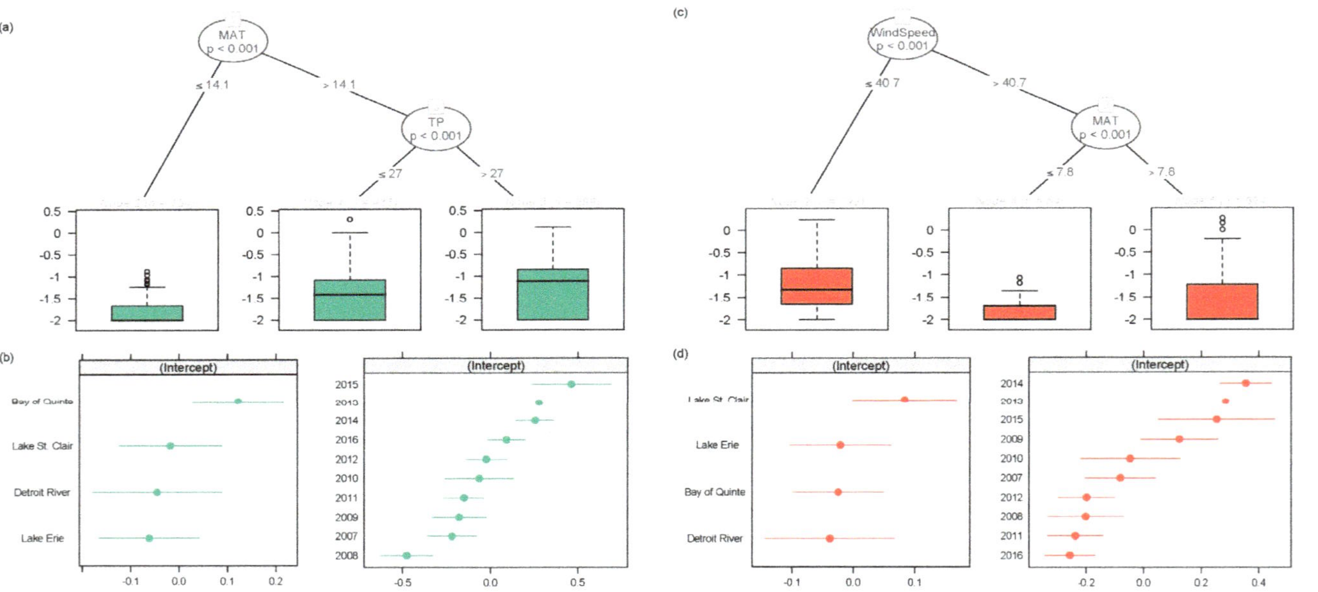

Figure 5. Mixed effect regression tree analysis for MC-LR and MC-LA concentrations from the most frequently sampled water bodies of the Laurentian Great Lakes basin, where fixed effects relationships are shown in (**a**,**c**) and random intercept coefficients are shown in (**b**,**d**) for MC-LR and -LA, respectively. MAT = monthly averaged maximum air temperature (°C), TP = Total Phosphorus (µg/L), and WindSpeed = monthly averaged maximum wind speed (km/hr).

2.2. Global Analysis of the Peer-Reviewed Literature

At the spatial scale of the Great Lakes basin and the surrounding region, we detected significant heterogeneity in MC congener dominance (notably between MC-LA and MC-LR, -RR) due in part to environmental variability among water bodies and over time. Within a global context, the synthesis of studies reporting MC congener data likewise showed a pattern in MC congener occurrence. In particular, MC-LA was most common in North and South American lakes (Figure 6a, Kruskal–Wallis test: $\chi^2 = 44.76$, $p < 0.0001$). The lakes sampled in this continent (predominantly the US and Canada) tended to have a smaller surface area, be shallower in depth, and have lower TP concentrations than the lakes sampled in the other five continents (Table 3, Figure S7a). MC-LR was more evenly distributed (Figure 6b, Kruskal–Wallis test: $\chi^2 = 1.73$ $p = 0.943$), though most common in eutrophic lakes (i.e., intermediate TP range for these sites (Figure S7b)), which corresponded to lakes sampled in Africa, Asia, Europe, and North America (Figure S7a, Table 3). Lastly, MC-RR and -YR tended to divide the depth niche space (the former being relatively more abundant in lakes of intermediate depth, the latter being relatively more abundant in deeper lakes (Figure S7b)).

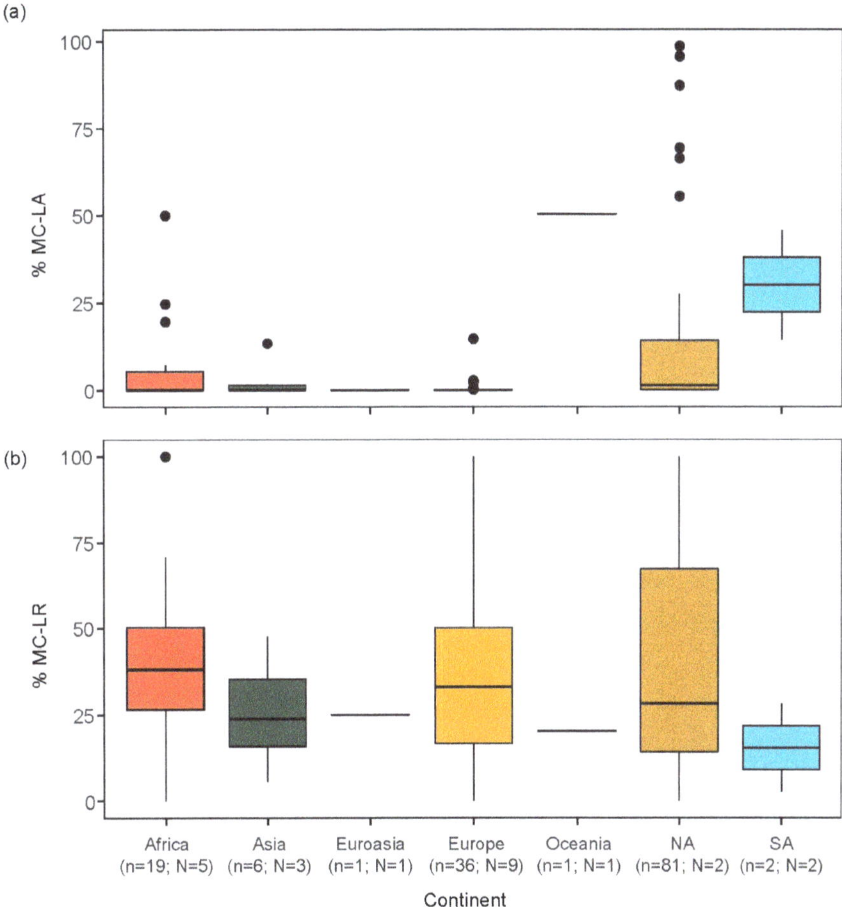

Figure 6. Summary of global meta-analysis of the peer-reviewed literature (OMECP raw intake sites were omitted). Boxplot of the percentage of (**a**) MC-LA and (**b**) MC-LR in different continents, where n = number of lakes and N = number of countries in each continent.

3. Discussion

3.1. Regional Relationship between Congener Occurrence and Environmental Conditions

Our regional analysis of the Laurentian Great Lakes showed that the variability in MC congener composition across raw water intake sites of Southern Ontario, Canada could be due to differences in meteorological conditions and lake nutrient status. MC-LA and MC-LR were the most commonly observed MC congeners in the region, though their dominance was distinct across the landscape (Figure 1). As environmental conditions changed (lower wind speed, decreased precipitation and higher temperatures in most recent years (Figures S3,S4)), the relative abundance of MC-LA and MC-LR also changed (increase proportion of -LA in western locations (Figure 2)), and the concentration of both MC congeners became increasingly variable among sampling dates from 2013 onwards (Figure S3). This observation led us to further assess whether environmental factors influenced MC congener prevalence. Our findings showed that when intermediate winds (average monthly wind speeds ranging from 36 to 51 km/hr) were coupled with wetter conditions, MC-LA tended to dominate. These conditions were typical of the raw water intake sites in Lake St. Clair (Figure 5c,d), as well as along the Detroit River and Lake Erie (Figure 1). In contrast, either weak wind (<36 km/hr), or stronger winds (>51 km/hr) coupled with warm conditions (>14 °C) and nutrient-rich waters (TP > 26 μg/L) were related to the dominance of MC-LR (and co-dominance of MC-RR). These conditions were typical of the Bay of Quinte (Figures 1 and 5a,b).

The overriding effect of monthly-averaged wind speed on MC composition in this regional dataset is noteworthy (Figure 4). Wind speed, through changes in turbulent mixing and water temperature, is known to affect phytoplankton and cyanobacteria species composition due to differences in buoyancy regulation and temperature optima [48]. That is, gas-vacuolated species can regulate their buoyancy and overcome settling to optimize resource acquisition during low wind speeds, strong stratification and high irradiance [41,49]. Michalak et al. [50] found that the combined effect of calm wind conditions, reduced lake mixing, increased nutrient loading and increased precipitation may have facilitated a record-breaking *Microcystis* bloom in Lake Erie in 2011, and furthermore, predicted increases in the frequency of such events in the future (50% increase in large storms with precipitation >30 mm under future climate models). Recently, Kelly et al. [51], found that lower wind speed (≤37 km/hr), combined with an increase in eutrophication indicators (Chl *a*, nutrients) and temperature, were associated with the increased probability of total MC concentrations exceeding drinking water standards (1.5 μg/L) in the Bay of Quinte. Given the decline in wind speed observed in recent years in the Laurentian Great Lakes region (Figure S4) and predictions of higher total annual precipitation [52], the increase in MC-LA dominating blooms may very well continue. However, MC-LA dominance was also observed in a small Ontario lake with no flushing (i.e., a closed-basin lake with no inflow or outflow [33]). Thus, the cumulative effect of environmental stressors (wind, precipitation, and nutrients) may vary across the landscape and interact with other factors such as lake morphometry. Furthermore, threshold wind speeds for complete mixing of the water column are likely to vary among water bodies and regions. In some lakes, winds greater than 29 km/hr were required to favour non-buoyant species [53,54], whereas other systems required stronger winds (>72 km/hr, [41]).

The likely reason for the observed relationship between environmental factors and MC congener dominance is that different congeners are produced by different cyanobacterial species or strains, which themselves vary along their ecological niche [55]. MC congener composition has been shown to vary between and within species (i.e., among strains), and to a lesser extent within strains. For example, within-strain variation in congener composition has been observed in response to changes in temperature [56], light availability [57,58] and nitrogen concentrations [59,60]. Different species may also produce many MC congeners concurrently. For instance, *Planktothrix* blooms have been dominated by both MC-LA (North American lakes [61]) and the -RR variant (European lakes [62–65]). Similarly, *Microcystis aeruginosa* has been associated with MC-LA dominance in some lakes (North America [33]) but -LR dominance in others (European lakes [66]).

A caveat of this regional study is the limitation of raw water intake data in terms of extrapolating to the surface water conditions. There may be times of the year (e.g., during storm events) when toxins measured at intake sites are partially or wholly derived from resuspended sediments, and may be less representative of surface water conditions. This is especially an issue given that MC-LA is more resilient to degradation and has lower sediment adsorption than MC-LR [32,67]. The differential resuspension and removal efficiency of MC congeners may thus vary with the depth of intake (greater exchange between surface and bottom waters in shallow sites). For the sites examined here however, we failed to detect any significant effect of depth of intake on MC congener composition, and all intake sites were relatively shallow (<12 m). Instead, we found that the distance from shore was more important (Figure 2), which may be related to the difference in water currents.

Chlorine treatment and dreissenid mussel occurrence could further decouple MC congener composition between surface and intake waters. Indeed, many water treatment plants chlorinate the raw water to serve as a chemical barrier to prevent dreissenid mussel veligers from clogging up intakes pipes as well as to prevent them from moving into the water treatment plant infrastructure. In this regional analysis, we detected a weak effect of chlorination on MC congener composition, which may be driven by the secondary effects of zebra mussel presence. There are reports, for instance in the US Midwest, of higher MC levels in dreissenid-infested water bodies compared to those without mussel [47]. This, in turn, may be linked to the changes in nutrient ratios and concentrations created by the mussels towards N:P ratios more suitable for toxic cyanobacteria growth [47,68–71]. Interestingly, distance from shore and chlorination were not selected by the MRT analysis. Wind speed and air temperature (>12 °C) were selected instead, though many water treatment plants only chlorinate when water temperatures exceeded 12 °C (i.e., during increased zebra mussel reproduction). Furthermore, although the effect of distance from shore and chlorine treatment on MC congener composition was less clear (Figure S8), the concentration of total MC within the four most frequently monitored water bodies tended to decrease with depth and distance from shoreline (notably so among the intake sites along Lakes Ontario and Erie (Figure S9a,b)) and increase with chlorination (or presence of zebra mussels Figure S5d). From a water treatment perspective, we suggest that more work is needed to determine the effect of chlorination and intake location on total MC concentration and on the relative composition of MC congeners, which may guide treatment optimization (such as using additional treatment methods at MC-LA dominated sites).

3.2. Global Relationships between Congener Occurrence and Environmental Conditions

Our meta-analysis of the peer-reviewed literature on MC congener dynamics identified significant differences in dominance patterns among continents. In general, the relative proportion of MC-LA was low for the lakes sampled in Africa, Europe, and Asia, but notably greater in lakes sampled in North and South America (Figure 6a). In contrast, MC-LR was more uniformly observed across sample sites (Figure 6b). In light of the relationships identified at the regional scale (Great Lakes and the surrounding region), we explored whether the difference in MC-LA occurrence among continents and countries was due in part to differences in environmental conditions among locations. We noted that climate (wind speed) was the most important driver of MC congener composition in the Great Lakes regions, and concordantly, MC-LA was more common in lakes with smaller surface area (i.e., North and South American lakes (Figure 6a, Figure S7)), which may modulate climate signals such as wind exposure [72,73]. We also noted that the effect of nutrients on regional congener composition was weaker, but still an important driver of MC-LR, with the latter more common at higher nutrient concentrations (Figure 5a). Similarly, on the global scale, we found that MC-LR was more common in eutrophic lakes (Figure S7). In general, despite the lack of a standardised protocol among the studies synthesizes in our meta-analysis, the global-scale relationships between lake morphometry, nutrient concentrations, and MC congener composition in the water column echoed the regional patterns observed in the Great Lakes Basin raw water intake data.

4. Conclusions

At the two geographical scales (regional and global) examined here, we identified strong spatial heterogeneity in congener composition, explained in part by routinely monitored environmental conditions. In particular, we found that:

- In the Laurentian Great Lakes, MC-LA was often more prevalent than MC-LR, and its concentration has increased over the last decade at several sites.
- The more toxic congeners (MC-LA and -LR) occurred under both low nutrient (MC-LA) and high nutrient (MC-LR) concentrations, while meteorological conditions (wind speed and precipitation) determined the relative concentration of each.
- Meso-oligotrophic waters with intermediate winds and frequent rain events showed greater percentage of MC-LA, while low winds or high winds combined with warm, nutrient-rich conditions showed greater percentage of MC-LR and -RR.
- Environmental conditions and related MC congener dominance were geographically distinct, with conditions that favoured MC-LA in the western part of our regional study (Lake St. Clair, Detroit River, and parts of Lake Erie), but windy, warm, eutrophic conditions that favoured MC-LR observed to the east in Lake Ontario's Bay of Quinte.
- Globally, MC-LA and MC-LR also showed geographically distinct patterns, with a smaller percentage of MC-LA in Africa, Europe and Asia, compared to North and South America. These patterns of MC congener dominance were associated with differences in lake morphometry and nutrient concentrations: MC-LA tended to be more prevalent in smaller lakes while MC-LR peaked in eutrophic lakes.

5. Materials and Methods

5.1. Great Lakes Regional Analysis

5.1.1. Analytical Data

Raw water samples were collected by water treatment plant personnel and sent to the Ontario Ministry of the Environment, Conservation and Parks (OMECP) for chemical analysis. For each raw water sample, the concentration of total MC (free and intracellular) was quantified by the OMECP laboratory technicians by isolation on silica gel and analysis by liquid chromatography-electrospray ionisation tandem mass spectrometry [74]. The concentrations of MC congeners (MC-LR, -RR, -LA, -YR, -LY, -WR, -HtyR, -HilR, -LW, -LF, and desmethyl-MC-LR, desmethyl-MC-RR) were measured using an internal standard quantification with multi-point calibration approach [74].

5.1.2. Statistical Analysis

The environmental factors considered as potential drivers of congener composition were meteorological condition (monthly averages of total precipitation (mm), wind direction (tens of degrees), wind speed (km/hr), and minimum, maximum and mean air temperature (°C) from nearby weather stations, Canadian Daily Climate Database) and nutrient concentrations (total phosphorus (TP) and total nitrogen (TN) from the raw water samples collected for MC analysis). Due to the proximity of Lake St. Clair and the Detroit River raw water intake sites (Figure 1), we used the same meteorological station for both water sources. We also considered the depth and distance from shore of the intake sites as these variables may track differences in water currents, light, and nutrient availability. Finally, we tested for any detectable effect of raw water pre-chlorination (used for dreissenid mussel control) as this may affect the concentration of total MC and MC congener composition (both via the direct effect of chlorine on the cyclic peptide structure and indirectly via secondary effects on dreissenid mussels [46,47]). Although the total chlorine concentration in raw water samples was measured by the OMECP, concentrations were often below detection (only detected in sites along Lake Ontario,

Lake Erie, and the Detroit River (Figure S5a)). To provide chlorine treatment data for all sample dates and sites, we thus used a categorical classification to indicate whether chlorination treatment was confirmed by the water treatment plants (Yes, No or turned Off prior to sample collection (Figure S5b)). This classification was an accurate indication of chlorine concentrations when detected (i.e., in Lake Ontario, Lake Erie, and the Detroit River (Figure S5c)). To create a variable that more broadly indicated chlorination treatment as well as the presence of dreissenid mussels, we grouped the one site where chlorination was turned off prior to sample collection (Union Water Treatment Plant along Lake Erie) with sites having continued chlorination.

To examine how the concentrations of all four congeners (MC-LR, -RR, -LA, and -YR) varied with changes in the environmental factors over time and space, we conducted a multivariate canonical ordination (redundancy analysis (RDA)) using the *rda* function from the {vegan} package in R [75]. To test for changes in the relative concentrations of MC congeners, we transformed the response matrix into relative abundances using the argument "total" of the *decostand* function. Although RDA examines the relationship between the multivariate response matrix and the suite of environmental factors, relationships are restricted to linear regressions. In addition, the method does not allow for missing data. We thus removed observations with missing environmental data prior to running the RDA.

Given that the RDA only tests for linear relationships between the multivariate response matrix and environmental drivers, we coupled this analysis with a multivariate regression tree (MRT, function *mvpart* in R [76,77]), which allows for non-linear and threshold responses [78]. In addition, MRTs allow for missing data. We used the cross-validation relative error (CVRE), which is the ratio of the variation unexplained by the tree to the total variation in the response, as the criterion for selecting the most parsimonious tree (i.e., the tree with the least splits whose CVRE value is within one standard error of the tree with the lowest CVRE [79]).

To further test whether similarity in meteorological and nutrient parameters among sites of closer proximity (Figure 1) could lead to the dominance of any particular MC congener, we ran a linear mixed effect regression tree (using the {glmertree} package in R [80]) with a random effect for year and site (main water body). By teasing apart the effect of co-location, this additional analysis evaluated whether the relationships observed with the multivariate regression tree on all sites and years were biased by the sampling design.

Lastly, to identify which congener was most often associated with which water body, we conducted an indicator species analysis using the indicator value index (function *indval* of the {labdsv} package [81]). Briefly, *indval* measures the fidelity and specificity of a "species" (congener) to a group (water body). Where specificity is defined as the mean abundance of the species within the targeted group compared to its mean abundance across all groups, and fidelity is the proportion of sites of the targeted group where the species is present [82]. The index is thus maximized when a species (congener) is observed at all sites belonging to a single group (water body), and not elsewhere [76].

5.2. Global Analysis

To assess the patterns in MC congener composition across many regions, we conducted a synthesis of published literature that provided MC congeners concentration data from freshwater lakes and reservoirs. Specifically, we conducted an ISI Web of Science and Google Scholar search using the keywords "microcystin", "congener", "cyanotoxin", "cyanobacteria", "lake", "eutrophication", "microcystin-LA", "MC-LA", "microcystin-LR" and/or "MC-LR" published between 2000 and 2017. We screened the studies using the criteria that: 1) Concentrations of MC congeners were measured (i.e., not just total MCs), 2) MC-LR, -LA, -RR and -YR were quantified (i.e., the use of the relevant standards was mentioned), and 3) data tables or graphs from which the raw data could be digitized were provided in each publication. A total of 146 sites which covered all continents (to the exception of Antarctica) met these search criteria (Table S1). Most studies presented their data as concentrations in the water column (μg/L), while some presented their data as seston content (g/dry weight). To analyze both types of concentration data, we calculated the composition as percentages of total MCs reported.

Supplementary Materials: The following are available online at http://www.mdpi.com/2072-6651/11/11/620/s1, Figure S1: Bar plot of average MC-congener concentrations across each main water body, Figure S2: Bar plots of yearly averaged (2006-2016) MC congener concentrations across each main water body, Figure S3: Boxplots of concentrations of MC–LA, MC–LR, total phosphorus (TP) and total nitrogen (TN) from the most frequently sampled water bodies, Figure S4: Boxplots of weather conditions from weather stations nearest to the most frequently sampled water bodies, Figure S5: Summary of chlorination treatment and chlorine concentration of the OMECP raw water intake sites, Figure S6: Multivariate regression tree (MRT) of congener relative abundance across the most frequently sampled water bodies of the Laurentian Great Lakes basin (excluding Total Phosphorus), Figure S7: Summary of microcystin data (boxplots and scatterplots) from the global dataset, Figure S8: Relationship between congener composition and raw water intake location and water treatment plant chlorination status, Figure S9: Relationship between total MC concentration and location of raw water intake sites, Table S1: Summary of microcystin literature review.

Author Contributions: Conceptualization, Z.E.T. and F.R.P.; statistical analysis, Z.E.T.; writing—Z.E.T., F.R.P., I.F.C., S.B.W., and A.Z.; supervision, F.R.P. and I.F.C.; project administration, F.R.P.; funding acquisition, Z.E.T., F.R.P. and I.F.C.

Funding: This research was funded by an NSERC postdoctoral grant, NSERC CREATE Algal Bloom Assessment through Technology & Education (ABATE) grant and a Best in Science grant from the Ontario Ministry of the Environment, Conservation and Parks (OMECP).

Acknowledgments: We thank Gillian Kingston, Patrick Cheung, Michelle Palmer, Claire Holeton and Jenny Kwong from the Environmental Monitoring and Reporting Branch of the Ontario Ministry of the Environment, Conservation and Parks for providing the raw data for analysis, as well as Xavier Ortiz for details on the OMECP's analytical methods for microcystin congeners. We are also most grateful for the help and information provided on water treatment facilities by Kyle Davis, Carolyn de Groot, Ainslie Timmons, Lindsay Ariss, John Armour, Bill Anderson, Susan Andrews, Sangeeta Chopra, Warren Higgins, Karen Burgess, Ryan Peterson, Amy Russell, Dean Walker, Sarah Clarke, Todd Harvey, Dale Dillen, John Hemingway, Paul Dyrda, Mike Purcell, John Hoos, Kayla Beach, Monica Reid, and Christa Paquette.

Conflicts of Interest: The authors declare no conflict of interest.

References

1. Paerl, H.W.; Paul, V.J. Climate change: Links to global expansion of harmful cyanobacteria. *Water Res.* **2012**, *46*, 1349–1363. [CrossRef] [PubMed]
2. Rastogi, R.P.; Sinha, R.P.; Incharoensakdi, A. The cyanotoxin-microcystins: Current overview. *Rev. Environ. Sci. Biotechnol.* **2014**, *13*, 215–249. [CrossRef]
3. Sukenik, A.; Quesada, A.; Salmaso, N. Global expansion of toxic and non-toxic cyanobacteria: Effect on ecosystem functioning. *Biodivers. Conserv.* **2015**, *24*, 889–908. [CrossRef]
4. Taranu, Z.E.; Gregory-Eaves, I.; Leavitt, P.R.; Bunting, L.; Buchaca, T.; Catalan, J.; Moorhouse, H. Acceleration of cyanobacteria dominance in north temperate-subarctic lakes during the Anthropocene. *Ecol. Lett.* **2015**, *18*, 375–384. [CrossRef]
5. Huisman, J.M.; Codd, G.A.; Paerl, H.S.; Ibelings, B.W.; Verspagen, J.M.H.; Visser, P.M. Cyanobacterial blooms. *Nat. Rev. Microbiol.* **2018**, *16*, 471–483. [CrossRef]
6. Harke, M.J.; Davis, T.W.; Watson, S.B.; Gobler, C.J. Nutrient-controlled niche differentiation of western Lake Erie cyanobacterial populations revealed via metatranscriptomic surveys. *Environ. Sci. Technol.* **2016**, *50*, 604–615. [CrossRef]
7. Pick, F.R. Blooming algae: A Canadian perspective on the rise of toxic cyanobacteria. *Can. J. Fish. Aquat. Sci.* **2016**, *73*, 1149–1158. [CrossRef]
8. Taranu, Z.E.; Gregory-Eaves, I.; Steele, R.; Beaulieu, M.; Legendre, P. Predicting microcystin occurrences in US lakes and reservoirs: A new framework for modeling the drivers of an important health risk factor. *Glob. Ecol. Biogeog.* **2017**, *26*, 625–637. [CrossRef]
9. Du, H.; Liu, H.; Yuan, L.; Wang, Y.; Ma, Y.; Wang, R.; Chen, X.; Losiewicz, M.D.; Guo, H.; Zhang, H. The diversity of cyanobacterial toxins on structural characterization, distribution and identification: A systematic review. *Toxins* **2019**, *11*, 530. [CrossRef]
10. Spoof, L.; Catherine, A. Appendix 3: Tables of microcystins and nodularins. In *Handbook of Cyanobacterial Monitoring and Cyanotoxin Analysis*; John Wiley & Sons, Ltd.: West Sussex, UK, 2017; pp. 526–537.
11. Beversdorf, L.J.; Rude, K.; Weirich, C.A.; Bartlett, S.L.; Seaman, M.; Kozik, C.; Biese, P.; Gosz, T.; Suha, M.; Stempa, C.; et al. Analysis of cyanobacterial metabolites in surface and raw drinking waters reveals more than microcystin. *Water Res.* **2018**, *140*, 280–290. [CrossRef]

12. Mantzouki, E.; Lürling, M.; Fastner, J.; de Senerpont Domis, L.; Wilk-Woźniak, E.; Koreivienė, J.; Walusiak, E. Temperature effects explain continental scale distribution of cyanobacterial toxins. *Toxins* **2018**, *10*, 156. [CrossRef] [PubMed]
13. Huisman, J.M.; Matthijs, H.C.P.; Visser, P.M. *Harmful Cyanobacteria*; Springer: Dordrecht, The Netherlands, 2005.
14. Miller, M.A.; Kudela, R.M.; Mekebri, A.; Crane, D.; Oates, S.C.; Tinker, M.T.; Hardin, D. Evidence for a novel marine harmful algal nloom: Cyanotoxin (microcystin) transfer from land to sea otters. *PLoS ONE* **2010**, *5*, e12576. [CrossRef] [PubMed]
15. Lévesque, B.; Gervais, M.C.; Chevalier, P.; Gauvin, D.; Anassour-Laouan-Sidi, E.; Gingras, S.; Fortin, N.; Brisson, G.; Greer, C.; Bird, D. Prospective study of acute health effects in relation to exposure to cyanobacteria. *Sci. Total Environ.* **2013**, *466*, 397–403. [CrossRef] [PubMed]
16. Carmichael, W.W. Cyanobacterial secondary metabolites–the cyanotoxins. *J. Appl. Bacteriol.* **1992**, *72*, 445–459. [CrossRef]
17. Ueno, T.; Nagata, S.; Tsutsumi, T.; Hasegawa, A.; Yoshida, F.; Suttajit, M.; Mebs, D.; Putsch, M.; Vasconcelos, V. Detection of microcystins, a blue-green algal hepatotoxin, in drinking water sampled in Haimen and Fusui, endemic areas of primary liver cancer in China, by highly senstive immunoassay. *Carcinogenesis* **1996**, *17*, 1317–1321. [CrossRef]
18. Carmichael, W.W. The cyanotoxins. *Adv. Bot. Res.* **1997**, *27*, 211–256.
19. Zhang, F.; Lee, J.; Shum, C.K. Cyanobacteria blooms and non-alcoholic liver disease: Evidence from a county level ecological study in the United States. *Environ. Health* **2015**, *14*, 41. [CrossRef]
20. Sotton, B.; Guillard, J.; Anneville, O.; Maréchal, M.; Savichtcheva, O.; Domaizon, I. Trophic transfer of microcystins through the lake pelagic food web: Evidence for the role of zooplankton as a vector in fish contamination. *Sci. Total Environ.* **2014**, *466*, 152–163. [CrossRef]
21. Ettoumi, A.; El Khalloufi, F.; El Ghazali, I.; Oudra, B.; Amrani, A.; Nasri, H.; Bouaïcha, N. Bioaccumulation of cyanobacterial toxins in aquatic organisms and its consequences for public health. In *Zooplankton and Phytoplankton: Types, Characteristics and Ecology*; Kattel, G., Ed.; Nova Science Publishers, Inc.: New York, NY, USA, 2011.
22. Hardy, F.J.; Johnson, A.; Hamel, K.; Preece, E.P. Cyanotoxin bioaccumulation in freshwater fish, Washington state, USA. *Environ. Monit. Assess.* **2015**, *187*, 667. [CrossRef]
23. Chorus, I. Cyanotoxin occurrence in freshwaters: A summary of survey results from different countries. In *Cyanotoxins: Occurrence, Causes, Consequences*; Springer: Berlin, Germany, 2001.
24. Graham, J.L.; Loftin, K.A.; Meyer, M.T.; Ziegler, A.C. Cyanotoxin mixtures and taste and odour compounds in cyanobacterial blooms from the Midwestern United States. *Environ. Sci. Technol.* **2010**, *44*, 7361–7368. [CrossRef]
25. Jones, G.J.; Falconer, I.F.; Wilkins, R.M. Persistence of cyclic peptide toxins in dried cyanobacterial crusts from Lake Mokoan, Australia. *Environ. Toxicol. Water Qual.* **1995**, *10*, 19–24. [CrossRef]
26. Falconer, I.; Bartram, J.; Chorus, I.; Kuiper-Goodman, T.; Utkilen, H.; Burch, M.; Codd, G. Safe levels and practices. In *Toxic Cyanobacteria in Water, A Guide to Their Public Health Consequences, Monitoring and Management, Chorus, I., Bartam, J., Eds.*; E & FN Spon: London, UK, 1999; pp. 155–178.
27. Srivastava, A.; Choi, G.G.; Ahn, C.Y.; Oh, H.M.; Ravi, A.K.; Asthana, R.K. Dynamics of microcystin production and quantification of potentially toxigenic Microcystis sp. using real-time PCR. *Water Res.* **2012**, *46*, 817–827. [CrossRef] [PubMed]
28. Monchamp, M.-E.; Pick, F.R.; Beisner, B.E.; Maranger, R. Nitrogen forms influence microcystin concentration and composition via changes in cyanobacterial community structure. *PLoS ONE* **2014**, *9*, e85573. [CrossRef] [PubMed]
29. Zastepa, A.; Taranu, Z.E.; Kimpe, L.E.; Blais, J.M.; Gregory-Eaves, I.; Pick, F.R. Reconstructing a long-term record of microcystins from the analysis of lake sediments. *Sci. Total Environ.* **2017**, *579*, 893–901. [CrossRef] [PubMed]
30. Health Canada. Guidelines for Canadian Drinking Water Quality: Guideline Technical Document-Cyanobacterial Toxins. 2018. Available online: https://www.canada.ca/en/health-canada/services/publications/healthy-living/guidelines-canadian-drinking-water-quality-guideline-technical-document-cyanobacterial-toxins-document.html (accessed on 25 October 2019).
31. Chorus, I.; Bartram, J. *Toxic Cyanobacteria Water: A Guide Public Health Conseq. Monitoring. Manag.*; E.&FN Spon: London, UK, 1999.

32. Newcombe, G.; Cook, D.; Brooke, S.; Ho, L.; Slyman, N. Treatment options for microcystin toxins: Similarities and differences between variants. *Environ. Technol.* **2003**, *24*, 299–308. [CrossRef] [PubMed]
33. Zastepa, A.; Pick, F.R.; Blais, J. Fate and persistence of particulate and dissolved microcystin-LA from *Microcystis* blooms. *Hum. Ecol. Risk Assess.* **2014**, *20*, 1670–1686. [CrossRef]
34. Namikoshi, M.; Rinehart, K.L.; Sakai, R.; Stotts, R.R.; Dahlem, A.M.; Beasley, V.R.; Carmichael, W.W.; Evans, W.R. Identification of 12 hepatotoxins from a Homer Lake bloom of the cyanobacteria *Microcystis aeruginosa*, *Microcystis viridis*, and *Microcystis wesenbergii*: Nine new microcystins. *J. Org. Chem.* **1992**, *57*, 866–872. [CrossRef]
35. Namikoshi, M.; Sun, F.; Choi, B.W.; Rinehart, K.L.; Carmichael, W.W.; Evans, W.R.; Beasley, V.R. Seven more microcystins from Homer Lake cells: Application of the general method for structure assignment of peptides containing α,β-dehydroamino acid unit(s). *J. Org. Chem.* **1995**, *60*, 3671–3679. [CrossRef]
36. Puddick, J.; Prinsep, M.R.; Wood, S.A.; Kaufononga, S.A.; Cary, S.C.; Hamilton, D.P. High levels of structural diversity observed in microcystins from *Microcystis* CAWBG 11 and characterization of six new microcystin congeners. *Mar. Drug.* **2014**, *12*, 5372–5395. [CrossRef]
37. Ibelings, B.W.; Backerb, L.C.W.; Kardinaa, E.A.; Chorus, I. Current approaches to cyanotoxin risk assessment and risk management around the globe. *Harmful Algae* **2014**, *40*, 63–74. [CrossRef]
38. Fischer, A.; Hoeger, S.J.; Stemmer, K.; Feurstein, D.J.; Knobeloch, D.; Nussler, A.; Dietrich, D.R. The role of organic anion transporting polypeptides (OATPs/SLCOs) in the toxicity of different microcystin congeners in vitro: A comparison of primary human hepatocytes and OATP-transfected HEK293 cells. *Toxicol. Appl. Pharm.* **2010**, *245*, 9–20. [CrossRef] [PubMed]
39. Jetoo, S.; Grover, V.I.; Krantzberg, G. The Toledo drinking water advisory: Suggested application of the water safety planning approach. *Sustainability* **2015**, *7*, 9787–9808. [CrossRef]
40. Jones, G.J.; Orr, P. Release and degradation of microcystin following algicide treatment of a *Microcystis aeruginosa* bloom in a recreational lake, as determined by HPLC and protein phosphatase inhibition assay. *Water Res.* **1994**, *28*, 871–876. [CrossRef]
41. Walsby, A.E.; Hayes, P.K.; Boje, R.; Stal, L.J. The selective advantage of buoyancy provided by gas vesibles for planktonic cyanobacteria in the Baltic Sea. *New Phytol.* **1997**, *136*, 407–417. [CrossRef]
42. Wagner, C.; Adrian, R. Cyanobacteria dominance: Quantifying the effects of climate change. *Limnol. Oceanogr.* **2009**, *54*, 2460–2468. [CrossRef]
43. Posch, T.; Köster, O.; Salcher, M.M.; Pernthaler, J. Harmful filamentous cyanobacteria favoured by reduced water turnover with lake warming. *Nat. Clim. Chang.* **2012**, *2*, 809–813. [CrossRef]
44. Downing, J.A.; Watson, S.B.; McCauley, E. Predicting cyanobacteria dominance in lakes. *Can. J. Fish. Aquat. Sci.* **2001**, *58*, 1905–1908. [CrossRef]
45. Dolman, A.M.; Rücker, J.; Pick, F.R.; Fastner, J.; Rohrlack, T.; Mischke, U.; Wiedner, C. Cyanobacteria and cyanotoxins: The influence of nitrogen versus phosphorus. *PLoS ONE* **2012**, *7*, e38757. [CrossRef]
46. Rosenblum, L.; Zaffiro, A.; Adams, W.A.; Wendelken, S.C. Effect of chlorination by-producs on the quantitation of microcystins in finished drinking water. *Toxicon* **2017**, *138*, 138–144. [CrossRef]
47. Knoll, L.B.; Sarnelle, O.; Hamilton, S.K.; Kissman, C.E.H.; Wilson, A.E.; Rose, J.B.; Morgan, M.R. Invasive zebra mussels (*Dreissena polymorpha*) increase cyanobacterial toxin concentrations in low-nutrient lakes. *Can. J. Fish. Aquat. Sci.* **2008**, *65*, 448–455. [CrossRef]
48. Jöhnk, K.D.; Huisman, J.; Sharples, J.; Sommeijer, B.; Visser, P.M.; Strooms, J.M. Summer heatwaves promote blooms of harmful cyanobacteria. *Glob. Chang. Biol.* **2008**, *14*, 495–512. [CrossRef]
49. Dokulil, M.T.; Teubner, K. Cyanobacterial dominance in lakes. *Hydrobiol.* **2000**, *438*, 1–12. [CrossRef]
50. Michalak, A.M.; Anderson, E.J.; Beletsky, D.; Boland, S.; Bosch, N.S. Record-setting algal bloom in Lake Erie caused by agricultural and meteorological trends consistent with expected future conditions. *Proc. Natl. Acad. Sci. USA* **2013**, *110*, 6448–6452. [CrossRef] [PubMed]
51. Kelly, N.E.; Javed, A.J.; Shimoda, Y.; Zastepa, A.; Watson, S.; Mugalingam, S.; Arhonditsis, G.B. A Bayesian risk assessment framework for microcystin violations of drinking water and recreational standards in the Bay of Quinte, Lake Ontario, Canada. *Water Res.* **2019**, *162*, 288–301. [CrossRef] [PubMed]
52. Bartolai, A.M.; He, L.; Hurst, A.E.; Mortsch, L.; Paehlke, R.; Scavia, D. Climate change as a driver of change in the Great Lakes St. Lawrence River basin. *J. Great Lakes Res.* **2015**, *41*, 45–58. [CrossRef]
53. Webster, I.T.; Hutchinson, P.A. Effects of wind on the distribution of phytoplankton cells in lakes-revisited. *Limnol. Oceanogr.* **1994**, *39*, 365–373. [CrossRef]

54. Šejnohová, L.; Maršálek, B.; Microcystin. *Ecology of Cyanobacteria II: Their Diversity in Space and Time*; Whitton, B.A., Ed.; Springer: Berlin/Heidelberg, Germany, 2012; pp. 195–228.
55. Tromas, N.; Taranu, Z.E.; Martin, B.D.; Willis, A.; Fortin, N.; Greer, C.W.; Shapiro, B.J. Niche separation increases with genetic distance among bloom-forming cyanobacteria. *Front. Microbiol.* **2018**. [CrossRef]
56. Amé, M.; Wunderlin, D. Effects of iron, ammonium and temperature on microcystin content by a natural concentrated *Microcystis aeruginosa* population. *Water Air Soil Pollut.* **2005**, *168*, 235–248. [CrossRef]
57. Tonk, L.; Visser, P.M.; Christiansen, G.; Dittmann, E.; Snelder, E.O.F.M.; Wiedner, C.; Mur, L.R.; Huisman, J. The microcystin composition of the cyanobacterium *Planktothrix agardhii* changes toward a more toxic variant with increasing light intensity. *Appl. Environ. Microbiol.* **2005**, *71*, 5177–5181. [CrossRef]
58. Agha, R.; Cirés, S.; Wömer, L.; Quesada, A. Limited stability of microcystins in oligopeptide compositions of *Microcystis aeruginos* (Cyanobacteria): Implications in the definition of chemotypes. *Toxins* **2013**, *5*, 1089–1104. [CrossRef]
59. Van de Waal, D.B.; Ferreruela, G.; Tonk, L.; Van Donk, E.; Huisman, J.; Visser, P.M.; Matthijs, H.C. Pulsed nitrogen supply induces dynamic changes in the amino acid composition and microcystin production of the harmful cyanobacterium *Planktothrix agardhii. FEMS Microbiol. Ecol.* **2010**, *74*, 430–438. [CrossRef] [PubMed]
60. Puddick, J.; Prinsep, M.R.; Wood, S.A.; Cary, S.C.; Hamilton, D.P. Modulation of microcystin congener abundance following nitrogen depletion of a *Microcystis* batch culture. *Aquat. Ecol.* **2016**, *50*, 235–246. [CrossRef]
61. Rinta-Kanto, J.M.; Wilhelm, S.W. Diversity of microcystin-producing cyanobacteria in spatially isolated regions of Lake Erie. *Appl. Environ. Microbiol.* **2006**, *72*, 5083–5085. [CrossRef] [PubMed]
62. Kutovaya, O.A.; McKay, R.M.L.; Beall, B.F.N.; Wihlem, S.W.; Kane, D.D.; Chaffin, J.D.; Bridgeman, T.B.; Bullerjahn, G.S. Evidence against fluvial seeding of recurrent toxic blooms of *Microcystis* spp. in Lake Erie's Western basin. *Harmful Algae* **2012**, *15*, 71–77. [CrossRef]
63. Rohrlack, T.; Bente, E.; Skulberg, R.; Halstvedt, C.B.; Utkilen, H.C.; Ptacnik, R.; Skulberg, O.M. Oligopeptide chemotypes of the toxic freshwater cyanobacterium *Planktothrix* can form sub-populations with dissimilar ecological traits. *Limnol. Oceanogr.* **2008**, *53*, 1279–1293. [CrossRef]
64. Bogialli, S.; di Gregorio, F.N.; Lucentini, L.; Ferretti, E.; Ottaviani, M.; Ungaro, N.; Cannarozzi de Grazia, M. Management of a toxic cyanobacterium bloom (*Planktothrix rubescens*) affecting an Italian drinking water basin: A case study. *Environ. Sci. Technol.* **2013**, *47*, 574–583. [CrossRef] [PubMed]
65. Ferranti, P.; Fabbrocino, S.; Chiaravalle, E.; Bruno, M.; Adriana Basile, A.; Serpe, L.; Gallo, P. Profiling microcystin contamination in a water reservoir by MALDI-TOF and liquid chromatography coupled to Q/TOF tandem mass spectrometry. *Food Res. Int.* **2013**, *54*, 1321–1330. [CrossRef]
66. Van de Waal, D.B.; Verspagen, J.M.; Lürling, M.; Van Donk, E.; Visser, P.M.; Huisman, J. The ecological stoichiometry of toxins produced by harmful cyanobacteria: An experimental test of the carbon-nutrient balance hypothesis. *Ecol. Lett.* **2009**, *12*, 1326–1335. [CrossRef]
67. Zastepa, A.; Pick, F.; Blais, J.M. Distribution and flux of microcystin congeners in lake sediments. *Lake Res. Manag.* **2017**, *33*, 444–451. [CrossRef]
68. Conroy, J.D.; Edwards, W.J.; Pontius, R.A.; Kane, D.D.; Zhang, H.; Shea, J.F.; Richey, J.N.; Culver, D.A. Soluble nitrogen and phosphorus excretion of exotic freshwater mussels (*Dreissena* spp.): Potential impacts for nutrient remineralization in western Lake Erie. *Freshw. Biol.* **2005**, *50*, 1146–1162. [CrossRef]
69. Bykova, O.; Laursen, A.; Bostan, V.; Bautista, J.; McCarthy, L. Do zebra mussels (*Dreissena polymorpha*) alter lake water chemistry in a way that favors *Microcystis* growth? *Sci. Total Environ.* **2006**, *371*, 362–372. [CrossRef] [PubMed]
70. Sarnelle, O.; White, J.D.; Horst, G.P.; Hamilton, S.K. Phosphorus addition reverses the positive effect of zebra mussels (*Dreissena polymorpha*) on the toxic cyanobacterium, *Microcystis aeruginosa. Water Res.* **2011**, *46*, 3471–3478. [CrossRef] [PubMed]
71. Harris, T.D.; Wilhelm, F.; Graham, J.L.; Loftin, K.A. Experimental manipulation of TN:TP ratios suppress cyanobacterial biovolume and microcystin concentration in large-scale in situ mesocosms. *Lake Res. Menag.* **2014**, *30*, 72–83. [CrossRef]
72. Kalff, J. *Limnology: Inland Water Ecosystems*; Upper: Saddle River, NJ, USA, 2002.
73. Håkanson, L.A. *Manual of Lake Morphometry*; Springer: Berlin, Germany, 1981.
74. Ortiz, X.; Korenkova, E.; Jobst, K.J.; MacPherson, K.A.; Reiner, E.J. A high throughput targeted and non-targeted method for the analysis of microcystins and anatoxin-A using on-line solid phase extraction

coupled to liquid chromatography-quadrupole time-of-flight high resolution mass spectrometry. *Anal. Bioanal. Chem.* **2017**, *409*, 4959–4969. [CrossRef]

75. Oksanen, J.; Blanchet, F.G.; Friendly, M.; Kindt, R.; Legendre, P.; McGlinn, D.; Minchin, P.R.; O'Hara, R.B.; Simpson, G.L.; Solymos, P.; et al. vegan: Community Ecology Package. R package, version 2.5-4. 2019. Available online: https://CRAN.R-project.org/package=vegan (accessed on 25 October 2019).
76. Legendre, P.; Legendre, L.F. *Numerical Ecology*; Elsevier: Amsterdam, The Netherlands, 2002; pp. 209–213.
77. Therneau, T.M.; Atkinson, B. mvpart: Multivariate partitioning. R package, version 1.6-2. 2014. Available online: https://CRAN.R-project.org/package=mvpart (accessed on 25 October 2019).
78. Ouellette, M.-H.; Legendre, P. MVPARTwrap: Additional features for package mvpart. R package, version 0.1-9.2. 2013. Available online: https://CRAN.R-project.org/package=MVPARTwrap (accessed on 25 October 2019).
79. De'Ath, G. Multivariate regression trees: A new technique for modeling species–environment relationships. *Ecology* **2009**, *83*, 1105–1117.
80. Fokkema, M.; Smits, N.; Zeileis, A.; Hothorn, T.; Kelderman, H. Detecting Treatment-Subgroup Interactions in Clustered Data with Generalized Linear Mixed-Effects Model Trees. Working Papers in Economics and Statistics, Research Platform Empirical and Experimental Economics, Universitaet Innsbruck. Available online: http://EconPapers.RePEc.org/RePEc:inn:wpaper:2015-10 (accessed on 25 October 2019).
81. Dufrêne, M.; Legendre, P. Species assemblages and indicator species: The need for a flexible asymmetrical approach. *Ecol. Monogr.* **1997**, *67*, 345–366. [CrossRef]
82. Borcard, D.; Gillet, F.; Legendre, P. *Numerical Ecology with R*, 2nd ed.; Springer: Berlin/Heidelberg, Germany, 2018; p. 453.

Review

Microcystin Incidence in the Drinking Water of Mozambique: Challenges for Public Health Protection

Isidro José Tamele [1,2,3] and Vitor Vasconcelos [1,4,*]

1 CIIMAR/CIMAR—Interdisciplinary Center of Marine and Environmental Research, University of Porto, Terminal de Cruzeiros do Porto, Avenida General Norton de Matos, 4450-238 Matosinhos, Portugal; isitamele@gmail.com

2 Institute of Biomedical Science Abel Salazar, University of Porto, R. Jorge de Viterbo Ferreira 228, 4050-313 Porto, Portugal

3 Department of Chemistry, Faculty of Sciences, Eduardo Mondlane University, Av. Julius Nyerere, n 3453, Campus Principal, Maputo 257, Mozambique

4 Faculty of Science, University of Porto, Rua do Campo Alegre, 4069-007 Porto, Portugal

* Correspondence: vmvascon@fc.up.pt; Tel.: +351-223-401-817; Fax: +351-223-390-608

Received: 6 May 2020; Accepted: 31 May 2020; Published: 2 June 2020

Abstract: Microcystins (MCs) are cyanotoxins produced mainly by freshwater cyanobacteria, which constitute a threat to public health due to their negative effects on humans, such as gastroenteritis and related diseases, including death. In Mozambique, where only 50% of the people have access to safe drinking water, this hepatotoxin is not monitored, and consequently, the population may be exposed to MCs. The few studies done in Maputo and Gaza provinces indicated the occurrence of MC-LR, -YR, and -RR at a concentration ranging from 6.83 to 7.78 $\mu g \cdot L^{-1}$, which are very high, around 7 times above than the maximum limit (1 $\mu g \cdot L^{-1}$) recommended by WHO. The potential MCs-producing in the studied sites are mainly *Microcystis* species. These data from Mozambique and from surrounding countries (South Africa, Lesotho, Botswana, Malawi, Zambia, and Tanzania) evidence the need to implement an operational monitoring program of MCs in order to reduce or avoid the possible cases of intoxications since the drinking water quality control tests recommended by the Ministry of Health do not include an MC test. To date, no data of water poisoning episodes recorded were associated with MCs presence in the water. However, this might be underestimated due to a lack of monitoring facilities and/or a lack of public health staff trained for recognizing symptoms of MCs intoxication since the presence of high MCs concentration was reported in Maputo and Gaza provinces.

Keywords: drinking water quality; microcystin; Mozambique; public health

Key Contribution: This review will contribute to the implementation of an operational monitoring program of MCs in order to reduce or avoid the possible cases of intoxications since the drinking water quality monitoring protocol recommended by the Ministry of Health does not include MC tests.

1. Introduction

Mozambique (Figure 1) is a country located in southeastern Africa (10°30′–26°52′ S and 40°50′–30°31′ E) covering a total land area of 800,000 km^2. It is bathed by the Indian Ocean in the east and makes borders with Tanzania in the north; Malawi and Zambia in the northwest; Zimbabwe in the west and Swaziland and South Africa in the southwest.

Figure 1. Map of Mozambique. Red points or lines indicate the sites where Microcystin (MC) or MC producers were detected in Mozambique, and in near sites or in the shared rivers with Mozambique, Green points indicate the water sources or water treatment centers.

According to IV populational census carried out in 2017, this country has 29.67 million habitants distributed in 11 provinces [1]. The climate of Mozambique varies from subtropical climates (north and center) to dry arid (south) [2]. Like many African countries, Mozambique is highly vulnerable to climate variability and extreme weather events (droughts, floods, and tropical cyclones) [3]. Droughts are the most frequent natural disaster that have a negative impact on the population that reside in these rural areas [4]. The location of Mozambique in the coastal area makes it vulnerable to floods since many transnational river basins end [2]. Unfortunately, only 50% of the population has access to "safe drinking water". Urban areas are the most favored, with 80%, while rural and most of the population have only 35% coverage and consume untreated water daily from rivers, lakes, and small puddles that form after or during the rain [5–7], putting at risk public health.

Eutrophication of freshwater resources may lead to the occurrence of cyanobacterial blooms and the presence of cyanotoxins, being microcystins (MC), the most common toxins worldwide [8]. The presence of MC in untreated drinking water is a major threat to public health because this potent cyanotoxin causes hepatotoxicity in humans. Thus, this review evaluates the incidence of MC and its producers in drinking water bodies of Mozambique, based on reported and available data and the estimated human illness case numbers and associated economic damage caused by Microcystin both overall globally, and in Mozambique—Africa. Recommendations for routine control and monitoring of MC will also be done since this hepatotoxin is not included in water control tests data in Mozambique.

2. Microcystin-Producing Species and Toxicology

2.1. Microcystin-Producing Species

MCs are secondary metabolites produced by cyanobacteria species that occur naturally (but it can be increased severely by human activities) in freshwater environments. The most reported cyanobacteria species, which produce MCs are listed in Table 1 and include species of the families *Microcystaceae, Nostocaceae, Microcoleaceae, Oscillatoriaceae, Pseudanabaenaceae* (Table 1). The occurrence and development of a particular genus and species of cyanobacteria and cyanotoxins production worldwide seem to be conditioned to water chemistry and climate conditions [8]. In a temperate climate, *Microcystis* and *Anabaena* blooms occur widely while *Cylindrospermopsis* develops in tropical regions [9]. There are toxic and non-toxic cyanobacteria of the same species, which may be found together [8,10,11]. Toxic cyanobacteria can produce several toxins with different toxicity making it uncertain to assess the overall toxicity of bloom due to the variations of toxins concentration spatially and seasonally [12]. To distinguish toxic and non-toxic cyanobacteria species is very complicated, and consequently, the methods used are also complex. It implicates that the prevention of cyanobacteria bloom development is a suitable way to control toxic blooms [13,14].

Table 1. Microcystin-producing species detected in freshwater bodies.

Order	Family	Species
Chroococcales	*Microcystaceae*	*Microcystis* sp. [15], M. *aeruginosa* [16–26], *M. viridis* [20,24], *M. wesenbergii* [25,27], *M.* spp. [28,29], *M. ichthyoblabe* [25] and *Synechocystis* sp. [25]
Nostocales	*Nostocaceae*	*Anabaena* spp. [29,30], *A. flos-aquae* [23,27], *A.* sp. [15,31], *A. subcylindrica* [32], *A, variables* [32], *Nostoc* sp. [27], *Aphanizomenon flos-aquae* [23,29,33] and *A. circinalis* [34]
Oscillatoriales	*Microcoleaceae*	*Planktothrix prolifica* [24] and *P. agardhii* [29]
	Oscillatoriaceae	*Oscillatoria agardhii* [35], *O. limosa* [36], *O. chlorina* [25], *Phormidium konstantinosum* (*O. tenuis*) [36], *P. corium* [32] and *Plectonema boryanum* [32]
Synechococcales	*Merismopediaceae*	*Synechocystis aquatilis f. salina* [37] and *Aphanocapsa cumulus* [38]
	Pseudanabaenaceae	*Pseudanabaena mucicola* [25] and *P. galeata* [25]

The factors that promote the MC synthesis are not yet clearly understood, however, the optimal growth of MC-producing species and toxicity seem influenced by light intensity, nutrients, and temperature, among other factors. For example, the higher toxicity of *M. aeruginosa* extracts was verified in extreme pH values [39,40], and heavy metals such as Zinc and Iron did not influence the *M. aeruginosa* toxicity [41]. The content of nitrogen and phosphorus influenced the toxicity of *M. aeruginosa* extracts. Low nitrogen content reduces the *M. aeruginosa* toxicity, while low phosphorous increased the toxicity in the natural population [42,43] and reduced in lab experiments [16,21,44,45]. Another lab conclusion was the correlation of colony size and content of toxic cyclic heptapeptide of the non-axenic strain of *M. viridis* and axenic *M. viridis* was also verified [20,46,47]. In general, the optimal temperature for which MC-producing species produce MC ranged from 20 to 25 °C [21,40,48,49]. This range of optimal temperature suggests that cyanobacteria blooms are most toxic during periods with warm weather and in areas with warm climates [8].

2.2. Toxicology

Microcystins (Figure 2) are the largest diverse group of cyanobacterial toxins, and to date, more than 240 MCs analogs are known, and they vary structurally in terms of the degree of methylation, hydroxylation, epimerization, peptide sequence, and consequently in their toxic effects [50–52]. Chemically, MC is a group of monocyclic heptapeptides (numbered in Figure 2) containing both D-

and L-amino acids plus N-methyldehydroalanine (Mdha) and a unique β-amino acid side-group, 3-amino-9-methoxy-2-6,8-trymethyl-10-phenyldeca-4,6-dienoic acid (Adda) and their analogs differ among them, at the two L-amino acids and on the methyl groups on D-erythro-β-methylaspartic acid (D-MeAsp) and Mdha with molecular weight varying from 900 to 1100 Daltons. MC-LR, MC-RR, and MC-YR are common MC variants, the letters L, R, and Y represent the aminoacids leucine, arginine, and tyrosine, which appear on the MC molecule in different combinations [50,53–58] being MC-LR the most studied. The biosynthesis of this group of cyanotoxin is regulated by non-ribosomal peptide synthetase and polyketide synthase domains, being *MCyS* the gene cluster, which has been sequenced and partially characterized in several cyanobacterial species of the family *Microcystaceae, Nostocaceae, Microcoleaceae, Oscillatoriaceae, Merismopediacea,* and *Pseudanabaenacea* [16–38,59,60]

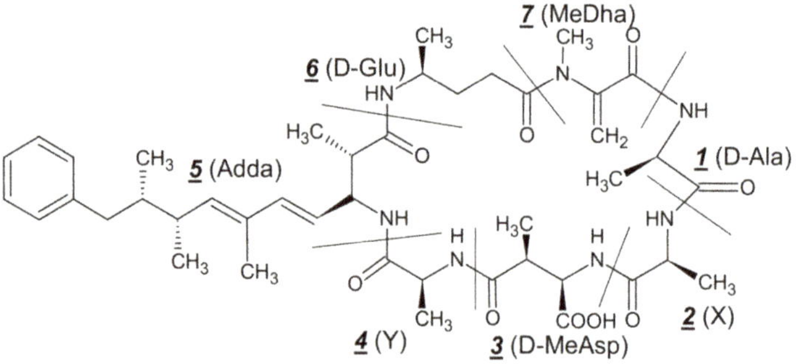

Figure 2. General chemical structure of microcystins. The common MC variant is MC-LR when X and Y correspond to L-Leu and L-Arg.

The mechanism of MCs toxicity seems to be well understood. They bind to serine/threonine-specific protein phosphatases (PPs) such as PP1 and PP2A, inhibiting their activity [61–63]. Adda moiety (Figure 1) plays an important role in the MC toxicity group since its isomerization and/or oxidation reduces the toxicity [64,65]. The inhibition of PP1and PP2A as a result of MC acute exposure causes excessive protein phosphorylation, alterations in the cytoskeleton, loss of cell shape, and consequently destruction of liver cells leading to intrahepatic hemorrhage or hepatic insufficiency [58]. Oxidative stress increasing in cells and consequent apoptosis, which can cause tumor promotion, is another mechanism of MC toxicity [66–68].

3. Effects of Microcystin in Humans, Symptoms, and Treatment

Microcystin effects in humans depend on the time of exposure and concentration ingested [69], and the studies are based on epidemiologic data but the reported studies on laboratory animals. Human health problems are mostly caused by chronic exposure by consumption of contaminated water or food, dermal exposure, or inhalation [57]. MC human poisoning episodes were reported in different parts of the world after the consumption of contaminated water or during sport or recreational activities [59,70–72]. Some examples of MC human poisoning cases are described; America—in 1996, an episode of human intoxications by MC was reported in Brazil with more than 76 deaths of patients at two dialysis centers in Caruaru. The municipal water supplied to the dialysis centers was the source of MC [71,73–75]. In Argentina, a human poisoning caused by MC involving a young man after immersion in an intense bloom *Microcystis* sp lake during sport and recreational activities were recorded. Four hours after exposure, the patient showed nausea, abdominal pain, and fever, and 48.6 $\mu g \cdot L^{-1}$ of microcystin-LR was detected in the water samples [76]. Other cases were recorded in Uruguay (January 2015) involving a 20-month-old child and her family during recreational activities. These victims were admitted to the hospital with diarrhea, vomiting, fatigue, and jaundice and the analysis confirmed the presence of MC-LR (2.4 $ng \cdot g^{-1}$ tissue) and [D-Leu1]MC-LR (75.4 $ng \cdot g^{-1}$ tissue) explanted liver [77]. Africa—toxic cyanobacteria suspected intoxication cases were reported in

Zimbabwe involving children that were hospitalized in the Hospital of Harare with gastroenteritis symptoms [78]. In Europe—121 people presented abdominal pain, nausea, vomiting, diarrhea, fever, headaches, and muscle pain after consumption of untreated water from the River Kavlingean in Sweden. In this case a bloom of MC—producing such as *Planktothrix agardhii* and *Microcystis* spp. was observed in the river [79]. The most affected human organ is the liver [57]. However, in vivo and in vitro studies indicated that the kidney and colon are also affected [80–85]. The symptoms generally reported in humans due to the MC intoxication include gastroenteritis and related diseases, allergic and irritation reactions, liver diseases, tumors, and primary liver cancer and colorectal cancers, and massive hepatic hemorrhage. MC human poisoning treatment is very complicated due to the rapid, irreversible, and severe liver damage [86], however, gastric lavage [87], administration of monoclonal antibodies against MC-LR [88], immunosupressant Cyclosporine A, antibiotic rifampin [89], and membrane-active antioxidant vitamin E, taken as a dietary supplement [90] are recommended.

4. Microcystin Detection and Monitoring in Freshwater

According to the World Health Organization (WHO) guideline, the permitted limit of MC-LR for drinking water is 1.0 $\mu g \cdot L^{-1}$, and the tolerable daily intake is 0.04 $\mu g Kg^{-1}$ [91]. There are several MC detection methods, the most reported are listed in Table 2. Immunoassays (IA) are suitable methods of MC detection in Mozambique because they do not require sophisticated laboratory equipment and have a limit of detection below the maximum limit (1 $\mu g \cdot L^{-1}$). Additionally, IA can be used in both laboratory and field studies.

Table 2. MC detection methods in drinking water. IA—immunoassays, HPLC—high-performance liquid chromatographic, PAD—photodiode-array detector, LC—liquid chromatography, MS—mass spectroscopy, MALDI-TOF MS—matrix-assisted laser desorption/ionization time-of-flight mass spectrometry, UV—ultraviolet detector.

MC Variant	Detection	LOD	LOQ	Reference
-LR: -LY: -LW: -LF: -LA: Asp^3(Z)-Dhb^7-HtyR: -DAsp^3-RR	IA	50–20,000 $pg \cdot mL^{-1}$		[88,92–102]
-RR: -LR: -LY: -LF	HPLC-UV			[102]
-RR: -LR: -LY: -LW: -LF: -FR; -WR	HPLC-PAD	5 ng		[103,104]
3-demethyl-MC-LR: -LR: -LY: -LA: -LW: -LF: 3-demethyl-MC-RR: -RR: 3-demethyl-MC-YR: -YR	LC-MS (/MS)	0.2 pg–2057 pg	1pg–15 $\mu g \cdot L^{-1}$	[93,102,104–108]
D-MC-LR; -LR: D-MC-RR: D-MC-YR: -RR: -YR: [H4]MC-YR: -WR	MALDI-TOF MS			[109]

5. The Occurrence of Microcystin in Mozambican Drinking Water

5.1. The Drinking Water Scenario in Mozambique

The drinking water supply scenario in Mozambique still faces major challenges because a majority of the population still consumes untreated drinking water and consequently is exposed to many water-borne diseases. Only 50% of the population has access to "safe drinking water". Urban areas are the most favored, with 80%, while rural and most of the population have only 35% coverage and consume untreated water daily from rivers, lakes, and small puddles that form after or during the raining season [5,7]. The low water supply cover in Mozambique is inconceivable due to several reasons, among them, the existence of natural water cover (rivers) in the whole country and the presence of excessive fragmentation of governmental organisms for water management (Figure 1). The water management is led by the Ministry of Public Works, Habitation and Hydric Resources (MOPHRH), which operates, among others, with the National Direction of Water Supply and Sanitation, National Direction of Hydric Resource Management, Water Regional Administrations, Sanitation, and Water Supply Infra-structure Administration, Water Regulation Council, Fund for Investment and Patrimony of Water Supply, and other private institutions, which provide goods and services. In order to improve the water management and expand the coverage, different projects funded by the Mozambican government and non-governmental organizations such as Plataforma Moçambicana de Água [6], Greater Maputo Water Supply Expansion Project [110], Integrated Water Supply and Sanitation Project for the provinces of Niassa and Nampula [111], National Rural Water Supply and Sanitation Program (PRONASAR) in Nampula and Zambezia Provinces [112], Inhambane Rural Water Supply and Sanitation Program [113], and others were implemented involving all the MOPHRH, civil society, and private organisms. However, still to date, the national water supply does not cover enough, with the population still consuming untreated water.

The water policy in Mozambique was approved in 1995, revised in 2007 and 2016, which, in the scope of water supply and sanitation, has the following relevant goals [114]:

- Achieve the sustainable development goals, universal access to water supply, and sanitation.
- Meeting of the basic needs of the poorest population, to reduce poverty, always looking for a sustainability situation.
- Water valuing, not only as a social and environmental asset but also with the economic value it holds.
- Government's concentration on the definition of priorities, standards, regulation, and promotion of the private sector.
- Development of an institutional framework that contributes to the management of water as a resource and provision of decentralized and autonomous water supply and sanitation, where the private sector is called upon to participate.

The Ministry of Health (MH) is the legal organism responsible for water quality control and follows the regulation of the WHO, which sets the parameters of the quality of water intended for human consumption and the methods of carrying out their checks in order to protect human health. The water quality control recommended by WHO include MC among other biological parameter and the provisional guideline value is 1 $\mu g \cdot L^{-1}$ for drinking water [115]. The challenge is enormous in Mozambique for control or monitoring of this hepatotoxin due to the lack of adequate laboratories for the detection of MC in drinking water, even for 50% of the population that consumes treated water. This scenario shows clearly that all the Mozambican population is very vulnerable to MC exposure.

5.2. Microcystin in Mozambican Drinking Water

The drinking water is supplied by private (autonomous systems) and governmental operators. In Table 3, are listed the main sources of drinking water in Mozambique and includes underground and water river. The drinking water treatment is performed mainly by disinfection with chlorine, but is some regions such as Pemba and Niassa, the water treatment system includes the removal of iron by aeration. Not only is there no drinking water treatment for MC removal, but also MC incidence data in Mozambique are very limited. However, according to the WHO, more than 500,000 cases of diarrhea were reported, which 100 and 7 cases correspond to dysentery and cholera, respectively, and others are unknown [116].

Table 3. Treatment and drinking water supply in Mozambique. Gov—Government system. HTH—High test hypochlorite [7,114].

Province	Water System	Water Treatment Center	Capacity, $m^3 \cdot dia^{-1}$	Water Source	Supplied Sites
Maputo	Gov-Umbeluzi	Umbeluzi	240,000	Umbeluzi river and Pequenos Limbobo Dam	Maputo, Matola and Boane
	Ka Tembe Autonomous	Ka Tembe	760	Underground – Ka Tembe	Ka Tembe
	Vila Olimpia Autonomous		-	Underground - Maputo	Vila Olimpia
	The Small		6500	Underground -Maputo	Zona Verde, Kongolote, Matola Gare na Matola, Magoanine and Albazine
Gaza	Gov-Xai-Xai	Xai-Xai	22,790	Limpopo river	Bairro 11, Bairro 13, Hospital, Patrice Lumumba, Inhamissa 6, CFPP, Marieny Gouaby, Chinuguine and Praia
	Gov-Limpopo				
	Gov-Chongoene				
	Xai-Xai Autonomous			Underground – Xai-Xai	Chicumbane, Julius Nyerere, Muahetane e Chongoene
	Gov-Chókwè		10,056	Limpopo river and underground - Chokwe	Lionde, Conhane, Massavassa, Nwachicoluane, Xilembene, Hókwe, Mapapa
	Chókwè Autonomous		6816		
	Gov-Guija			Underground - Guija	vila-sede do distrito de Guijá
Inhambane	Gov-Inhambane		11,176		Inhambane City, Salela, Nhamua e Josina Machel
	Gov-Maxixi		9120	Inhanombe river	Chambone, Rumbana, Nhambiho, Bato, Habana, Malalane, Macupula, Macuamene, Maquetela, Eduardo Mandlane, Nhamaxaxa, Matadouro, Mabil, Barrane and Bembe
	Mangapana and Mabil Autonomous				Mangapana and Mabil
Sofala	Beira and Dondo	Mutua	50,000	Pungué river	Beira and Dondo
Manica	Gov-Manica	Chicamba	38,600		Manica, Chimoio and Gondola and Messica and Bandula village
Tete	Gov-Tete	Tete: Aeration through a cascade, followed by two decantation tanks and then filtration and finally disinfection with granular chlorine	38,495	Zambeze river	Tete city
	The Degué small				Degué
Zambezia	Gov-Zambezia	Licuar: Disinfection with HTH	19,512	Underground - Licuar	Quelimane, Nicoadala and Licuar

Table 3. *Cont.*

Province	Water System	Water Treatment Center	Capacity, $m^3 \cdot dia^{-1}$	Water Source	Supplied Sites
Nampula	Gov-Nampula	Nampula: Pre-chlorination, flocculation, decanting and filtration	20,000	Monapo dam	Nampula city
	Gov-Nacala	Nacala: A mixture of flocculation, decantation, filtration, and disinfection	6000	Nacala dam	Nacala city
Cabo Delgado	Gov-Pemba	Pemba: Removal of iron by aeration and filtration	12,000	Underground-Metuge	Pemba city
	Gov-Angoche	Angoche: Disinfection with HTH	1800	Underground-Malatane	Angoche
Niassa	Gov-Lichinga	Locumué	2400	Locumué dam	Lichinga
	Chiuaula Autonomous			Underground - Chiuala	Chiuaula
	Cuamba	Cuamba: Disinfection with HTH	960	Mpopole dam	Cuamba

These data indicate that many people of Mozambique consume food and water unsafely. Few studies (Figure 1) were done by Pedro et al. [117–119] and Bojcevska and Jergil [120] in Pequenos Libombos dam, Nhambavale lake, Chòkwé irrigation channels and Chidenguele sites in the South of Mozambique during 2003, 2008, and 2009 and their studies indicated the occurrence of MC-LR, -YR, and -RR produced by *Microcystis* sp. (*M. novacekii, botrys* and other) and *Cylindrospermopsis raciborskii* [117–120] (Table 4). MC concentration varies from less than 0.01 (below quantification levels) to 0.02 in Pequenos Libombos dam, less to 0.01 to 0.68 in Chòkwé irrigation channels, 0.86 to 7.82 in Nhambavale lake and 0.57 to 6.83 $\mu g{\cdot}L^{-1}$ in Chidenguele. Higher MC concentration values than the maximum limit ranging from 6.83 to 7.78 $\mu g{\cdot}L^{-1}$ (around 7 times above) were found in the Nhambavale lake and Chidenguele sites. These data highlight (suggest) the need to implement an operational monitoring program of MCs since the tests recommended by MH do not include the MC test [121]. Neighboring countries published other data, which support the need for MC monitoring in Mozambique (Figure 1), namely:

- **South Africa**: MC Producers: *Synechocystis* sp. *Microcystis aeruginosa, Microcystis panniformis, Nostoc* sp., *Planktothrix* sp., *Phormidium* sp., in the Limpopo river basin [122–126], Hartbeespoort dam [127–131], Kruger National Park [132], Sand, Mawoni, Lephalale, Mokolo, Crocodile, Nzhelele ivers [126] MC -YR, -LR, -FR, -YA, -LA, -LAba (0.156–0.270, 0.059–0.18, 0.09, 0.02–0.044, 0.051–0.241, 0.080 $mg.g^{-1}$) in Natal and Transvaal dams [133], 8.6 $\mu g{\cdot}L^{-1}$ in Hartbeespoort dam [134], 12,300 $\mu g{\cdot}L^{-1}$ in Hartbeespoort lake [135].
- **Tanzania**: MC-LR and -RR in different tissues of dead flamingos (*Phoeniconaias minor*) from Empakai Crater, Lake Natron and Lake Manyara (0.165–1.16 $ng.g^{-1}$) [136–138], MC-RR (0.4–13 μgL^{-1}) in Victoria lake [139,140], MC producers: *Aphanocapsa* sp., *Anabaena* sp., *Microcystis* sp. In Victoria lake [139,140].
- **Zimbabwe**: MC producers: *Microcystis aeruginosa* in Mzingwane river, Shashe River [126], *Microcystis wesenbergii* [141,142]. MC-LR (1.62–22 μgL^{-1}) in Chivero lake [141,142].
- **Malawi**: MC producers: *Anabaena* sp. In Malawi lake [143].

Table 4. The Incidence of Microcystin and its producers in the aquatic environments of Mozambique. PL—Pequenos Libombos dam, NL—Nhambavale lake, CH—Chòkwé irrigation channels, RFLP—restriction fragment length polymorphism, MC—microcystins, ELISA—enzyme-linked immunosorbent assay, CG—Chidengule, LM—light microscope, PCR— polymerase chain reaction, ML—Malawi lake, NL— Niassa lake.

Local	Date	Producer		MC			Reference
		Species	Detection	MC Variant	Detection	Conc.	
PL	2008–2009	*Microcystis* sp.	PC gene	LR and YR	LC-MS	3.9 ng·g^{-1}	[118]
		Microcystis sp.	MCyA-MISY gene				
		Microcystis sp.	MCyB gene				
		Microcystis sp.	RFLP				
NL		*Microcystis* sp.	PC gene	LR, YR and RR		159.4 ng·g^{-1}	
		Microcystis sp.	MCyA-MISY gene				
		Microcystis sp.	MCyB gene				
		Microcystis sp.	RFLP				
CH		*Microcystis* sp.	PC gene	LR		2.7 ng·g^{-1}	
		negative	MCyA-MISY gene				
		negative	MCyB gene				
		negative	RFLP				
PL	2002		LM	MC	ELISA	0.22 μg·L^{-1}	[120]
CH		*Cylindrospermopsis raciborskii*				< 0.01 μg·L^{-1}	
CG		*Microcystis novacekii* and *M. botrys*				6.83 μg·L^{-1}	
PL	2008–2009			LR	LC-MS	< 0.01 μg·L^{-1}	[117]
				YR		0.01 μg·L^{-1}	
CH				LR		0.68 μg·L^{-1}	
				YR		0.06 μg·L^{-1}	
NL				LR		7.78 μg·L^{-1}	
				YR		0.07 μg·L^{-1}	
				RR		< 0.01 μg·L^{-1}	

Table 4. *Cont.*

Local	Date	Producer		MC			Reference
		Species	Detection	MC Variant	Detection	Conc.	
PL	2008–2009	*Microcystis aeruginosa*	PC gene				[119]
			MCyB-Taq-Nuclease assay				
NL			PC gene				
			MCyB-Taq-Nuclease assay				
CH			PC gene				
			MCyB-Taq-Nuclease assay				
ML/NL	2002	*Anabaena* sp.	LM				[143]
NKP	2007	*Microcystis aeruginosa*	PCR	LR	ELISA	23718 $\mu g \cdot L^{-1}$	[132]

For example, the drinking water in the Xai Xai district (Gaza province) (Figure 1) is supplied from the Limpopo river. This river contains different MC producers such as *Synechocystis* sp. *Microcystis aeruginosa*, *Microcystis panniformis*, *Nostoc* sp., *Planktothrix* sp., *Phormidium* sp., which were detected in South Africa areas [122–126]. The presence of a potentially toxic algae is not an indication of MC production but is an indication of the need for MC screening in order to confirm the MC presence.

5.3. *Removal of Microscystin from Drinking Water in Mozambique*

MCs can be removed from drinking water using several rapid and low-cost. The most are reported in laboratory studies, and they are not adaptable to economic conditions in Mozambique [144–150]. However, the following techniques seem to be useful in Mozambique and can be implemented in both rural and urban zones: Photolysis at 254 and 185 nm [147], use of wood-based and coconut-based activated carbons [148], use of bamboo-based charcoal adsorbent modified with chitosan [149], hydrophyte filter bed [150], biological activated carbon process [151], aquatic vegetable bed [152], and activated carbon from the seed husks of the pan-tropical tree, *Moringa oleifera* [153], among others.

6. Final Considerations and Recommendations

The drinking water supply scenario in Mozambique still faces major challenges because the majority of the population still consumes untreated drink water (from rivers, lakes, and small puddles that form after or during the raining season) and consequently exposed to many water diseases [5,7] including, for example, gastroenteritis, which is caused by hepatotoxins MCs. To date, no data of water poisoning episodes recorded were associated with MCs presence in the water. However, this might be underestimated due to a lack of monitoring facilities and/or a lack of public health staff trained for recognizing symptoms of MCs intoxication since the presence of high MCs concentration was reported in Maputo and Gaza provinces. Few studies done in Maputo and Gaza provinces indicated the occurrence of MC-LR, -YR, and -RR at a concentration ranging from 6.83 to 7.78 $\mu g \cdot L^{-1}$ [117–120], which are very high, around 7 times above the maximum limit (1 $\mu g \cdot L^{-1}$) recommended by WHO [59]. The potential MC-producing in the studied sites is mostly *Microcystis* sp. [117–120]. However, MC distribution in Mozambique is unknown, and a monitoring program would help to understand the dimension of the problem. To date, no water MC poisoning episodes data recorded in Mozambique. The absence of MC intoxication episodes might be underestimated due to the absence of MC monitoring plan and/or a lack of public health staff trained in recognizing symptoms of MC intoxication. MC monitoring may be implemented according to recommendations of WHO [59] (1 $\mu g \cdot L^{-1}$), and the respective MC analysis can be done in the existing water treatment centers in each province (Table 3). Rapid tests for MC detection, such as ELISA, can be used in each center. In the case of higher MCs content, some suitable techniques for MC removal may be used. The recommended techniques include photolysis at 254 and 185 [147], the use of wood-based and coconut-based activated carbons [148], use of bamboo-based charcoal adsorbent modified with chitosan [149], hydrophyte Filter Bed [150], Biological Activated Carbon Process [151], and aquatic vegetable bed [152], among others. These techniques can be used in both rural and urban areas due to their low-cost implementation and local access.

Author Contributions: Conceptualization, I.J.T.; introduction, I.J.T.; microcystin-producing species and toxicology, I.J.T.; effects of microcystin in humans, symptoms, and treatment, I.J.T.; microcystin detection and monitoring in freshwater, I.J.T.; the occurrence of Microcystin in Mozambican drinking water, I.J.T.; orientation, supervision and corrections, V.V. All authors have read and agreed to the published version of the manuscript.

Funding: This research was funded by Fundação para a Ciência e a Tecnologia (FCT) projects UIDB/04423/2020 and UIDP/04423/2020.

Acknowledgments: The authors acknowledge the Fundação Calouste Gulbenkian for the partial scholarship of Isidro José Tamele and the project EMERTOX [grant 734748], funded by H2020-MSCA-RISE 2016.

Conflicts of Interest: The authors declare no conflict of interest

References

1. Moçambique INE Destaques. IV Censo 2017—Instituto Nacional de Estatistica—INE-Moçambique. Available online: http://www.ine.gov.mz/ (accessed on 26 November 2019).
2. Arndt, C.; Strzepeck, K.; Tarp, F.; Thurlow, J.; Fant, C.; Wright, L. Adapting to climate change: An integrated biophysical and economic assessment for Mozambique. *Sustain. Sci.* **2011**, *6*, 7–20. [CrossRef]
3. Strzepek, K.; Arndt, C.; Chinowsky, P.; Kuriakose, A.; Neumann, J.; Nicholls, R.; Thurlow, J.; Wright, L. *Economics of Adaptation to Climate Change: Mozambique*; World Bank: Washington, DC, USA, 2010.
4. Arndt, C.; Benfica, R.; Maximiano, N.; Nucifora, A.M.; Thurlow, J.T. Higher fuel and food prices: Impacts and responses for Mozambique. *Agric. Econ.* **2008**, *39*, 497–511. Available online: https://www.unicef.org/evaldatabase/files/283729_B85_IOB_360_BW_WEB_Mozambique-Final.pdf (accessed on 22 May 2020). [CrossRef]
5. UNICEF; Government of the Netherlands Partnership for Water Supply S.a.H. *Impact Evaluation of Drinking Water Supply and Sanitation Interventions in Rural Mozambique: More than Water*; Printed Report; UNICEF: New York, NY, USA; Government of the Netherlands Partnership for Water Supply: The Hague, The Netherlands, 2011; Available online: https://www.unicef.org/evaldatabase/files/283729_B85_IOB_360_BW_WEB_Mozambique-Final.pdf (accessed on 22 May 2020).
6. Ministério das Obras Públicas, Habitação e Recursos Hídricos. *O Sector de Água em Moçambique. PLASMA-PLATAFORMA MOCAMBICANA DA AGUA*; Ministério das Obras Públicas, Habitação e Recursos Hídricos: Maputo, Moçambique, 2016. Available online: https://plama.org.mz/index.php (accessed on 1 June 2020).
7. UNICEF. *Situação da Água, Saneamento e Higiene em Moçambique*; UNICEF: Maputo, Mozambique, 2017; Available online: https://www.unicef.org/mozambique/%C3%A1gua-saneamento-e-higiene (accessed on 1 June 2020).
8. World Health Organization. *Cyanobacterial Toxins: Microcystin-LR in Drinking-Water: Background Document for Development of WHO Guidelines for Drinking-Water Quality*; World Health Organization: Geneva, Switzerland, 2003.
9. Yoo, R.S. *Cyanobacterial (Blue-Green Algal) Toxins: A Resource Guide*; American Water Works Association: Denver, CO, USA, 1995.
10. Lawton, L.A.; Codd, G. Cyanobacterial (blue-green algal) toxins and their significance in UK and European waters. *Water Environ. J.* **1991**, *5*, 460–465. [CrossRef]
11. Skulberg, O.M.; Carmichael, W.W.; Codd, G.A.; Skulberg, R. Taxonomy of toxic Cyanophyceae (cyanobacteria). In *Algal Toxins in Seafood and Drinking Water*; Academic Press: New York, NY, USA, 1993; pp. 145–164.
12. Falconer, I.R.; Humpage, A.R. Health risk assessment of cyanobacterial (blue-green algal) toxins in drinking water. *Int. J. Environ. Res. Public Health* **2005**, *2*, 43–50. [CrossRef]
13. FerrÃo-Filho, A.S.; Azevedo, S.M.; DeMott, W.R. Effects of toxic and non-toxic cyanobacteria on the life history of tropical and temperate cladocerans. *Freshw. Biol.* **2000**, *45*, 1–19. [CrossRef]
14. Neilan, B.A.; Jacobs, D.; Blackall, L.L.; Hawkins, P.R.; Cox, P.T.; Goodman, A.E. rRNA sequences and evolutionary relationships among toxic and nontoxic cyanobacteria of the genus Microcystis. *Int. J. Syst. Evol. Microbiol.* **1997**, *47*, 693–697. [CrossRef]
15. Vaitomaa, J.; Rantala, A.; Halinen, K.; Rouhiainen, L.; Tallberg, P.; Mokelke, L.; Sivonen, K. Quantitative real-time PCR for determination of microcystin synthetase E copy numbers for Microcystis and Anabaena in lakes. *Appl. Environ. Microbiol.* **2003**, *69*, 7289–7297. [CrossRef]
16. Oh, H.-M.; Lee, S.J.; Jang, M.-H.; Yoon, B.-D. Microcystin production by Microcystis aeruginosa in a phosphorus-limited chemostat. *Appl. Environ. Microbiol.* **2000**, *66*, 176–179. [CrossRef]
17. Kotak, B.G.; Lam, A.K.Y.; Prepas, E.E.; Kenefick, S.L.; Hrudey, S.E. Variability of the hepatotoxin microcystin - LR in hypereutrophic drinking water lakes 1. *J. Phycol.* **1995**, *31*, 248–263. [CrossRef]
18. Zhao, G.; Wu, D.; Cao, S.; Du, W.; Yi, Y.; Gu, H. Efects of CeO2 Nanoparticles on Microcystis aeruginosa Growth and Microcystin Production. *Bull. Environ. Contam. Toxicol.* **2020**, *104*, 834–839. [CrossRef] [PubMed]
19. Utkilen, H.; Gjølme, N. Toxin production by Microcystis aeruginosa as a function of light in continuous cultures and its ecological significance. *Appl. Environ. Microbiol.* **1992**, *58*, 1321–1325. [CrossRef]
20. Watanabe, M.F.; Harada, K.-I.; Matsuura, K.; Watanabe, M.; Suzuki, M. Heptapeptide toxin production during the batch culture of two Microcystis species (Cyanobacteria). *J. Appl. Phycol.* **1989**, *1*, 161–165. [CrossRef]

21. Watanabe, M.F.; Oishi, S. Effects of environmental factors on toxicity of a cyanobacterium (Microcystis aeruginosa) under culture conditions. *Appl. Environ. Microbiol.* **1985**, *49*, 1342–1344. [CrossRef] [PubMed]
22. Wicks, R.J.; Thiel, P.G. Environmental factors affecting the production of peptide toxins in floating scums of the cyanobacterium Microcystis aeruginosa in a hypertrophic African reservoir. *Environ. Sci. Technol.* **1990**, *24*, 1413–1418. [CrossRef]
23. Eynard, F.; Mez, K.; Walther, J.-L. Risk of cyanobacterial toxins in Riga waters (Latvia). *Water Res.* **2000**, *34*, 2979–2988. [CrossRef]
24. Willén, E.; Ahlgren, G.; Söderhielm, A.-C. Toxic cyanophytes in three Swedish lakes. *Int. Ver. Für Theor. Und Angew. Limnol. Verh.* **2000**, *27*, 560–564. [CrossRef]
25. Oudra, B.; Loudiki, M.; Vasconcelos, V.; Sabour, B.; Sbiyyaa, B.; Oufdou, K.; Mezrioui, N. Detection and quantification of microcystins from cyanobacteria strains isolated from reservoirs and ponds in Morocco. *Environ. Toxicol. Int. J.* **2002**, *17*, 32–39. [CrossRef]
26. Mankiewicz-Boczek, J.; Gągała, I.; Jurczak, T.; Urbaniak, M.; Negussie, Y.Z.; Zalewski, M. Incidence of microcystin-producing cyanobacteria in Lake Tana, the largest waterbody in Ethiopia. *Afr. J. Ecol.* **2015**, *53*, 54–63. [CrossRef]
27. Vasconcelos, V.; Sivonen, K.; Evans, W.; Carmichael, W.; Namikoshi, M. Hepatotoxic microcystin diversity in cyanobacterial blooms collected in Portuguese freshwaters. *Water Res.* **1996**, *30*, 2377–2384. [CrossRef]
28. Frank, C.A. Microcystin-producing cyanobacteria in recreational waters in southwestern Germany. *Environ. Toxicol. Int. J.* **2002**, *17*, 361–366. [CrossRef] [PubMed]
29. Henriksen, P. Toxic freshwater cyanobacteria in Denmark. In *Cyanotoxins*; Springer: Berlin/Heidelberg, Germany, 2001; pp. 49–56.
30. Rapala, J.; Sivonen, K. Assessment of environmental conditions that favor hepatotoxic and neurotoxic Anabaena spp. strains cultured under light limitation at different temperatures. *Microb. Ecol.* **1998**, *36*, 181–192. [CrossRef] [PubMed]
31. Rapala, J.; Sivonen, K.; Lyra, C.; Niemelä, S.I. Variation of microcystins, cyanobacterial hepatotoxins, in Anabaena spp. as a function of growth stimuli. *Appl. Environ. Microbiol.* **1997**, *63*, 2206–2212. [CrossRef]
32. Mohamed, Z.A.; El-Sharouny, H.M.; Ali, W.S. Microcystin production in benthic mats of cyanobacteria in the Nile River and irrigation canals, Egypt. *Toxicon* **2006**, *47*, 584–590. [CrossRef] [PubMed]
33. Maatouk, I.; Bouaïcha, N.; Fontan, D.; Levi, Y. Seasonal variation of microcystin concentrations in the Saint-Caprais reservoir (France) and their removal in a small full-scale treatment plant. *Water Res.* **2002**, *36*, 2891–2897. [CrossRef]
34. Vezie, C.; Brient, L.; Sivonen, K.; Bertru, G.; Lefeuvre, J.-C.; Salkinoja-Salonen, M. Variation of microcystin content of cyanobacterial blooms and isolated strains in Lake Grand-Lieu (France). *Microb. Ecol.* **1998**, *35*, 126–135. [CrossRef]
35. Sivonen, K. Effects of light, temperature, nitrate, orthophosphate, and bacteria on growth of and hepatotoxin production by Oscillatoria agardhii strains. *Appl. Environ. Microbiol.* **1990**, *56*, 2658–2666. [CrossRef]
36. Mez, K.; Beattie, K.A.; Codd, G.A.; Hanselmann, K.; Hauser, B.; Naegeli, H.; Preisig, H.R. Identification of a microcystin in benthic cyanobacteria linked to cattle deaths on alpine pastures in Switzerland. *Eur. J. Phycol.* **1997**, *32*, 111–117. [CrossRef]
37. Nascimento, S.M.; De Oliveira e Azevedo, S.M.F. Changes in cellular components in a cyanobacterium (Synechocystis aquatilis f. salina) subjected to different N/P ratios—An ecophysiological study. *Environ. Toxicol. Int. J.* **1999**, *14*, 37–44. [CrossRef]
38. Domingos, P.; Rubim, T.; Molica, R.; Azevedo, S.; Carmichael, W. First report of microcystin production by picoplanktonic cyanobacteria isolated from a northeast Brazilian drinking water supply. *Environ. Toxicol. Int. J.* **1999**, *14*, 31–35. [CrossRef]
39. Carmichael, W. *The Water Environment: Algal Toxins and Health*; Springer Science & Business Media: Berlin/Heidelberg, Germany, 2013.
40. Song, L.; Sano, T.; Li, R.; Watanabe, M.M.; Liu, Y.; Kaya, K. Microcystin production of Microcystis viridis (cyanobacteria) under different culture conditions. *Phycol. Res.* **1998**, *46*, 19–23. [CrossRef]
41. Lukač, M.; Aegerter, R. Influence of trace metals on growth and toxin production of Microcystis aeruginosa. *Toxicon* **1993**, *31*, 293–305. [CrossRef]

42. Rinta-Kanto, J.M.; Konopko, E.A.; DeBruyn, J.M.; Bourbonniere, R.A.; Boyer, G.L.; Wilhelm, S.W. Lake Erie Microcystis: Relationship between microcystin production, dynamics of genotypes and environmental parameters in a large lake. *Harmful Algae* **2009**, *8*, 665–673. [CrossRef]
43. Te, S.H.; Gin, K.Y.-H. The dynamics of cyanobacteria and microcystin production in a tropical reservoir of Singapore. *Harmful Algae* **2011**, *10*, 319–329. [CrossRef]
44. Codd, G. Cyanobacterial toxins. In *Biochemistry of the Algae and Cyanobacteria*; Oxford University Press: Oxford, UK, 1988; pp. 283–296.
45. Orr, P.T.; Jones, G.J. Relationship between microcystin production and cell division rates in nitrogen-limited Microcystis aeruginosa cultures. *Limnol. Oceanogr.* **1998**, *43*, 1604–1614. [CrossRef]
46. Jungmann, D.; Ludwichowski, K.U.; Faltin, V.; Benndorf, J. A field study to investigate environmental factors that could effect microcystin synthesis of a Microcystis population in the Bautzen reservoir. *Ienternationale Rev. Der Gesamten Hydrobiol. Und Hydrogr.* **1996**, *81*, 493–501. [CrossRef]
47. Kaya, K.; Watanabe, M.M. Microcystin composition of an axenic clonal strain of Microcystis viridis and Microcystis viridis-containing water blooms in Japanese freshwaters. *J. Appl. Phycol.* **1990**, *2*, 173–178. [CrossRef]
48. Van der Westhuizen, A.; Eloff, J. Effect of temperature and light on the toxicity and growth of the blue-green alga Microcystis aeruginosa (UV-006). *Planta* **1985**, *163*, 55–59. [CrossRef] [PubMed]
49. Gorham, P.R. Toxic algae. In *Algae and Man*; Springer: Berlin/Heidelberg, Germany, 1964; pp. 307–336.
50. Bartram, J.; Chorus, I. *Toxic Cyanobacteria in Water: A Guide to Their Public Health Consequences, Monitoring and Management*; CRC Press: Boca Raton, FL, USA, 1999.
51. Welker, M.; Von Döhren, H. Cyanobacterial peptides—nature's own combinatorial biosynthesis. *FEMS Microbiol. Rev.* **2006**, *30*, 530–563. [CrossRef] [PubMed]
52. Svirčev, Z.; Drobac, D.; Tokodi, N.; Mijović, B.; Codd, G.A.; Meriluoto, J. Toxicology of microcystins with reference to cases of human intoxications and epidemiological investigations of exposures to cyanobacteria and cyanotoxins. *Arch. Toxicol.* **2017**, *91*, 621–650. [CrossRef] [PubMed]
53. Botes, D.P.; Wessels, P.L.; Kruger, H.; Runnegar, M.T.; Santikarn, S.; Smith, R.J.; Barna, J.C.; Williams, D.H. Perkin Transactions 1. Structural studies on cyanoginosins-LR,-YR,-YA, and-YM, peptide toxins from Microcystis aeruginosa. *J. Chem. Soc. Perkin Trans. 1* **1985**, 2747–2748. [CrossRef]
54. Namikoshi, M.; Yuan, M.; Sivonen, K.; Carmichael, W.W.; Rinehart, K.L.; Rouhiainen, L.; Sun, F.; Brittain, S.; Otsuki, A. Seven new microcystins possessing two L-glutamic acid units, isolated from Anabaena sp. strain 186. *Chem. Res. Toxicol.* **1998**, *11*, 143–149. [CrossRef]
55. Rinehart, K.L.; Namikoshi, M.; Choi, B.W. Structure and biosynthesis of toxins from blue-green algae (cyanobacteria). *J. Appl. Phycol.* **1994**, *6*, 159–176. [CrossRef]
56. Sivonen, K. Cyanobacterial toxins and toxin production. *Phycologia* **1996**, *35*, 12–24. [CrossRef]
57. Campos, A.; Vasconcelos, V. Molecular mechanisms of microcystin toxicity in animal cells. *Int. J. Mol. Sci.* **2010**, *11*, 268–287. [CrossRef]
58. Prieto, A.I.; Jos, A.; Pichardo, S.; Moreno, I.; de Sotomayor, M.Á.; Moyano, R.; Blanco, A.; Cameán, A.M. Time-dependent protective efficacy of Trolox (vitamin E analog) against microcystin-induced toxicity in tilapia (Oreochromis niloticus). *Environ. Toxicol. Int. J.* **2009**, *24*, 563–579. [CrossRef]
59. World Health Organization. *Guidelines for Drinking-Water Quality [Electronic Resource]: Incorporating First Addendum. Vol. 1, Recommendations*; Springer: Geneva, Switzerland, 2006.
60. Tillett, D.; Dittmann, E.; Erhard, M.; Von Döhren, H.; Börner, T.; Neilan, B.A. Structural organization of microcystin biosynthesis in Microcystis aeruginosa PCC7806: An integrated peptide–polyketide synthetase system. *Chem. Biol.* **2000**, *7*, 753–764. [CrossRef]
61. Van Apeldoorn, M.E.; Van Egmond, H.P.; Speijers, G.J.; Bakker, G.J. Toxins of cyanobacteria. *Mol. Nutr. Food Res.* **2007**, *51*, 7–60. [CrossRef]
62. MacKintosh, C.; Beattie, K.A.; Klumpp, S.; Cohen, P.; Codd, G.A. Cyanobacterial microcystin-LR is a potent and specific inhibitor of protein phosphatases 1 and 2A from both mammals and higher plants. *FEBS Lett.* **1990**, *264*, 187–192. [CrossRef]
63. Gulledge, B.; Aggen, J.; Huang, H.; Nairn, A.; Chamberlin, A. The microcystins and nodularins: Cyclic polypeptide inhibitors of PP1 and PP2A. *Curr. Med. Chem.* **2002**, *9*, 1991–2003. [CrossRef] [PubMed]

64. Song, W.; De La Cruz, A.A.; Rein, K.; O'Shea, K.E. Ultrasonically induced degradation of microcystin-LR and-RR: Identification of products, effect of pH, formation and destruction of peroxides. *Environ. Sci. Technol.* **2006**, *40*, 3941–3946. [CrossRef] [PubMed]
65. Tsuji, K.; Naito, S.; Kondo, F.; Ishikawa, N.; Watanabe, M.F.; Suzuki, M.; Harada, K.-I. Stability of microcystins from cyanobacteria: Effect of light on decomposition and isomerization. *Environ. Sci. Technol.* **1994**, *28*, 173–177. [CrossRef] [PubMed]
66. Zhang, H.; Zhang, J.; Chen, Y.; Zhu, Y. biochemistry. Microcystin-RR induces apoptosis in fish lymphocytes by generating reactive oxygen species and causing mitochondrial damage. *Fish Physiol.* **2008**, *34*, 307–312.
67. Fujiki, H.; Suganuma, M. Carcinogenic aspects of protein phosphatase 1 and 2A inhibitors. In *Marine Toxins as Research Tools*; Springer: Berlin/Heidelberg, Germany, 2009; pp. 221–254.
68. Nishiwaki-Matsushima, R.; Ohta, T.; Nishiwaki, S.; Suganuma, M.; Kohyama, K.; Ishikawa, T.; Carmichael, W.W.; Fujiki, H. Liver tumor promotion by the cyanobacterial cyclic peptide toxin microcystin-LR. *J. Cancer Res. Clin. Oncol.* **1992**, *118*, 420–424. [CrossRef]
69. Herfindal, L.; Selheim, F. Microcystin produces disparate effects on liver cells in a dose dependent manner. *Mini Rev. Med. Chem.* **2006**, *6*, 279–285. [CrossRef]
70. Carmichael, W.W.; Azevedo, S.M.; An, J.S.; Molica, R.J.; Jochimsen, E.M.; Lau, S.; Rinehart, K.L.; Shaw, G.R.; Eaglesham, G.K. Human fatalities from cyanobacteria: chemical and biological evidence for cyanotoxins. *Environ. Health Perspect.* **2001**, *109*, 663–668. [CrossRef]
71. Yuan, M.; Carmichael, W.W.; Hilborn, E.D. Microcystin analysis in human sera and liver from human fatalities in Caruaru, Brazil 1996. *Toxicon* **2006**, *48*, 627–640. [CrossRef]
72. Texeira, M.D.G.L.C.; Costa, M.D.C.N.; Carvalho, V.L.P.D.; Pereira, M.D.S.; Hage, E. Gastroenteritis epidemic in the area of the Itaparica Dam, Bahia, Brazil. *Bull. Paho* **1993**, *27*, 1993.
73. Jochimsen, E.M.; Carmichael, W.W.; An, J.; Cardo, D.M.; Cookson, S.T.; Holmes, C.E.; Antunes, M.B.; de Melo Filho, D.A.; Lyra, T.M.; Barreto, V.S.T. Liver failure and death after exposure to microcystins at a hemodialysis center in Brazil. *N. Engl. J. Med.* **1998**, *338*, 873–878. [CrossRef]
74. Pouria, S.; de Andrade, A.; Barbosa, J.; Cavalcanti, R.; Barreto, V.; Ward, C.; Preiser, W.; Poon, G.K.; Neild, G.; Codd, G. Fatal microcystin intoxication in haemodialysis unit in Caruaru, Brazil. *Lancet* **1998**, *352*, 21–26. [CrossRef]
75. Azevedo, S.M.; Carmichael, W.W.; Jochimsen, E.M.; Rinehart, K.L.; Lau, S.; Shaw, G.R.; Eaglesham, G.K. Human intoxication by microcystins during renal dialysis treatment in Caruaru—Brazil. *Toxicology* **2002**, *181*, 441–446. [CrossRef]
76. Giannuzzi, L.; Sedan, D.; Echenique, R.; Andrinolo, D. An acute case of intoxication with cyanobacteria and cyanotoxins in recreational water in Salto Grande Dam, Argentina. *Mar. Drugs* **2011**, *9*, 2164–2175. [CrossRef]
77. Vidal, F.; Sedan, D.; D'Agostino, D.; Cavalieri, M.L.; Mullen, E.; Parot Varela, M.M.; Flores, C.; Caixach, J.; Andrinolo, D. Recreational exposure during algal bloom in Carrasco Beach, Uruguay: A liver failure case report. *Toxins* **2017**, *9*, 267. [CrossRef] [PubMed]
78. Zilberg, B. Gastroenteritis in Salisbury European children-a five-year study. *Cent. Afr. J. Med.* **1966**, *12*, 164–168. [PubMed]
79. Annadotter, H.; Cronberg, G.; Lawton, L.; Hansson, H.-B.; Göthe, U.; Skulberg, O. An extensive outbreak of gastroenteritis associated with the toxic cyanobacterium Planktothrix agardhii (Oscillatoriales, Cyanophyceae) in Scania, South Sweden. In *Cyanotoxins*; Springer: Berlin, Germany, 2001; pp. 200–208.
80. Milutinović, A.; Živin, M.; Zorc-Pleskovič, R.; Sedmak, B.; Šuput, D. Nephrotoxic effects of chronic administration of microcystins-LR and-YR. *Toxicon* **2003**, *42*, 281–288. [CrossRef]
81. Milutinović, A.; Sedmark, B.; Horvat-Žnidaršić, I.; Šuput, D. Renal injuries induced by chronic intoxication with microcystins. *Cell Mol. Biol. Lett.* **2002**, *7*, 139–141.
82. Nobre, A.; Jorge, M.; Menezes, D.; Fonteles, M.; Monteiro, H. Effects of microcystin-LR in isolated perfused rat kidney. *Braz. J. Med. Biol. Res.* **1999**, *32*, 985–988. [CrossRef]
83. Botha, N.; van de Venter, M.; Downing, T.G.; Shephard, E.G.; Gehringer, M.M. The effect of intraperitoneally administered microcystin-LR on the gastrointestinal tract of Balb/c mice. *Toxicon* **2004**, *43*, 251–254. [CrossRef]
84. Žegura, B.; Volčič, M.; Lah, T.T.; Filipič, M. Different sensitivities of human colon adenocarcinoma (CaCo-2), astrocytoma (IPDDC-A2) and lymphoblastoid (NCNC) cell lines to microcystin-LR induced reactive oxygen species and DNA damage. *Toxicon* **2008**, *52*, 518–525. [CrossRef]

85. Dias, E.; Andrade, M.; Alverca, E.; Pereira, P.; Batoréu, M.; Jordan, P.; Silva, M.J. Comparative study of the cytotoxic effect of microcistin-LR and purified extracts from Microcystis aeruginosa on a kidney cell line. *Toxicon* **2009**, *53*, 487–495. [CrossRef] [PubMed]
86. De Figueiredo, D.R.; Azeiteiro, U.M.; Esteves, S.M.; Gonçalves, F.J.; Pereira, M.J. Microcystin-producing blooms—a serious global public health issue. *Ecotoxicol. Environ. Saf.* **2004**, *59*, 151–163. [CrossRef] [PubMed]
87. Gorham, P.; Carmichael, W. Hazards of freshwater blue-green algae (cyanobacteria). In *Algae and Human Affairs*; Lembi, C.A., Waaland, J.R., Eds.; Cambridge University Press: Cambridge, UK, 1988; Sponsored by the Phycological Society of America, Inc.
88. Nagata, S.; Soutome, H.; Tsutsumi, T.; Hasegawa, A.; Sekijima, M.; Sugamata, M.; Harada, K.I.; Suganuma, M.; Ueno, Y. Novel monoclonal antibodies against microcystin and their protective activity for hepatotoxicity. *Nat. Toxins* **1995**, *3*, 78–86. [CrossRef]
89. Dawson, R. The toxicology of microcystins. *Toxicon* **1998**, *36*, 953–962. [CrossRef]
90. Gehringer, M.M.; Govender, S.; Shah, M.; Downing, T.G. An investigation of the role of vitamin E in the protection of mice against microcystin toxicity. *Environ. Toxicol. Int. J.* **2003**, *18*, 142–148. [CrossRef]
91. World Health Organization. *Guidelines for Drinking-Water Quality. Volume 2, Health Criteria and Other Supporting Information: Addendum*; World Health Organization: Geneva, Switzerland, 1998; Available online: https://apps.who.int/iris/bitstream/handle/10665/63844/WHO_EOS_98.1.pdf?sequence=1&isAllowed=y (accessed on 22 May 2020).
92. Metcalf, J.; Bell, S.; Codd, G. Production of novel polyclonal antibodies against the cyanobacterial toxin microcystin-LR and their application for the detection and quantification of microcystins and nodularin. *Water Res.* **2000**, *34*, 2761–2769. [CrossRef]
93. Zweigenbaum, J.; Henion, J.; Beattie, K.; Codd, G.; Poon, G. Direct analysis of microcystins by microbore liquid chromatography electrospray ionization ion-trap tandem mass spectrometry. *J. Pharm. Biomed. Anal.* **2000**, *23*, 723–733. [CrossRef]
94. Ward, C.J.; Beattie, K.A.; Lee, E.Y.; Codd, G.A. Colorimetric protein phosphatase inhibition assay of laboratory strains and natural blooms of cyanobacteria: Comparisons with high-performance liquid chromatographic analysis for microcystins. *FEMS Microbiol. Lett.* **1997**, *153*, 465–473. [CrossRef]
95. Carmichael, W.W.; An, J.J. Using an enzyme linked immunosorbent assay (ELISA) and a protein phosphatase inhibition assay (PPIA) for the detection of microcystins and nodularins. *Nat. Toxins* **1999**, *7*, 377–385. [CrossRef]
96. Ueno, Y.; Nagata, S.; Tsutsumi, T.; Hasegawa, A.; Yoshida, F.; Suttajit, M.; Mebs, D.; Pütsch, M.; Vasconcelos, V. Survey of microcystins in environmental water by a highly sensitive immunoassay based on monoclonal antibody. *Nat. Toxins* **1996**, *4*, 271–276. [CrossRef]
97. Zeck, A.; Weller, M.G.; Bursill, D.; Niessner, R. Generic microcystin immunoassay based on monoclonal antibodies against Adda. *Analyst* **2001**, *126*, 2002–2007. [CrossRef] [PubMed]
98. Lindner, P.; Molz, R.; Yacoub-George, E.; Dürkop, A.; Wolf, H. Development of a highly sensitive inhibition immunoassay for microcystin-LR. *Anal. Chim. Acta* **2004**, *521*, 37–44. [CrossRef]
99. Lindner, P.; Molz, R.; Yacoub-George, E.; Wolf, H. Rapid chemiluminescence biosensing of microcystin-LR. *Anal. Chim. Acta* **2009**, *636*, 218–223. [CrossRef]
100. Fontal, O.; Vieytes, M.; de Sousa, J.B.; Louzao, M.; Botana, L.J.A.b. A fluorescent microplate assay for microcystin-LR. *Anal. Biochem.* **1999**, *269*, 289–296. [CrossRef]
101. Tippkötter, N.; Stückmann, H.; Kroll, S.; Winkelmann, G.; Noack, U.; Scheper, T.; Ulber, R. A semi-quantitative dipstick assay for microcystin. *Anal. Bioanal. Chem.* **2009**, *394*, 863–869. [CrossRef]
102. Spoof, L.; Vesterkvist, P.; Lindholm, T.; Meriluoto, J. Screening for cyanobacterial hepatotoxins, microcystins and nodularin in environmental water samples by reversed-phase liquid chromatography–electrospray ionisation mass spectrometry. *J. Chromatogr. A* **2003**, *1020*, 105–119. [CrossRef]
103. Lawton, L.A.; Edwards, C.; Codd, G.A. Extraction and high-performance liquid chromatographic method for the determination of microcystins in raw and treated waters. *Analyst* **1994**, *119*, 1525–1530. [CrossRef]
104. Douma, M.; Ouahid, Y.; Del Campo, F.; Loudiki, M.; Mouhri, K.; Oudra, B. Identification and quantification of cyanobacterial toxins (microcystins) in two Moroccan drinking-water reservoirs (Mansour Eddahbi, Almassira). *Environ. Monit. Assess.* **2010**, *160*, 439. [CrossRef]
105. Neffling, M.-R.; Spoof, L.; Meriluoto, J. Rapid LC–MS detection of cyanobacterial hepatotoxins microcystins and nodularins—Comparison of columns. *Anal. Chim. Acta* **2009**, *653*, 234–241. [CrossRef]

106. Barco, M.; Rivera, J.; Caixach, J. Analysis of cyanobacterial hepatotoxins in water samples by microbore reversed-phase liquid chromatography–electrospray ionisation mass spectrometry. *J. Chromatogr. A* **2002**, *959*, 103–111. [CrossRef]
107. Cong, L.; Huang, B.; Chen, Q.; Lu, B.; Zhang, J.; Ren, Y. Determination of trace amount of microcystins in water samples using liquid chromatography coupled with triple quadrupole mass spectrometry. *Anal. Chim. Acta* **2006**, *569*, 157–168. [CrossRef]
108. Mekebri, A.; Blondina, G.; Crane, D. Method validation of microcystins in water and tissue by enhanced liquid chromatography tandem mass spectrometry. *J. Chromatogr. A* **2009**, *1216*, 3147–3155. [CrossRef] [PubMed]
109. Via-Ordorika, L.; Fastner, J.; Kurmayer, R.; Hisbergues, M.; Dittmann, E.; Komarek, J.; Erhard, M.; Chorus, I. Distribution of microcystin-producing and non-microcystin-producing Microcystis sp. In European freshwater bodies: Detection of microcystins and microcystin genes in individual colonies. *Syst. Appl. Microbiol.* **2004**, *27*, 592–602. [CrossRef]
110. Miguel, M. *Mozambique - AFRICA- P125120- Greater Maputo Water Supply Expansion Project - Procurement Plan (English)*; World Bank: Washington, DC, USA, 2019.
111. Republic, M. *Integrated Water Supply and Sanitation Project for the Provinces of Niassa and Nampula-MOZ/PWWS/2000/01*; African Development Fund: Maputo, Mozambique, 2000.
112. Housing, M.o.P.W.a. *National Rural Water Supply and Sanitation Program (PRONASAR) in Nampula and Zambezia Provinces*; African Development Bank Group: Maputo, Mozambique, 2010.
113. Canada, G.o.C.-G.A. *Inhambane Rural Water Supply and Sanitation Program*; Global Affairs Canada: Maputo, Mozambique, 2017.
114. Uandela, A. Mecanismos e Instrumentos de Planificação e Orçamentação no Sector de águas em Moçambique. Folheto Informativo Moç. M03. 2012, pp. 1–11. Available online: https://www.ircwash.org/sites/default/files/m03_mecanismos_e_instrumentos_de_planificacao_e_oramentacao_no_sector_de_aguas_em_mocambique.pdf (accessed on 22 May 2020).
115. Water Sanitation; World Health Organization. Guidelines for Drinking-Water Quality [Electronic Resource]: Incorporating First Addendum. Vol. 1, Recommendations; ONU News, Maputo, Mozambique 2006. Available online: https://news.un.org/pt/story/2019/06/1675251 (accessed on 22 May 2020).
116. Pota, O. Em Moçambique, mais de 500 mil pessoas tiveram doenças causadas por consumo de alimentos inseguros. *ONU News*, 7 June 2019.
117. Pedro, O.; Rundberget, T.; Lie, E.; Correia, D.; Skaare, J.U.; Berdal, K.G.; Neves, L.; Sandvik, M. Occurrence of microcystins in freshwater bodies in Southern Mozambique. *J. Res. Environ. Sci. Toxicol.* **2012**, *1*, 58–65.
118. Pedro, O.; Correia, D.; Lie, E.; Skåre, J.U.; Leão, J.; Neves, L.; Sandvik, M.; Berdal, K.G. Polymerase chain reaction (PCR) detection of the predominant microcystin-producing genotype of cyanobacteria in Mozambican lakes. *Afr. J. Biotechnol.* **2011**, *10*, 19299–19308.
119. Pedro, O.; Lie, E.; Correia, D.; Neves, L.; Skaare, J.U.; Sandvik, M.; Berdal, K.G. Quantification of microcystin-producing microcystis in freshwater bodies in the Southern Mozambique using quantitative real time polymerase chain reaction. *Afr. J. Biotechnol.* **2013**, *12*. [CrossRef]
120. Bojcevska, H.; Jergil, E. Removal of cyanobacterial toxins (LPS endotoxin and microcystin) in drinking-water using the BioSand household water filter. *Minor Field Study* **2003**, *91*, 1–44.
121. MD. SAUDE. Regulamento sobre a Qualidade da Água para o Consumo Humano. In *Diploma Ministerial 180/2004 de 15 de Setembro*; Boletim da Republica, I serie, numero 37; MD. SAUDE: Maputo, Mozambique, 2004; Volume Diploma Ministerial 180/2004.
122. Magonono, M.; Oberholster, P.J.; Shonhai, A.; Makumire, S.; Gumbo, J.R. The presence of toxic and non-toxic Cyanobacteria in the sediments of the Limpopo River Basin: Implications for human health. *Toxins* **2018**, *10*, 269. [CrossRef]
123. Fosso-Kankeu, E.; Jagals, P.; Du Preez, H. Exposure of rural households to toxic cyanobacteria in container-stored water. *Water SA* **2008**, *34*, 631–636. [CrossRef]
124. Falconer, I.R.; Runnegar, M.T.; Beresford, A.M. Evidence of liver damage by toxin from a bloom of the blue-green alga, Microcystis aeruginosa. *Med J. Aust.* **1983**, *1*, 511–514. [CrossRef] [PubMed]
125. Kirumba, W.; Shushu, D.; Masundire, H.; Oyaro, N. Diversity of Algae and Potentially Toxic Cyanobacteria in a River Receiving Treated Sewage Effluent: A case of Notwane River (Gaborone, Botswana). *Int. Res. J. Environ. Sci.* **2014**.

126. Mbukwa, E.A.; Boussiba, S.; Wepener, V.; Leu, S.; Kaye, Y.; Msagati, T.A.; Mamba, B.B. PCR amplification and DNA sequence of mcyA gene: The distribution profile of a toxigenic Microcystis aeruginosa in the Hartbeespoort Dam, South Africa. *J. Water Health* **2013**, *11*, 563–572. [CrossRef]
127. Robarts, R.; Zohary, T. Microcystis aeruginosa and underwater light attenuation in a hypertrophic lake (Hartbeespoort Dam, South Africa). *J. Ecol.* **1984**, *72*, 1001–1017. [CrossRef]
128. Jarvis, A.C.; Hart, R.C.; Combrink, S. Zooplankton feeding on size fractionated Microcystis colonies and Chlorella in a hypertrophic lake (Hartbeespoort Dam, South Africa): Implications to resource utilization and zooplankton succession. *J. Plankton Res.* **1987**, *9*, 1231–1249. [CrossRef]
129. Zohary, T. Hyperscums of the cyanobacterium Microcystis aeruginosa in a hypertrophic lake (Hartbeespoort Dam, South Africa). *J. Plankton Res.* **1985**, *7*, 399–409. [CrossRef]
130. Scott, W.E.; Barlow, D.J.; Hauman, J.H. Studies on the ecology, growth and physiology of toxic Microcystis aeruginosa in South Africa. In *The Water Environment*; Springer: Berlin/Heidelberg, Germany, 1981; pp. 49–69.
131. Ballot, A.; Sandvik, M.; Rundberget, T.; Botha, C.J.; Miles, C. Diversity of cyanobacteria and cyanotoxins in Hartbeespoort Dam, South Africa. *Mar. Freshw. Res.* **2014**, *65*, 175–189. [CrossRef]
132. Oberholster, P.J.; Myburgh, J.G.; Govender, D.; Bengis, R.; Botha, A.-M. Identification of toxigenic Microcystis strains after incidents of wild animal mortalities in the Kruger National Park, South Africa. *Ecotoxicol. Environ. Saf.* **2009**, *72*, 1177–1182. [CrossRef]
133. Scott, W. Occurrence and significance of toxic cyanobacteria in Southern Africa. *Water Sci. Technol.* **1991**, *23*, 175–180. [CrossRef]
134. Mokoena, M.; Mukhola, M.; Okonkwo, O. Hazard assessment of microcystins from the household's drinking water. *Appl. Ecol. Environ. Res.* **2016**, *14*, 695–710. [CrossRef]
135. Oberholster, P.; Cloete, T.; van Ginkel, C.; Botha, A.; Ashton, P. The use of remote sensing and molecular markers as early warning indicators of the development of cyanobacterial hyperscum crust and microcystin producing genotypes in the hypertrophic Lake Hartebeespoort, South Africa. *Pretoria Counc. Sci. Ind. Res.* 2008. Available online: https://pdfs.semanticscholar.org/5ff3/c3915fbb680ddcd6a9f5dff1fd247892d541.pdf (accessed on 22 May 2020).
136. Nonga, H.; Sandvik, M.; Miles, C.; Lie, E.; Mdegela, R.; Mwamengele, G.; Semuguruka, W.; Skaare, J. Possible involvement of microcystins in the unexplained mass mortalities of Lesser Flamingo (Phoeniconaias minor Geoffroy) at Lake Manyara in Tanzania. *Hydrobiologia* **2011**, *678*, 167–178. [CrossRef]
137. Fyumagwa, R.D.; Bugwesa, Z.; Mwita, M.; Kihwele, E.S.; Nyaki, A.; Mdegela, R.H.; Mpanduji, D.G. Cyanobacterial toxins and bacterial infections are the possible causes of mass mortality of lesser flamingos in Soda lakes in northern Tanzania. *Res. Opin. Anim. Vet. Sci.* **2013**, *3*, 1–6.
138. Lugomela, C.; Pratap, H.B.; Mgaya, Y.D. Cyanobacteria blooms—a possible cause of mass mortality of Lesser Flamingos in Lake Manyara and Lake Big Momela, Tanzania. *Harmful Algae* **2006**, *5*, 534–541. [CrossRef]
139. Sekadende, B.C.; Lyimo, T.J.; Kurmayer, R. Microcystin production by cyanobacteria in the Mwanza Gulf (Lake Victoria, Tanzania). *Hydrobiologia* **2005**, *543*, 299–304. [CrossRef]
140. Mbonde, A.S.; Sitoki, L.; Kurmayer, R. Phytoplankton composition and microcystin concentrations in open and closed bays of Lake Victoria, Tanzania. *Aquat. Ecosyst. Health Manag.* **2015**, *18*, 212–220. [CrossRef]
141. Ndebele, M.R.; Magadza, C.H. The occurrence of microcystin-LR in Lake Chivero, Zimbabwe. *Lakes Reserv. Res. Manag.* **2006**, *11*, 57–62. [CrossRef]
142. Mhlanga, L.; Day, J.; Cronberg, G.; Chimbari, M.; Siziba, N.; Annadotter, H. Cyanobacteria and cyanotoxins in the source water from Lake Chivero, Harare, Zimbabwe, and the presence of cyanotoxins in drinking water. *Afr. J. Aquat. Sci.* **2006**, *31*, 165–173. [CrossRef]
143. Gondwe, M.J.; Guildford, S.J.; Hecky, R.E. Planktonic nitrogen fixation in Lake Malawi/Nyasa. *Hydrobiologia* **2008**, *596*, 251–267. [CrossRef]
144. Liu, I.; Lawton, L.A.; Cornish, B.; Robertson, P.K. Mechanistic and toxicity studies of the photocatalytic oxidation of microcystin-LR. *J. Photochem. Photobiol. A Chem.* **2002**, *148*, 349–354. [CrossRef]
145. Yuan, B.; Li, Y.; Huang, X.; Liu, H.; Qu, J. Fe (VI)-assisted photocatalytic degradating of microcystin-LR using titanium dioxide. *J. Photochem. Photobiol. A Chem.* **2006**, *178*, 106–111. [CrossRef]
146. Pavagadhi, S.; Tang, A.L.L.; Sathishkumar, M.; Loh, K.P.; Balasubramanian, R. Removal of microcystin-LR and microcystin-RR by graphene oxide: Adsorption and kinetic experiments. *Water Res.* **2013**, *47*, 4621–4629. [CrossRef]

147. Chintalapati, P.; Mohseni, M. Degradation of cyanotoxin microcystin-LR in synthetic and natural waters by chemical-free UV/VUV radiation. *J. Hazard. Mater.* **2020**, *381*, 120921. [CrossRef] [PubMed]
148. Pendleton, P.; Schumann, R.; Wong, S.H. Microcystin-LR adsorption by activated carbon. *J. Colloid Interface Sci.* **2001**, *240*, 1–8. [CrossRef] [PubMed]
149. Zhang, H.; Zhu, G.; Jia, X.; Ding, Y.; Zhang, M.; Gao, Q.; Hu, C.; Xu, S. Removal of microcystin-LR from drinking water using a bamboo-based charcoal adsorbent modified with chitosan. *J. Environ. Sci.* **2011**, *23*, 1983–1988. [CrossRef]
150. Song, H.-L.; Lv, X.-L.; Li, X.-L. Safety Improvement for Drinking Water in Rural Region by Hydrophyte Filter Bed. *China Water Wastewater* **2006**, *22*, 17–20.
151. Guang-can, Z.; Xi-wu, L. Removal of Microcystins by Biological Activated Carbon Process. *China Water Wastewater* **2005**, *2*, 14–17.
152. Song, H.-L.; Li, X.-N.; Lu, X.-W.; Inamori, Y. Investigation of microcystin removal from eutrophic surface water by aquatic vegetable bed. *Ecol. Eng.* **2009**, *35*, 1589–1598. [CrossRef]
153. Warhurst, A.; Raggett, S.; McConnachie, G.; Pollard, S.; Chipofya, V.; Codd, G. Adsorption of the cyanobacterial hepatotoxin microcystin-LR by a low-cost activated carbon from the seed husks of the pan-tropical tree, Moringa oleifera. *Sci. Total Environ.* **1997**, *207*, 207–211. [CrossRef]

Article

Effect of Zinc on *Microcystis aeruginosa* UTEX LB 2385 and Its Toxin Production

Jose L. Perez and Tinchun Chu *

Department of Biological Sciences, Seton Hall University, South Orange, NJ 07079, USA
* Correspondence: tin-chun.chu@shu.edu

Received: 1 January 2020; Accepted: 28 January 2020; Published: 30 January 2020

Abstract: Cyanobacteria harmful algal blooms (CHABs) are primarily caused by man-made eutrophication and increasing climate-change conditions. The presence of heavy metal runoff in affected water systems may result in CHABs alteration to their ecological interactions. Certain CHABs produce by-products, such as microcystin (MC) cyanotoxins, that have detrimentally affected humans through contact via recreation activities within implicated water bodies, directly drinking contaminated water, ingesting biomagnified cyanotoxins in seafood, and/or contact through miscellaneous water treatment. Metallothionein (MT) is a small, metal-sequestration cysteine rich protein often upregulated within the stress response mechanism. This study focused on zinc metal resistance and stress response in a toxigenic cyanobacterium, *Microcystis aeruginosa* UTEX LB 2385, by monitoring cells with (0, 0.1, 0.25, and 0.5 mg/L) $ZnCl_2$ treatment. Flow cytometry and phase contrast microscopy were used to evaluate physiological responses in cultures. Molecular assays and an immunosorbent assay were used to characterize the expression of MT and MC under zinc stress. The results showed that the half maximal inhibitory concentration (IC_{50}) was 0.25 mg/L $ZnCl_2$. Flow cytometry and phase contrast microscopy showed morphological changes occurred in cultures exposed to 0.25 and 0.5 mg/L $ZnCl_2$. Quantitative PCR (qPCR) analysis of selected cDNA samples showed significant upregulation of *Mmt* through all time points, significant upregulation of *mcyC* at a later time point. ELISA MC-LR analysis showed extracellular MC-LR (μg/L) and intracellular MC-LR (μg/cell) quota measurements persisted through 15 days, although 0.25 mg/L $ZnCl_2$ treatment produced half the normal cell biomass and 0.5 mg/L treatment largely inhibited growth. The 0.25 and 0.5 mg/L $ZnCl_2$ treated cells demonstrated a ~40% and 33% increase of extracellular MC-LR(μg/L) equivalents, respectively, as early as Day 5 compared to control cells. The 0.5 mg/L $ZnCl_2$ treated cells showed higher total MC-LR (μg/cell) quota yield by Day 8 than both 0 mg/L $ZnCl_2$ control cells and 0.1 mg/L $ZnCl_2$ treated cells, indicating release of MCs upon cell lysis. This study showed this *Microcystis aeruginosa* strain is able to survive in 0.25 mg/L $ZnCl_2$ concentration. Certain morphological zinc stress responses and the upregulation of *mt* and *mcy* genes, as well as periodical increased extracellular MC-LR concentration with $ZnCl_2$ treatment were observed.

Keywords: cyanobacteria; cyanotoxins; microcystin; metal; zinc; *Microcystis aeruginosa*

Key Contribution: Zinc stress causes cyanotoxin production increase in *M. aeruginosa* UTEX LB 2385.

---

1. Introduction

Cyanobacteria harmful algal blooms (CHABs) are described as toxigenic or irritating biomasses of mostly oxygenic, photosynthetic bacteria on the rise worldwide due to anthropogenic eutrophication (via excessive P and N loading) and increasing climate-change conditions [1–6]. Oftentimes, metal pollutant runoff in water systems may also affect the ecological interaction of a given CHAB population [7]. Certain essential element processes, such as iron's regulation by (FUR) uptake regulators, may act as growth-limiting factors in established cyanobacteria populations [1,8]. Additionally, the effects of different heavy metal compounds and concentrations on varying cyanobacteria populations and cyanotoxin production have been demonstrated in a few studies [9–11].

The rising occurrence of global CHABs may lead to a greater probability of human exposure and animal detrimental effects through environmental interactions [12,13]. Generally, humans may be exposed to CHABs cyanotoxins or irritants via recreation activities in contaminated water sources, imbibement or infusion (via medical dialysis) of contaminated water, ingestion of biomagnified cyanotoxins through food sources, and possible long-term chronic exposure through ingestion [14–16]. Cyanotoxins may structurally present as alkaloids, polyketides, cyclic peptides, and amino acid complexes, and are potently classified as neurotoxins, hepatotoxins, or cytotoxins [17,18]. The toxic health effects of cyanotoxins in humans are both varied and dependent on the class, but effects may display from mild symptoms to severe and fatal [16,19].

Aside from escalating incidences of detrimental effects to organisms, CHABs will most likely play an increasing role in global economic dynamics. The greater occurrence of global CHABs have reported some monitoring and contingency plan estimate costs of 10,000 s–1,000,000 s USD per country per year [20]. While these numbers do not often reflect loss of recreation revenue, cost of treatment actions, or socio-economic strategies (for anthropogenic P-N load reduction), an estimate report that took these components into account (within the United States) calculated potential losses in the billions USD per year [21].

Microcystis spp. and the hepatotoxins microcystins (MCs), have been identified as several of the most commonly encountered freshwater CHABs species and cyanotoxins, respectively, on a global scale and within the U.S. [22,23]. As such, studies evaluating both *Microcystis* and MCs have increased in importance in the recent years. Adding to the gravity of these considerations, examples of toxigenic *Microcystis* outcompeting non-toxigenic strains within elevated surface water temperature parameters have been observed [24,25]. This indicates a potentially competitive selection of toxigenic strains over non-toxigenic strains under climate change conditions. *Microcystis aeruginosa* is a commonly identified species of *Microcystis* within many CHABs, and have been found on all continents except for the Antarctic [26]. *M. aeruginosa* are typically 2–8 μm wide, unicellular planktonic cyanobacteria that possess variable, intracellular gas vesicles and often form colonies in natural and eutrophic conditions [27–29]. MCs (~995 Da for MC-LR) are water soluble, monocyclic heptapeptides that can be produced by freshwater, terrestrial, and benthic cyanobacteria. The most common genus and species producing a given variant(s) of the ~ 200 identified MCs are: *Microcystis, Chroococcus, Planktothrix, Anabaena, Nostoc, Oscillatoria, Hapalosiphon,* and *Phormidium* [18,30,31]. MCs are biosynthesized by a non-ribosomal peptide synthetase/polyketide synthase complex (known as microcystin synthetase) [18], and vary in toxicity depending on the L-amino acid components and binding capacity to receptor sites [32]. The role of MCs in the extracellular environment remains largely undetermined, but several studies have proposed an intercell signal-like characteristic where environmental conditions and introduced MCs enhanced *mcy* gene and toxin production in established cell populations [33,34]. Therefore, further evaluation of stimulatory MC release to the extracellular environment and their possible extracellular functions with environmental factors is of great importance. Zinc (Zn) is an important mineral integral to the physiological functions of all organisms. Zn naturally occurs as a trace element in world average river waters (0.27–27 μg/L), as ionic complexes in continental crust and world soil averages (70 mg/kg, respectively), and as precipitating minerals, organic, and inorganic compounds in water (~3.25 μg/L in worldwide, clean drinking water), but organic and inorganic zinc compounds are oftentimes identified as metal runoff contaminants within aqueous environments [35–37]. A comparison of the average Zn levels originating from a continental crustal average (52 ppm) was found to be significantly different to an anthropogenic impacted region, New York Harbor, United States (188–244 ppm), showing a further trend in heavy metal accumulation within water environments proximal to urbanized and industrial sources [38,39]. The toxicity of $ZnCl_2$ has been known in many organisms [37,40,41] and can cause external irritation, severe inflammation, and gastrointestinal toxicity dependent on percent ingestion [42]. Cyanobacteria sensitivity, resistance or adaptive sequestration of heavy metal concentrations has been documented within both colonial CHABs and unicellular species [7,43–45]. Aside from the observed zinc metal-complexing potential of MCs (MC-LR-Zn = −617 ± 7 kcal mol^{-1}; MC-RR-Zn =−777 ± 9 kcal mol^{-1}) [46], metallothioneins (MTs) are well documented metal-cation chelating, cysteine-rich proteins (<10 kDa) ubiquitously found in prokaryotes and eukaryotes [47,48]. MTs have been shown to be upregulated in different species of cyanobacteria when exposed to Zn^{2+} or Cd^{2+} concentrations while remaining relatively constant at basal levels [49,50]. Because of the variability of the MC synthetase gene cluster (encoding *mcyABC–mcyD–J*) in different cyanobacteria clades and within strains [51] and a relative conservation of MT cysteine domain sequences and motifs across cyanobacteria and bacteria [48], these genes may be possible quantification method candidates involving heavy metal zinc response and resistance in identified MC producing *Microcystis* species and strains.

While the use of quantitative PCR (qPCR) has yielded both successes and noncorrelation in relating MC synthetase gene copy numbers or gene expressions with collection site MC concentrations [52], qPCR remains a very powerful and accessible technique for the study of gene regulation in known toxigenic or identified cyanobacteria species [53]. Along with other quantitative analysis (HPLC, LC/MS, ELISA) and sequencing profiles, it may lead to the development of known metal-response gene standards for important identified toxic cyanobacteria species and strains. These parameters may better assist in determining toxic vs. non-toxic cyanobacteria response and resistance to heavy metal pollution.

The aim of this study was (1) to study the growth and physiological effects of zinc concentrations on an established toxigenic *M. aeruginosa* strain; (2) to design *mcyC*, *mcyE*, and *Mmt* qPCR oligonucleotides to quantify *mcyC*, *mcyE* and *Mmt* relative gene expression profiles of this strain treated with varying Zn^{2+} concentrations; and (3) to determine relative quantitation of MC-LR equivalents within $ZnCl_2$-treated *M. aeruginosa* using intracellular and extracellular portions.

2. Results

2.1. Growth Response to $ZnCl_2$ in M. Aeruginosa UTEX LB 2385

To better evaluate the zinc metal resistance and response mechanisms of a globally important toxigenic cyanobacteria species, *M. aeruginosa* UTEX LB 2385 cultures were used for the observation of physiological responses to long-term $ZnCl_2$ concentrations exposure via growth monitoring (Figure 1). For *M. aeruginosa*—UTEX LB 2385 cells exposed to 0.1 mg/L $ZnCl_2$, the cell concentration was similar to 0 mg/L $ZnCl_2$ (control) through 15 days (Figure 1). However, the average turbidity for *M. aeruginosa* UTEX LB 2385 culture cells exposed to all $ZnCl_2$ concentrations decreased below the initial control measurement (optical density, $OD_{750\ nm}$ ≈ 0.47) for 0.1, 0.25, and 0.5 mg/L $ZnCl_2$ by Day 1 (Figure 1b). When compared to control cells, *M. aeruginosa* UTEX LB 2385 culture cell numbers exposed to 0.25 mg/L $ZnCl_2$ were reduced by ~28% by Day 15 ((4.73 ± 1.243/8.41 ± 0.122) × 100) (Figure 1a). At the highest concentration treatment (0.5 mg/L $ZnCl_2$), *M. aeruginosa* UTEX LB 2385 culture cells were almost completely inhibited in comparison to 0.25 mg/L $ZnCl_2$ cells by the end of the growth monitoring period. This concentration (0.5 mg/L $ZnCl_2$) may therefore present as the $ZnCl_2$ minimum inhibitory concentration (MIC) for *M. aeruginosa* UTEX LB 2385 within this study (Figure 1). Aside from these observations, *M. aeruginosa* UTEX LB 2385 culture cells treated with 0.25 and 0.5 mg/L $ZnCl_2$ were observed via hemocytometer to possess larger aggregate cell clusters and extracellular debris compared to control cells.

2.2. Phase Contrast Microscopy Imaging in $ZnCl_2$ Exposed Cells

M. aeruginosa UTEX LB 2385 culture cells exposed to 0.25 mg/L $ZnCl_2$ possessed morphological and size characteristics similar to the control cells through eight days as single cells (Figure 2), but also showed an increase in multi-paired cell aggregation and size by Days 5 and 8 when observed via phase contrast microscopy. Additionally, though 0.5 mg/L $ZnCl_2$ treatment greatly inhibited *M. aeruginosa*

UTEX LB 2385 culture cell numbers by Day 8 compared to the control, there was an observable number of cells possessing similar *M. aeruginosa* UTEX LB 2385 morphology to control cells (Figure 2). It also appeared a greater observable amount of multi-cell aggregation within 0.5 mg/L $ZnCl_2$ treated cells versus control cells. Flow cytometry was used to further evaluate the population sizes of $ZnCl_2$ treated *M. aeruginosa* UTEX LB 2385.

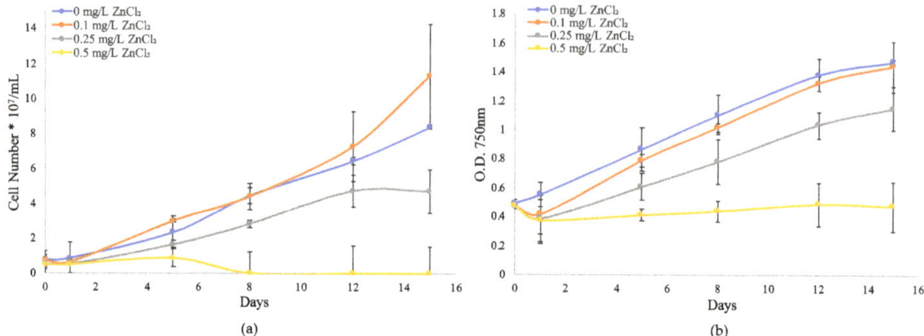

Figure 1. Growth curves of *Microcystis aeruginosa* UTEX LB 2385 treated with 0, 0.1, 0.25, and 0.5 mg/L $ZnCl_2$. (**a**) Direct count of cell number was made via hemocytometer through 15 days. (**b**) Turbidity was evaluated via optical density ($OD_{750\ nm}$) readings through 15 days. Data is presented as mean ± SD of three replicates.

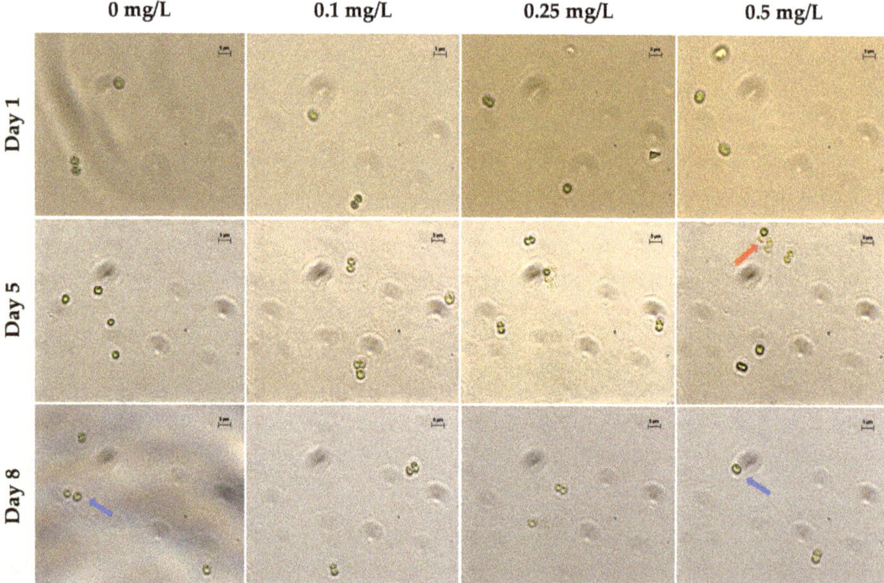

Figure 2. Phase-contrast microscopy of *Microcystis aeruginosa* UTEX LB 2385 treated with 0, 0.1, 0.25, and 0.5 mg/L $ZnCl_2$ at Days 1, 5, and 8, 1000× total magnification. Although greatly reduced in cell number, *M. aeruginosa* UTEX LB 2385 cells treated with 0.5 mg/L $ZnCl_2$ presented morphology similar to 0 mg/L cells through Day 8, as indicated by blue arrows. The red arrow indicates a small cluster of multi-cell aggregation at Day 5 for 0.5 mg/L $ZnCl_2$ cells. Scale bar: 5 μm.

2.3. Flow Cytometry

Flow cytometry was used over the eight day time-course to evaluate the relative morphology size of *M. aeruginosa* UTEX LB 2385 culture cell populations treated with the highest $ZnCl_2$ concentrations (0.25 and 0.5 mg/L). The flow cytometry measurement (FCM) histogram profiles were ungated for subpopulations to evaluate overall distribution of single cell + multi-cell aggregates. Figure 3 demonstrates a histogram profile of 0, 0.25, and 0.5 mg/L $ZnCl_2$ treated *M. aeruginosa* UTEX LB 2385 cultures through size (forward scatter: FSC-A) parameter. By Day 8, (Figure 3) the FSC-A histograms of *M. aeruginosa* UTEX LB 2385 cultures treated with 0.25 and 0.5 mg/L $ZnCl_2$ had positively shifted in comparison to the control cultures, with increasing FSC-A measurement (Figure 3). *M. aeruginosa* UTEX LB 2385 cultures treated with 0.5 mg/L $ZnCl_2$ showed a distinct, larger population size and two small peaks at Day 8 compared to the control—supporting both hemocytometer cell count and microscopic observations of greater numbers of multi-cell aggregates (Figure 3c(D8)). FlowJo software statistical analysis showed average 0.5 mg/L $ZnCl_2$ treated cells FSC-A measurement was approximately 91 versus 62 for 0 mg/L $ZnCl_2$ treated cells at Day 8.

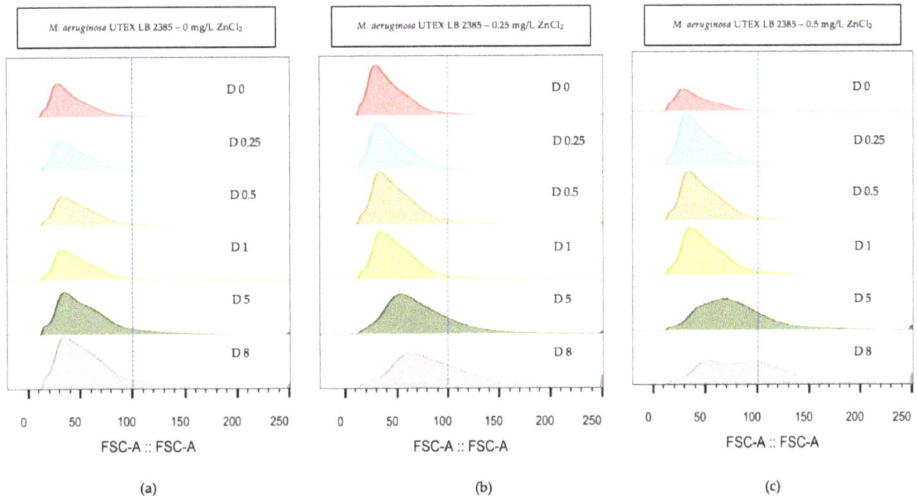

Figure 3. Flow cytometry analysis of population size (forward scatter: FSC-A) histograms within *Microcystis aeruginosa* UTEX LB 2385 treated with 0, 0.25, and 0.5 mg/L $ZnCl_2$. (**a**) 0 mg/L $ZnCl_2$ cells showed relative uniform FSC-A size through eight days. (**b**) 0.25 mg/L $ZnCl_2$ cells and (**c**) 0.5 mg/L $ZnCl_2$ cells showed increased FSC-A population size at Day 8 compared to control cells. Histograms were generated via FlowJo software analysis. (D represent days).

2.4. Quantitative Polymerase Chain Reaction (qPCR) Analysis of M. Aeruginosa UTEX LB 2385

The qPCR analysis results were calculated as fold change gene expressions of *Mmt*, *mcyC*, and *mcyE* relative to the *M. aeruginosa* UTEX LB 2385 16S ribosomal RNA internal control, within the specific time period and samples (Days 1, 5, 12, 15 for 0, 0.1, and 0.25 mg/L $ZnCl_2$) and the initial 0 mg/L $ZnCl_2$ Day 1 sample. Fold change gene expression values were expressed as $[2^{-\Delta\Delta Ct}]$ and evaluated as a combined $[\log_2(RQ)]$ heatmap within RStudio statistical software. The genes expressed in log2 > 0 were represented within an orange-red color scale and indicate upregulation as a response to $ZnCl_2$ treatment, and the genes in log2 < 0 were represented within a green color scale and indicated downregulation as a response to $ZnCl_2$ treatment. Welch two sample *t*-tests of 16S rRNA endogenous average C_t values control were found to not be statistically different within each sample condition (all p values > 0.05). The specific sample treatment and days were abbreviated as D1_0 mg/L $ZnCl_2$, D1_0.1 mg/L $ZnCl_2$,

D1_0.25 mg/L $ZnCl_2$, etc. Correlating to Day 1 treatments for 0, 0.1, and 0.25 mg/L $ZnCl_2$, respectively (Figure 4).

M. aeruginosa UTEX LB 2385 *Mmt* expression displayed an overall upregulation over the time course period of 15 days, with 0.25 mg/L $ZnCl_2$ treated cells showing > 3-fold expression by Day 1 (Figure 4). By Day 15 there was a > 5-fold *Mmt* gene expression within 0.25 mg/L $ZnCl_2$ treated cells, demonstrating a steady increase of *Mmt* gene expression within this study. Additionally, the overall trend throughout 15 days showed 0.1 and 0.25 mg/L $ZnCl_2$ treated cells to be comparable or greater than control (0 mg/L $ZnCl_2$) cells (Figure 4—top row).

M. aeruginosa UTEX LB 2385 *mcyE* gene expression showed the least change of expression between both *mcy* genes profiles, with an initial slight *mcyE* down-regulation by Day 1 in 0.1 and 0.25 mg/L $ZnCl_2$ treated cells. For 0.25 mg/L $ZnCl_2$ treated cells, *mcyE* gene expression showed slight up-regulation from Days 5 to 15 in comparison to the control cells (Figure 4—bottom row). The expression of *mcyC* was shown to be significantly upregulated at Day 12, with >4.5 fold-increase in 0.1 mg/L $ZnCl_2$ treated cells and >3 fold-increase in 0.25 mg/L $ZnCl_2$ treated cells (Figure 4—middle row). The 0.25 mg/L $ZnCl_2$ treated cells also showed a slight *mcyC* upregulation in comparison to both the control and 0.1 mg/L $ZnCl_2$ treated cells at Day 15 (Figure 4—middle row, right). The *mcyC* gene expression profile in $ZnCl_2$ treated cells showed an up-regulation trend at Days 5 and 12 in comparison to all other time periods (Figure 4—middle and bottom rows, center).

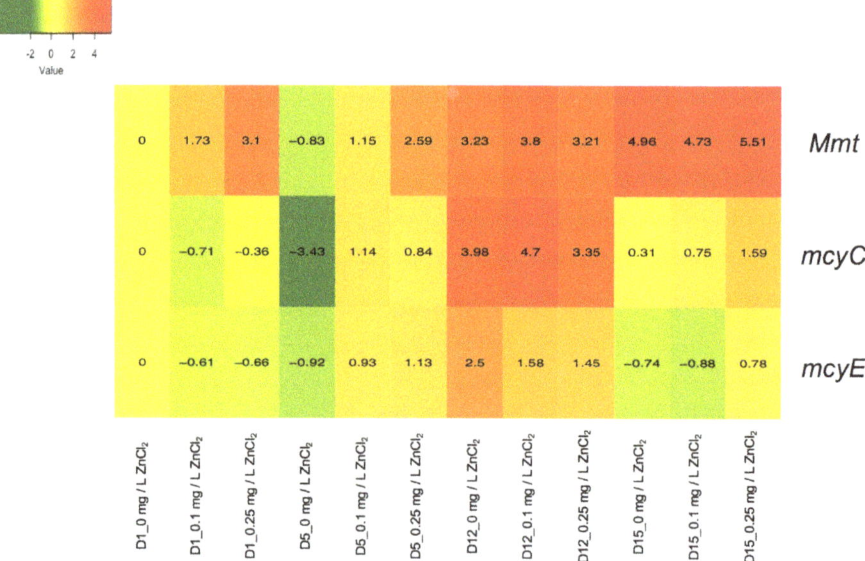

Figure 4. qPCR analysis after cDNA synthesis, of three genes: *Mmt*, *mcyC*, and *mcyE* as potential zinc metal response genes in *M. aeruginosa* UTEX LB 2385. *Mmt* (top row) overall expression was shown to be highest in 0.25 mg/L $ZnCl_2$ treated cells, with significant (Day 1) upregulation in comparison to 0 and 0.1 mg/L $ZnCl_2$ treatment. *mcyC* and *mcyE* (middle and bottom rows, respectively) were distinct in expression patterns, with *mcyC* gene significantly upregulating at Day 12, showing a fold-increase >2.5 in $ZnCl_2$ treated cells. For 0.1 and 0.25 mg/L $ZnCl_2$ treated cells, Day 12 showed gene expression increase for *Mmt* and *mcyC* genes. (D represent days; 0, 0.1, 0.25 mg/L represent $ZnCl_2$ concentrations; color scale of yellow refers to no change, green-downregulation, orange-red-upregulation).

2.5. ELISA Analysis

ELISA quantitative analysis of 0, 0.1, 0.25, and 0.5 mg/L $ZnCl_2$ treated *M. aeruginosa* UTEX LB 2385 cells was performed as MC-LR equivalents using an MC-LR standard curve generated with MC-LR standards (0–5 μg/L). A standard 2nd-order polynomial equation ($y = 7.5536x^2 - 54.927x + 96.06$) was used to calculate MC-LR equivalents below the 2.5 μg/L standard range. The ELISA analysis was performed on extracted MC-LR intracellular and extracellular samples at time periods (Day: 1, 5, 8, 12, and 15). Total MC-LR (μg/cell) quota was calculated by adding intracellular and extracellular MC-LR equivalents and dividing by average cells/L. The Pearson's *r* correlation showed that total MC-LR (μg/cell) quota was positively correlated with $ZnCl_2$ concentration ($p < 0.05$, $r = 0.6$).

The data was presented as intracellular-extracellular MC-LR (μg/L) equivalents to better evaluate relative concentrations of MC-LR within each sample portion. The extracellular MC-LR (μg/L) equivalents for Days 8 and 12 of *M. aeruginosa* UTEX LB 2385 cultures treated with 0.1 mg/L $ZnCl_2$ were comparable to 0 mg/L $ZnCl_2$ control cells (18% increase of MC-LR equivalents for Day 8: {(1.67/2.04) × 100}; 23% increase of MC-LR equivalents for Day 12: {(1.40/1.82) × 100}) (Figure 5). Day 12 extracellular MC-LR (μg/L) equivalent measurements for 0.25 mg/L $ZnCl_2$ treated cells decreased from Day 8, but once again increased comparable to control cells at Day 15 (Figure 5). This same time period saw a 40% increase in cell number/mL biomass within 0.25 mg/L $ZnCl_2$ treated cells (see Figure 1a above) and a greater amount of multicell aggregates at Day 8. The 0.25 and 0.5 mg/L $ZnCl_2$ treated cells possessed higher extracellular MC-LR (μg/L) equivalents at Day 5 compared to control cells {~ 40% increase for 0.25 mg/L $ZnCl_2$ cells (1.05/1.77) × 100; ~ 30% increase for 0.5 mg/L $ZnCl_2$ cells (1.05/1.57) × 100) (Figure 5). From Days 8 through 15 of this study, 0.5 mg/L $ZnCl_2$ treated cells yielded significantly higher total MC-LR (μg/cell) quota yield (student's t-test $p < 0.05$) than all other $ZnCl_2$ treatment. However, the cell concentration had decreased from Day 5 to Day 8 for 0.5 mg/L $ZnCl_2$ treated cells as stated above. This would indicate possible release of MCs into the extracellular matrix from lysed cells rather than increased MC production due to $ZnCl_2$ concentration.

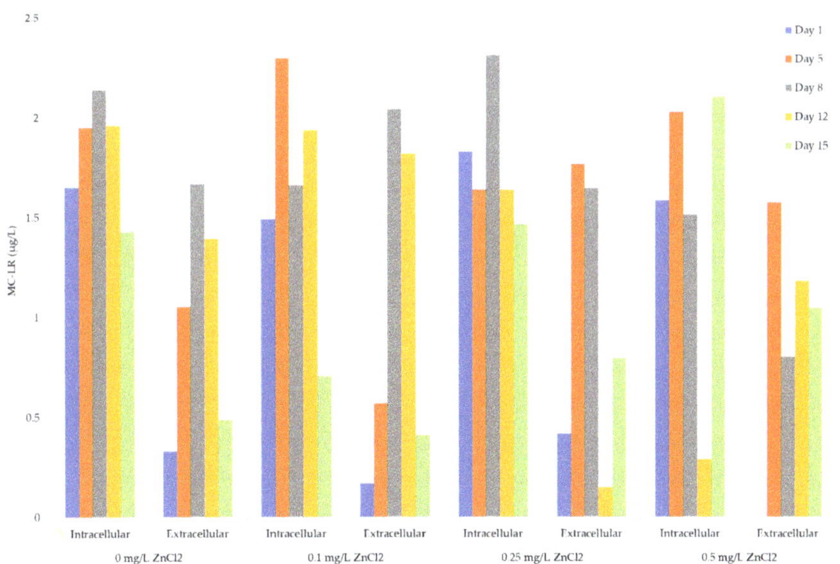

Figure 5. ELISA quantitative analysis of intracellular and extracellular MC-LR (μg/L) equivalents within *Microcystis aeruginosa* UTEX LB 2385 treated with 0, 0.1, 0.25, and 0.5 mg/L $ZnCl_2$ at Days: One, 5, 8, 12, 15. Extracellular MC-LR equivalents (μg/L) of 0.1 mg/L $ZnCl_2$ cells showed a 18% and 23% increase

for Days 8 and 12 compared to control; 0.25 and 0.5 mg/L $ZnCl_2$ treated cells showed a ~40% and 33% increase of extracellular MC-LR, respectively, as early as day 5. The 0.25 mg/L $ZnCl_2$ extracellular MC-LR (μg/L) equivalents decreased and increased from Days 8 to 15, and 0.5 mg/L $ZnCl_2$ intracellular MC-LR (μg/L) intracellular measurements decreased and increased from this same time point. All samples were performed as duplicate replicates.

3. Discussion

Zn is a classified essential, transitional metal ubiquitously involved in numerous cellular biochemical pathways, and is often found within enzyme cavities and protein infrastructures. Zn is often employed in galvanization processes, vulcanization procedures, automobile applications, coating alloy formulations, and various manufacturing components where it often enters the environment in a sequential fashion after ZnO is formed [54]. Environmental chloride ions and pH often form $ZnCl_2$ near man-made sources high in processed Zn [55].

The interactions of zinc and heavy metal pollution with organisms in naturally occurring and man-made aquatic environments presents its own immediate and long-term problems in our increasingly industrial societies. As an example, a surface river, estuary, and bay sediment study in highly industrial Jinzhou Bay, China found Cu, Zn, Pb, and Cd heavy metals concentrations to be correlated to man-made industrial activities, and has been identified as a polluted ecological risk [56]. Relatively, documented metal resistance and stress response mechanisms within cyanobacteria populations necessitates that focus shifts towards risk assessment, prevention, and treatment. *Synechococcus* sp. IU 625 cultures were shown to be highly tolerant to 25 mg/L $ZnCl_2$ treatment, and a similar strain showed comparable growth to control when treated with 0.1 mg/L and 0.5 mg/L $HgCl_2$ [43,50]. Relatively and conversely, a total-protein profile of an *Anabaena flos-aquae* isolate showed progressive decrease upon treatment with increasing concentrations of $CdCl_2$ and $CuCl_2$—with the highest concentrations presenting the most considerable biophysiological and biomass damage [57].

CHABs-associated species and strains often found in metal polluted environments must be given special attention due to their potential survivability and resultant detrimental effects (i.e., cyanotoxin release). Although distinct environmental factors have been observed to contribute to favorable conditions conducive to cyanobacteria growth, the specific combinations of effects that causes heavy metal tolerance or resistance in cyanobacteria population are not well understood, or in fact, remain unknown.

Our results showed that *M. aeruginosa* UTEX LB 2385 is tolerant to $ZnCl_2$ concentrations of 0.1 and 0.25 mg/L, with 0.1 mg/L $ZnCl_2$ treated cell numbers and turbidity measurements being similar to control cells through 15 days (Figure 1a,b). All cell concentrations were found to be statistically similar through five days (student's t-test $p > 0.05$ for 0 mg/L and 0.5 mg/L $ZnCl_2$ treated cells/mL). Further, 0.5 mg/L $ZnCl_2$ concentration was found to largely inhibit cell concentration (cell/mL) by Day 15 $\{(0.01/8.41) \times 100 \sim 99.9\%$ inhibition$\}$ and decreased the turbidity of the culture. This result is similar to gradual cell density decrease with increasing Zn^{2+} concentration [11]. Additionally, it was also similar to *M. aeruginosa* Kütz 854 cultures showing chlorophyll and phycobiliprotein decrease at $CdCl_2$ concentrations of 1 and 2 μM, and a bactericidal reaction at 4 μM $CdCl_2$ treatment [58]. Both a toxic *M. aeruginosa* (FACHB-905) and a non-toxic strain (FACHB-469) were shown to accumulate Zn^{2+} and Cd^{2+} intracellularly as an uptake metal concentration vs. four-hour time rate [59]. This previous study observed intracellular Zn concentration was a good predictor of Zn toxicity in *M. aeruginosa*. In a different study, a 10 nM treatment of Zn as a free ion concentration did not decrease microcystin/chlorophyll-*a* (μg/μg) in comparison to UVR and Cu^{2+} metal treatments of *M. aeruginosa* strains (UTEX LB 2385 and LE3), but total added Zn to lake water did affect the final biomass and growth rates [60]. Different *M. aeruginosa* chlorophyll concentrations were also found to be more stable to $CdNO_3$ treatment than to $Pb(NO_3)_2$, and only showed chlorophyll decrease at 20 mg/L Cd after 24 h [61]. Additionally, low levels of both Cd and Pb (1–5 mg/L) resulted in increases of chlorophyll fluorescence during 24-h incubation, and showed that specific *M. aeruginosa* strain was not inhibited at those concentrations. Conversely, Fe^{2+} and Fe^{3+} was shown to increase *M. aeruginosa* cell density with increasing concentrations up to

12 mg/L [11]. Lastly, *M. aeruginosa* (FACHB-905) cultures were observed to show levels of resistance and recovery to treatment with arsenic (III) concentrations (0.01, 0.1, and 1 mg/L) after 48 h incubation, and only showed marked damaging effects to growth and carotenoids production at 10 mg/L [62]. These observations indicate that heavy metal inhibition, possible resistance, or resultant growth rate in *M. aeruginosa* is likely dependent on strain and metal species.

Cell counts, phase-contrast microscopy, and flow cytometry (FSC-A) histogram profile showed that *M. aeruginosa* UTEX LB 2385 treated with 0.25 and 0.5 mg/L $ZnCl_2$ possessed larger amounts of multi-cell aggregates compared to 0 and 0.1 mg/L $ZnCl_2$ treated cells at Day 8 (Figure 3c(D8)). Progressive positive shifting of histograms occurred from Day 5 to Day 8 referenced from FCM of (100) FSC-A for 0.25 and 0.1 mg/L $ZnCl_2$ treated cells (Figure 3). Future experiments should concentrate on the dynamics of aggregation formation as it relates to intracellular molecular profile, in response to increasing $ZnCl_2$ concentrations. Aggregation of many *Microcystis* species is thought to contribute to possible natural bloom formation, and subsequent protection from environmental factors such as heavy metals [63]. *M. aeruginosa* were shown able to survive in low concentrations of variable heavy metal compounds, and *Microcystis* blooms showed a capacity to bioaccumulate and sequester heavy metal compounds within historically eutrophic lakes [7,10,58].

qPCR analysis of 0.1 and 0.25 mg/L $ZnCl_2$ treated *M. aeruginosa* UTEX LB 2385 showed that *Mmt* gene expression within this strain was significantly upregulated by $ZnCl_2$ concentration dependent and time period dependent parameters (Figure 4—top row). MTs are well documented divalent/monovalent metal sequestrating, intracellular proteins that are ubiquitously found from prokaryotes to humans. Our results are supported by the observations of cyanobacteria *Synechococcus* sp. IU 625, which showed a marked upregulation of *smtA* at 25 mg/L $ZnCl_2$ when analyzed via transcriptome analysis and qPCR analysis [50], and a relative, *Synechococcus elongatus* PCC 7942, which showed upregulation of *smtA* with Zn treatment [64]. The expression of *mcyC* was significantly upregulated at Day 12 for 0.1 and 0.25 mg/L $ZnCl_2$ treatment, while *mcyE* only showed slight upregulation for 0.1 and 0.25 mg/L $ZnCl_2$ treatment cells at this same time point (Figure 4—middle and bottom row). While there is an obvious correlation of *mcyC* expression levels and $ZnCl_2$ treatment within this study, *mcyE* expression level trends varied by both concentration and time period (0 and 0.1 mg/L $ZnCl_2$ showed slight downregulation at Day 15 vs. *mcyC*) and was not predictive of $ZnCl_2$ concentration response (Figure 4—middle and bottom row, right). Additional *mcy* gene expression, metal treatment studies should be performed to better assess possible intracellular MC functions. Though *mcyC* function in *Microcystis aeruginosa* PCC 7806 has been evaluated to be important in the final condensation reaction release of the entire MC heptapeptide [65], further analysis must be performed to evaluate *mcyC* gene relation to heavy metal stress response, MC-variant identification, and total MC release to the extracellular environment. The intracellular roles of MCs are still largely undetermined, but a possible siderophore-like function towards zinc and iron may exist under specific intracellular conditions [66]. Molecular modeling for MC-LR metal binding showed that the total potential energy ($Kcal{\cdot}mol^{-1}$) (relative stability) for MC-LR was: Zn > Cu ≥ Fe ≥ Mg > Ca [46].

ELISA quantitative analysis showed MC-LR (µg/L) were present in all $ZnCl_2$ treated intracellular and extracellular *M. aeruginosa* UTEX LB 2385 culture portions (Figure 5). There was a higher correlation of MC-LR (µg/cell) yield to 0.5 mg/L $ZnCl_2$ treatment cells from Day One to Day 15 versus 0, 0.1, and 0.2 mg/L $ZnCl_2$ treated cells ($p < 0.05$, $r = 0.6$). This correlation was inversely associated with the decreased cell biomass through 15 days, indicating release of MC-LR into the extracellular environment. It was observed that intracellular microcystin production was decreased with increasing multiple metal concentrations, indicating a possible distribution of microcystins (intracellularly to extracellularly) due to cell lysis and/or cellular damage [10]. The 0.25 and 0.5 mg/L $ZnCl_2$ treated cells showed a ~40% and 33% increase of extracellular MC-LR equivalents, respectively, as early as Day 5 (Figure 5). Although 0.5 mg/L $ZnCl_2$ cell concentration was largely inhibited by the end of the monitor period (99.9% inhibition), a number of multi-cell aggregates and a very low cell number of individual cells (1×10^5 cells/mL) were present at Day 15. The sum of all measured extracellular and intracellular

MC-LR (μg/L) concentration through 15 days was slightly higher for cultures treated with 0, 0.1, and 0.25 mg/L $ZnCl_2$ than for the 0.5 mg/L $ZnCl_2$ treated culture. These results are supported by observations that total MC presence is largely associated with overall cell biomass [67]. Extracellular MC-LR (μg/L) equivalent measurements for 0.25 mg/L $ZnCl_2$ treated cells fluctuated between Days 8, 12, and 15 (1.64, 0.15, 0.80 μg/L, respectively), but was decreased from Day 5 (Figure 5). Interestingly, the presence of observed multi-cell aggregates in both 0.25 and 0.5 mg/L $ZnCl_2$ treated cells from Days 8 through 15 coincided with this observation. These structures may have possibly affected MC interactions with higher $ZnCl_2$ concentrations (0.25 and 0.5 mg/L $ZnCl_2$) in the extracellular environment for the measured time period. A further study of multi-cell aggregation, Exopolysaccharides (EPS) production, and the MC role in the extracellular environment must be evaluated to determine interaction dynamics and possible association. These observations are supported by the following studies. The application of 0.5 mg·L $^{-1}$ Zn^{2+} was shown to increase the dissolve organic carbon (DOC) concentration of *M. aeruginosa* (FACHB 469) into the surrounding media by a 18-day study [68]. Furthermore, it was shown that the addition of MC-RR (0.25–10 μg·L^{-1}) significantly enlarged *Microcystis* colony size, specifically increasing EPS production [69]. There are a number of undetermined factors concerning the interaction of MCs-metal species in the extracellular environment, and these factors are complicated by the presence of EPS, amino acids, and miscellaneous molecules produced by stress-responsive cells.

Further associative evaluation of *Microcystis* aggregation and the associated multi-cell components to heavy metal stress is necessary to determine potential resistance mechanisms and survivability dynamics of this cyanobacteria. A future transcriptomic evaluation of Zn heavy metal response in *M. aeruginosa* is necessary to profile which intracellular and extracellular localizing genes are in regulation, and to distinguish this important CHAB-associated cyanobacteria species from other cyanobacteria species. Finally, the role of MCs in the extracellular environment must be further determined to understand cyanobacteria-cyanobacteria environment dynamics.

4. Conclusions

This study exhibits the $ZnCl_2$ metal stress response and resistance capabilities that a universal CHAB species, *Microcystis aeruginosa* (in this study with strain *M. aeruginosa* UTEX LB 2385) possesses, and the potential survivability this species demonstrates in increasing polluted aquatic environments. *M. aeruginosa* UTEX LB 2385 was observed to survive $ZnCl_2$ concentrations of up to 0.25 mg/L, with increasing biomass through 15 days. Though mostly inhibited, 0.5 mg/L $ZnCl_2$ treated cultures presented multi-cell aggregates and residual populations through 15 days. A persistent yield of the cyanotoxin MC-LR (μg/cell) was observed in all $ZnCl_2$ treated cells by 15 days, indicating that this cyanotoxin remains present in the environment even with low cell concentrations. This finding was supported by qPCR data gene expression profiles of *Mmt* and *mcyC*, suggesting that *M. aeruginosa* UTEX LB 2385 possesses several metal response mechanisms.

5. Materials and Methods

5.1. Growth Monitoring

Microcystis aeruginosa UTEX LB 2385 (NCBI Taxon ID: 1296356) was acquired from UTEX Culture Collection of Algae, TX, USA, and was grown to late exponential growth phase ($OD_{750\ nm}$ = 1.0; cells/mL = 6 × 10^7) in sterile 1X Cyanobacteria BG-11 Freshwater Medium (Sigma Life Sciences, St. Louis, MO, USA), within a sterile 250 mL Borosilicate Erlenmeyer flask at 21 ± 2 °C under constant 24 h cool-white fluorescent light (24 μmol·m^{-2} s^{-1} photons) at a constant agitation of 100 rpm (Innova 2000 Platform Shaker—New Brunswick Scientific, Edison, NJ, USA). The 1X Cyanobacteria BG-11 media was prepared with sterile deionized Milli-Q water (Milli-Q Plus ultra-pure water system—Millipore, Billerica, MA, USA) and the pH was adjusted to 8.0 ± 0.1 with 1 M NaOH. Duplicate sets of four 20 mL volumes of *M. aeruginosa* UTEX LB 2385 culture were centrifuged at 2900× *g* for 10 min, the supernatants were discarded, and the cell pellets were gently washed with sterile deionized

Milli-Q water. These cells were repelleted and constituted in 20 mL fresh sterile 1X Cyanobacteria BG-11 media. The 20 mL volumes were aseptically transferred and diluted with 80 mL fresh sterile 1X BG-11 Medium into eight new sterile 250 mL Borosilicate Erlenmeyer Flasks.

5.2. Experimental Design

A sterile 1% zinc chloride ($ZnCl_2$) (Sigma-Aldrich, St. Louis, MO, USA) solution was added to each *M. aeruginosa* UTEX LB 2385 culture to yield relative $ZnCl_2$ concentrations: 0 mg/L, 0.1 mg/L, 0.25 mg/L, and 0.5 mg/L (0 μM, 0.734 μM, 1.835 μM, 3.669 μM, respectively), as described previously [50,70]. These $ZnCl_2$ concentrations were targeted through multiple growth curve analysis from past experiments, with concentrations as high as 10 mg/L $ZnCl_2$ (data not shown). The rationale was predicated on a previously studied cyanobacterium, *Synechococcus* sp. IU 625 with $ZnCl_2$. All $ZnCl_2$-treated cultures were maintained at the same culture growth parameters described above and were monitored by turbidity observation at optical density (O.D.$_{750\ nm}$) with an UltraSpec III (Pharmacia LKB—Pfizer, New York, NY, USA) for a predetermined time course of Days: 0, 1, 5, 8, 12, and 15. All *M. aeruginosa* UTEX LB 2385 culture cell counts were performed via hemocytometer at 400× magnification using an Olympus BH2 BHS-312 Trinocular Microscope (Olympus Corp, Waltham, MA, USA), and all experiments were repeated in triplicates. The standard deviations of duplicate set means were used to generate the growth curves.

5.3. Phase Contrast Microscopy

To determine the physiological effects of $ZnCl_2$ concentration on *M. aeruginosa* UTEX LB 2385 cultures, cell cultures were collected and imaged. All $ZnCl_2$-treated *M. aeruginosa* UTEX LB 2385 cultures were collected during the set, predetermined time points (Days: 0, 1, 5, 8, 12, and 15) and immediately imaged with a Zeiss AxioLab A1 phase contrast microscope coupled with an AxioCam MrC camera (Carl Zeiss, Oberkochen, Germany) at 1000× total magnification.

5.4. Flow Cytometry

Sample cell size for $ZnCl_2$-treated *M. aeruginosa* UTEX LB 2385 cultures was measured (FCM) in 10,000 events per sample within a calibrated MACSQuant Analyzer 10 (Miltenyi Biotec, Auburn, CA, USA) containing 405 $_{nm}$, 488 $_{nm}$, and 638 $_{nm}$ lasers, using the forward scatter setting (FSC-A) for a set short term and long term course period of 8 days (0, 0.25, 0.5, 1, 5, 8). For each time point, 1 mL of each cell treatment was aseptically transferred into a 1.5 mL microcentrifuge tubes, and the parameters were set to gently resuspend before 100 μL was measured in FCM. Different cyanobacteria populations may exhibit distinct patterns of size, complexity, and autofluorescence in regards to colony formation and to their phycobilisome complex. Furthermore, the light harvesting components and phycobilisome complex has been shown to be affected by heavy metals in concentration dependent kinetics [71,72]. Allophycocyanin and phycoerythrin fluorescent intensities for *M. aeruginosa* UTEX LB 2385 cells were measured in a calibrated MACSQuant Analyzer 10 (Miltenyi Biotec, Auburn, CA, USA) for the set time course period of 8 days (data not shown). All flow cytometry histograms and statistical analysis of measurements were generated via FlowJo software analysis (FlowJo, Ashland, OR, USA) to compare the effects of varying concentrations of $ZnCl_2$ treatment within all cultures.

5.5. Total RNA Isolation and cDNA Synthesis

The $ZnCl_2$ treatment (0.5 mg/L) was found to inhibit *M. aeruginosa* UTEX LB 2385 cell number and decrease turbidity by Day 8. Therefore, total RNA (*M. aeruginosa* UTEX LB 2385—0, 0.1, and 0.25 mg/L $ZnCl_2$ for Days: 1, 5, 12, and 15) was isolated and purified using a modified Ambion® RiboPure™ Kit (Ambion, Austin, TX, USA) approach. Homogenization and sample disruption preparation was prepared by adding a volume of 48 mL of 100% EtOH to 60 mL of Wash Solution Concentrate. One mL TRI Reagent was then aseptically mixed with 250 μL of cultures within sterile 2 mL microcentrifuge tubes. The mixtures were homogenized (by vortexing), sonicated 15 times with 3 s pulses (20% power) using a

Branson Sonifier Cell Disruptor 200 (Emerson Industrial, St. Louis, MO, USA), and incubated for 5 min at room temperature. Homogenates were centrifuged at 12,000× *g* for 10 min at 4 °C, and supernatants transferred to new sterile 2 mL microcentrifuge tubes. RNA extraction was performed by aseptically adding a volume of 200 μL $CHCl_3$ solution to each sample, tightly capping and vortexing the sample tubes at 700× *g* for 15 s, and then incubating at room temperature for 5 min. Each sample was centrifuged at 12,000× *g* for 10 min at 4 °C before 400 μL of aqueous phase was aseptically transferred to sterile 2 mL microcentrifuge tubes. Then, 200 μL of 100% EtOH was aseptically added to the 400 μL of aqueous phase and immediately vortexed at 700× *g* for 5 s. The sample mixtures were aseptically transferred to a filter cartridge placed within a collection tube, capped, and centrifuged at 12,000× *g* for 30 s at room temperature. Each sample flow-through was discarded and the filter cartridge (with bound RNA) was replaced within the same collection tube. Next, 500 μL of wash solution was aseptically added to the filter cartridges, capped, and centrifuged at 12,000× *g* for 30 s at room temperature. Each sample flow-through was discarded and the filter cartridge was replaced within the same collection tube. The same wash solution step was repeated. Then, the filter cartridges were transferred to new sterile collection tubes and 100 μL elution buffer was added to each respective filter column. The samples were then incubated at RT for 2 min and centrifuged at 12,000× *g* for 30 s to elute RNA.

The concentration (μg/mL) and A260/280 ratios for total RNA samples were checked using a BioDrop UV/VIS Spectrophotometer (Denville Scientific, Metuchen, NJ, USA). The isolated RNA samples served as templates for cDNA synthesis with an ABI High Capacity cDNA Reverse Transcription Kit (Applied Biosystems-Life Technologies, Camarillo, CA, USA), using random hexamers as per the manufacturer's specification. Briefly for each sample, 10 μL DNase-treated RNA was carefully mixed with 10 μL 2X RT MasterMix I (2 μL RT Buffer; 0.8 μL 25X dNTP Mix; 2 μL 10X RT Random Primers; 1 μL MultiScribe™ Reverse Transcriptase; 4.2 μL sterile nuclease-free H_2O) within a 200 μL nuclease-free reaction tube. The sample tubes were placed and run in a Veriti 96 well Thermocycler (Applied Biosystems, Camarillo, CA, USA) via incubation at 25 °C for ten minutes, 37 °C for two hours, and RT inactivation at 85 °C for 5 m. cDNA sample concentrations and A260/280 ratios were checked using a BioDrop UV/VIS Spectrophotometer, and stored at −20 °C until prepared to use.

Microcystis aeruginosa Mmt, *mcyC*, and *mcyE* genes were chosen for primer design based upon phylogenetic analysis of 16S-23S rRNA ITS sequences, multiple alignment of *Microcystis aeruginosa* MC synthetase sequences, and multiple alignment of selected *Microcystis* MT amino acid sequences (T-Coffee Program) (data not shown).

5.6. Quantitative Polymerase Chain Reaction (qPCR)

Oligonucleotides for qPCR were designed using IDT qPCR PrimerQuest (IDT, IA, USA) based on *Microcystis aeruginosa mt* gene (*M. aeruginosa* PCC 7941 (NZ_HE973171), and qPCR *mcyC* and *mcyE* oligonucleotides were designed based on *M. aeruginosa mcy* genes (*M. aeruginosa* UTEX LB 2388: EU009881 and *M. aeruginosa* K-139: AB032549, respectively) using IDT qPCR PrimerQuest. The following primer pairs were generated: MmtRT2f: 5′-TTGTGAATCCTGTACGTGTCAA-3′ and MmtRT2r: 5′-GTCCACAGCCTTTCCCTTTA-3′; mcyC_RT1f: 5′-GGCTAAACCTGACGGGTATAAA-3′ and mcyC_RT1r: 5′-CGCAATATTGAGGGAACACAAG-3′; mcyE_RT1f: 5′-AAGTGGGACCAAGACCAATAC-3′ and mcyE_RT1r: 5′-TCTAAGCCACGATTGAGAGAAC-3′. An endogenous control was designed to normalize the relative gene expression of *mcyC*, *mcyE*, and *mt* genes using IDT PrimerQuest with *M. aeruginosa* UTEX LB 2385 (KF372572) 16S ribosomal RNA {MA8516S_RT1F: 5′-GTAGCAGGAATTCCCAGTGTAG-3′ and MA8516S_RT1R: 5′-TTCGTCCCTGAGTGTCAGATA-3′}.

The cDNA samples were diluted to a concentration of ~100 μg/mL and measured with a BioDrop UV/VIS Spectrophotometer to determine A260/280 ratios. The comparative C_T relative gene expression method was used with the final equation:

$$\text{Fold change} = 2^{-\Delta\Delta Ct} \quad (1)$$

Quantitative RT PCR reactions were performed within 96-well plate assays using an Applied Biosystems StepOnePlus™ Real Time PCR System (ThermoFisher Scientific, MA, USA) with a Luna® Universal qPCR Master Mix (New England BioLabs, Ipswich, MA, USA) (containing Hot Start *Taq* DNA Polymerase). SYBR green dye chemistry was used for reactions containing: 10 μL qPCR Master Mix, 7 μL nuclease-free H_2O, 1 μL of forward and 1 μL reverse primers for a total final 1 μM concentration, and 1 μL < 100 ng final mass of diluted cDNA. The qPCR reactions were performed in triplicates through an initial incubation of 50 °C for two minutes, an initial one-cycle denaturation step at 95 °C for 60 s, and 40 cycles of 95 °C denaturation for 15 s with 60 °C extension for 30 s.

5.7. Enzyme-Linked Immunosorbent Assay (ELISA)

$ZnCl_2$-treated *M. aeruginosa* UTEX LB 2385 cultures were collected for a predetermined course of 15 days (Days: 1, 5, 8, 12, and 15) by centrifuging at 2900× *g* for 10 min at RT. The supernatants were aseptically removed from the pellets and filtered into new sterile centrifuge tubes using sterile microfiltration apparatus (with sterile ≤ 0.22 μm membrane filters). All pellet and supernatant samples were frozen at −20 °C and stored for 15 days. Intracellular microcystin samples were extracted via modified extraction methods [22,73]. Briefly, frozen *M. aeruginosa* UTEX LB 2385 pellet samples were thawed, refrozen at −20 °C and thawed again, before 5 mL 75% methanol was added and samples were sonicated for 1 min at 20% power using a Branson Sonifier Cell Disruptor 200 (Emerson Industrial, St. Louis, MO, USA). The homogenized pellet samples were allowed to sit at RT for 5 min before they were transferred to a sterile microfiltration device/apparatus (with a sterile ≤ 0.22 μm membrane filter). For initial supernatant samples, 5 mL 75% methanol was added and mixed, before aseptically transferring to their respective sterile microfiltration device/apparatus. The total filtrates were diluted to final 1 mL 4% methanol working sample solutions.

ELISA quantitation for intracellular and extracellular microcystin-LR equivalents was performed using a Microcystin-LR ELISA kit (colorimetric) (Abnova, Taipei, Taiwan) as per manufacturer specifications. Final absorbances were read at 450 nm in duplicates using a Varioskan™ LUX multimode microplate reader (ThermoFisher Scientific, Waltham, MA, USA) with SkanIt Software.

5.8. Statistical Analysis

The growth analysis experiments were performed in triplicates and the standard deviation of means were used to generate growth curves. The statistical analysis of flow cytometry measurements (FCM) were generated via FlowJo software analysis (FlowJo, Ashland, OR, USA). The initial raw data was analyzed using the comparative C_t method in the ABI StepOne Software (Life Technologies, Camarillo, CA, USA). To ascertain if the endogenous 16S rRNA reference gene varied due to experimental conditions, student's *t*-tests were performed for all accumulated averaged C_t values organized as treatment groups [74]. The full form of the comparative C_t method equation (1) was used to evaluate relative quantifications of *mcyE*, *mcyC*, and *mt* genes using the MA_UTEX LB 2385-specific 16S rRNA endogenous control for the calculations. A Welch two sample t-test was used to evaluate possible statistical significance between different, pooled $ZnCl_2$ treatment groupings of 16S rRNA endogenous control, using RStudio statistical software (RStudio Team, 2018). A standard 2nd-order polynomial equation ($y = 7.5536x^2 - 54.927x + 96.06$) was used to calculate MC-LR equivalents below the 2.5 μg/L standard range. Total MC-LR (μg/cell) quota was calculated by adding intracellular-extracellular MC-LR equivalents (μg/L) and dividing by average cells/L. A Pearson's product-moment *r* correlation was used to evaluate increasing $ZnCl_2$ concentration.

Author Contributions: Conceptualization, T.C.; methodology, J.L.P., and T.C.; investigation, J.L.P.; formal analysis, J.L.P., and T.C.; resources, T.C.; data curation, T.C.; writing—original draft preparation, J.LP.; writing—review and editing, T.C.; visualization, J.L.P., and T.C.; supervision, T.C.; project administration, T.C.; funding acquisition, T.C. All authors have read and agreed to the published version of the manuscript.

Funding: This research was funded by Seton Hall University (SHU) Biological Sciences Department Annual Research Fund and William and Doreen Wong Foundation to T.C.

Conflicts of Interest: The authors declare no conflicts of interest.

References

1. Cassier-Chauvat, C.; Chauvat, F. Responses to Oxidative and Heavy Metal Stresses in Cyanobacteria: Recent Advances. *IJMS* **2014**, *16*, 871–886. [CrossRef] [PubMed]
2. O'Neil, J.M.; Davis, T.W.; Burford, M.A.; Gobler, C.J. The rise of harmful cyanobacteria blooms: The potential roles of eutrophication and climate change. *Harmful Algae* **2012**, *14*, 313–334. [CrossRef]
3. Paerl, H.W.; Fulton, R.S.; Moisander, P.H.; Dyble, J. Harmful Freshwater Algal Blooms, With an Emphasis on Cyanobacteria. *Sci. World J.* **2001**, *1*, 76–113. [CrossRef] [PubMed]
4. Paerl, H.W.; Hall, N.S.; Calandrino, E.S. Controlling harmful cyanobacterial blooms in a world experiencing anthropogenic and climatic-induced change. *Sci. Total Environ.* **2011**, *409*, 1739–1745. [CrossRef] [PubMed]
5. Patidar, S.K.; Chokshi, K.; George, B.; Bhattacharya, S.; Mishra, S. Dominance of cyanobacterial and cryptophytic assemblage correlated to CDOM at heavy metal contamination sites of Gujarat, India. *Environ. Monit. Assess* **2015**, *187*, 4118. [CrossRef] [PubMed]
6. Visser, P.M.; Verspagen, J.M.H.; Sandrini, G.; Stal, L.J.; Matthijs, H.C.P.; Davis, T.W.; Paerl, H.W.; Huisman, J. How rising CO2 and global warming may stimulate harmful cyanobacterial blooms. *Harmful Algae* **2016**, *54*, 145–159. [CrossRef]
7. Jia, Y.; Chen, W.; Zuo, Y.; Lin, L.; Song, L. Heavy metal migration and risk transference associated with cyanobacterial blooms in eutrophic freshwater. *Sci. Total Environ.* **2018**, *613–614*, 1324–1330. Available online: https://www.sciencedirect.com/science/article/pii/S0048969717325317?via%3Dihub (accessed on 20 December 2019). [CrossRef]
8. Ludwig, M.; Chua, T.T.; Chew, C.Y.; Bryant, D.A. Fur-type transcriptional repressors and metal homeostasis in the cyanobacterium Synechococcus sp. PCC 7002. *Front. Microbiol.* **2015**, *6*, 1217. [CrossRef]
9. Martínez-Ruiz, E.B.; Martínez-Jerónimo, F. How do toxic metals affect harmful cyanobacteria? An integrative study with a toxigenic strain of Microcystis aeruginosa exposed to nickel stress. *Ecotoxicol. Environ. Saf.* **2016**, *133*, 36–46. [CrossRef]
10. Polyak, Y.; Zaytseva, T.; Medvedeva, N. Response of Toxic Cyanobacterium Microcystis aeruginosa to Environmental Pollution. *Water Air Soil Pollut.* **2013**, *224*, 1494. [CrossRef]
11. Zhou, H.; Chen, X.; Liu, X.; Xuan, Y.; Hu, T. Effects and control of metal nutrients and species on Microcystis aeruginosa growth and bloom. *Water Environ. Res.* **2019**, *91*, 21–31. [CrossRef] [PubMed]
12. Cheung, M.Y.; Liang, S.; Lee, J. Toxin-producing cyanobacteria in freshwater: A review of the problems, impact on drinking water safety, and efforts for protecting public health. *J. Microbiol.* **2013**, *51*, 1–10. [CrossRef] [PubMed]
13. Lopez-Rodas, V.; Maneiro, E.; Lanzarot, M.P.; Perdigones, N.; Costas, E. Mass wildlife mortality due to cyanobacteria in the Donana National Park, Spain. *Vet Rec.* **2008**, *162*, 317–318. [CrossRef] [PubMed]
14. Codd, G.; Bell, S.; Kaya, K.; Ward, C.; Beattie, K.; Metcalf, J. Cyanobacterial toxins, exposure routes and human health. *Eur. J. Phycol.* **1999**, *34*, 405–415. [CrossRef]
15. Hitzfeld, B.C.; Hoger, S.J.; Dietrich, D.R. Cyanobacterial toxins: Removal during drinking water treatment, and human risk assessment. *Environ. Health Perspect.* **2000**, *108*, 10.
16. Otten, T.G.; Paerl, H.W. Health Effects of Toxic Cyanobacteria in U.S. Drinking and Recreational Waters: Our Current Understanding and Proposed Direction. *Curr. Envir. Health Rpt.* **2015**, *2*, 75–84. [CrossRef]
17. Carmichael, W.W. Cyanobacteria secondary metabolites-the cyanotoxins. *J. Appl. Bacteriol.* **1992**, *72*, 445–459. [CrossRef]
18. Dittmann, E.; Fewer, D.P.; Neilan, B.A. Cyanobacterial toxins: biosynthetic routes and evolutionary roots. *Fems. Microbiol. Rev.* **2013**, *37*, 23–43. [CrossRef]
19. Carmichael, W.W.; Azevedo, S.M.; An, J.S.; Molica, R.J.; Jochimsen, E.M.; Lau, S.; Rinehart, K.L.; Shaw, G.R.; Eaglesham, G.K. Human fatalities from cyanobacteria: chemical and biological evidence for cyanotoxins. *Envrion. Health Perspect* **2001**, *109*, 663–668. [CrossRef]

20. Sanseverino, I.; Conduto, D.; Pozzoli, L.; Dobricic, S.; Lettieri, T.; European, C.; Joint Research, C. *Algal Bloom and Its Economic Impact*; JRC Science Hub: Ispra, Italy, 2016; Volume 52. Available online: https://publications.jrc.ec.europa.eu/repository/bitstream/JRC101253/lbna27905enn.pdf. (accessed on 20 December 2019).
21. Dodds, W.K.; Bouska, W.W.; Eitzmann, J.L.; Pilger, T.J.; Pitts, K.L.; Riley, A.J.; Schloesser, J.T.; Thornbrugh, D.J. Eutrophication of U.S. Freshwaters: Analysis of Potential Economic Damages. *Environ. Sci. Technol.* **2009**, *43*, 12–19. [CrossRef]
22. Chorus, I.; Falconer, I.R.; Salas, H.J.; Bartram, J. Health risks caused by freshwater cyanobacteria in recreational waters. *J. Toxicol Environ. Health B Crit. Rev.* **2000**, *3*, 323–347. [CrossRef]
23. D'Anglada, L.V. Editorial on the special issue "harmful algal blooms (HABs) and public health: progress and current challenges". *Toxins (Basel)* **2015**, *7*, 4437–4441. [CrossRef] [PubMed]
24. Davis, T.W.; Berry, D.L.; Boyer, G.L.; Gobler, C.J. The effects of temperature and nutrients on the growth and dynamics of toxic and non-toxic strains of Microcystis during cyanobacteria blooms. *Harmful Algae* **2009**, *8*, 715–725. [CrossRef]
25. Dziallas, C.; Grossart, H.P. Increasing oxygen radicals and water temperature select for toxic Microcystis sp. *PloS ONE* **2011**, *6*, e25569. [CrossRef]
26. van Gremberghe, I.; Leliaert, F.; Mergeay, J.; Vanormelingen, P.; Van der Gucht, K.; Debeer, A.E.; Lacerot, G.; De Meester, L.; Vyverman, W. Lack of phylogeographic structure in the freshwater cyanobacterium Microcystis aeruginosa suggests global dispersal. *PLoS ONE* **2011**, *6*, e19561. [CrossRef]
27. Mlouka, A.; Comte, K.; Castets, A.M.; Bouchier, C.; Tandeau de Marsac, N. The gas vesicle gene cluster from Microcystis aeruginosa and DNA rearrangements that lead to loss of cell buoyancy. *J. Bacteriol.* **2004**, *186*, 2355–2365. [CrossRef]
28. Mlouka, A.; Comte, K.; Tandeau de Marsac, N. Mobile DNA elements in the gas vesicle gene cluster of the planktonic cyanobacteria Microcystis aeruginosa. *Fems. Microbiol Lett.* **2004**, *237*, 27–34. [CrossRef]
29. Wu, Z.-X.; Gan, N.-Q.; Huang, Q.; Song, L.-R. Response of Microcystis to copper stress – Do phenotypes of Microcystis make a difference in stress tolerance? *Environ. Pollut.* **2007**, *147*, 324–330. [CrossRef]
30. Hu, C.; Rzymski, P. Programmed Cell Death-Like and Accompanying Release of Microcystin in Freshwater Bloom-Forming Cyanobacterium Microcystis: From Identification to Ecological Relevance. *Toxins (Basel)* **2019**, *11*, 706. [CrossRef]
31. Oksanen, I.; Lohtander, K.; Sivonen, K.; Rikkinen, J. Repeat-type distribution in trnL intron does not correspond with species phylogeny: comparison of the genetic markers 16S rRNA and trnL intron in heterocystous cyanobacteria. *Int. J. Syst. Evol. Microbiol.* **2004**, *54*, 765–772. [CrossRef]
32. Atencio, L.; Moreno, I.; Jos, A.; Prieto, A.I.; Moyano, R.; Blanco, A.; Camean, A.M. Effects of dietary selenium on the oxidative stress and pathological changes in tilapia (Oreochromis niloticus) exposed to a microcystin-producing cyanobacterial water bloom. *Toxicon* **2009**, *53*, 269–282. [CrossRef]
33. Omidi, A.; Esterhuizen-Londt, M.; Pflugmacher, S. Still challenging: the ecological function of the cyanobacterial toxin microcystin – What we know so far. *Toxin Rev.* **2018**, *37*, 87–105. [CrossRef]
34. Yeung, A.C.Y.; D'Agostino, P.M.; Poljak, A.; McDonald, J.; Bligh, M.W.; Waite, T.D.; Neilan, B.A. Physiological and Proteomic Responses of Continuous Cultures of Microcystis aeruginosa PCC 7806 to Changes in Iron Bioavailability and Growth Rate. *Appl. Environ. Microbiol.* **2016**, *82*, 5918–5929. [CrossRef]
35. Gasperi, J.; Garnaud, S.; Rocher, V.; Moilleron, R. Priority pollutants in wastewater and combined sewer overflow. *Sci. Total Enviorn.* **2008**, *407*, 263–272. [CrossRef]
36. Sakson, G.; Brzezinska, A.; Zawilski, M. Emission of heavy metals from an urban catchment into receiving water and possibility of its limitation on the example of Lodz city. *Enviorn. Monit. Assess* **2018**, *190*, 281. [CrossRef]
37. Sandstead, H.H.; Au, W. Zinc. In *Handbook on the Toxicology of Metals*; Nordberg, G.F., Fowler, B.A., Nordberg, M., Friberg, L.T., Eds.; Academic Press: Cambridge, MA, USA, 2007; pp. 925–947. Available online: https://www.sciencedirect.com/book/9780444594532/handbook-on-the-toxicology-of-metals (accessed on 20 December 2019).
38. Martin, G.D.; George, R.; Shaiju, P.; Muraleedharan, K.R.; Nair, S.M.; Chandramohanakumar, N. Toxic Metals Enrichment in the Surficial Sediments of a Eutrophic Tropical Estuary (Cochin Backwaters, Southwest Coast of India). *Sci. World J.* **2012**, *2012*, 1–17. [CrossRef]

39. USEPA; USACE; USDOE-BNL. *Fast Track Dredged Material Decontamination Demonstration for the Port of New York and New Jersey*; USEPA: Springfield, VA, USA, 1999; Volume 65. Available online: https://www.nj.gov/dep/passaicdocs/docs/NJDOTSupportingCosts/DECON-REPORT-EPA-FastTrkDredgedMatDeconDemoPortNYNJ1999.pdf (accessed on 20 December 2019).
40. Hemalatha, S.; Banerjee, T.K. Histopathological analysis of sublethal toxicity of zinc chloride to the respiratory organs of the airbreathing catfish Heteropneustes fossilis (Bloch). *Biol. Res.* **1997**, *30*, 11–21.
41. Salvaggio, A.; Marino, F.; Albano, M.; Pecoraro, R.; Camiolo, G.; Tibullo, D.; Bramanti, V.; Lombardo, B.M.; Saccone, S.; Mazzei, V.; et al. Toxic Effects of Zinc Chloride on the Bone Development in Danio rerio (Hamilton, 1822). *Front Physiol.* **2016**, *7*, 153. [CrossRef]
42. Roney, N.; Smith, C.V.; Williams, M.; Osier, M.; Paikoff, S.J. Toxicological Profile for Zinc. ATSDR, 2005; p. 352. Available online: https://www.atsdr.cdc.gov/toxprofiles/tp60.pdf (accessed on 20 December 2019).
43. Chu, T.-C.; Murray, S.R.; Todd, J.; Perez, W.; Yarborough, J.R.; Okafor, C.; Lee, L.H. Adaption of Synechococcus sp. IU 625 to growth in the presence of mercuric chloride. *Acta Histochem.* **2012**, *114*, 6–11. [CrossRef]
44. Dudkowiak, A.; Olejarz, B.; Łukasiewicz, J.; Banaszek, J.; Sikora, J.; Wiktorowicz, K. Heavy Metals Effect on Cyanobacteria Synechocystis aquatilis Study Using Absorption, Fluorescence, Flow Cytometry, and Photothermal Measurements. *Int. J. Thermophys* **2011**, *32*, 762–773. [CrossRef]
45. Morby, A.P.; Turner, J.S.; Huckle, J.W.; Robinson, N.J. SmtB is a metal-dependent repressor of the cyanobacterial metallothionein gene smtA: identification of a Zn inhibited DNA-protein complex. *Nucleic Acids Res.* **1993**, *21*, 921–925. [CrossRef]
46. Pochodylo, A.L.; Klein, A.R.; Aristilde, L. Metal-binding selectivity and coordination dynamics for cyanobacterial microcystins with Zn, Cu, Fe, Mg, and Ca. *Environ. Chem Lett.* **2017**, *15*, 695–701. [CrossRef]
47. Capdevila, M.; Atrian, S. Metallothionein protein evolution: A miniassay. *J. Biol. Inorg. Chem.* **2011**, *16*, 977–989. [CrossRef]
48. Gutiérrez, J.-C.; de Francisco, P.; Amaro, F.; Díaz, S.; Martín-González, A. Structural and Functional Diversity of Microbial Metallothionein Genes. In *Microbial Diversity in the Genomic Era*; Academic Press: Cambridge, MA, USA, 2019; pp. 387–407. Available online: https://www.sciencedirect.com/science/article/pii/B9780128148495000228?via%3Dihub (accessed on 20 December 2019).
49. Liu, T.; Nakashima, S.; Hirose, K.; Uemura, Y.; Shibasaka, M.; Katsuhara, M.; Kasamo, K. A metallothionein and CPx-ATPase handle heavy-metal tolerance in the filamentous cyanobacterium Oscillatoria brevis. *Febs. Lett.* **2003**, *542*, 159–163. [CrossRef]
50. Newby, R.; Lee, L.H.; Perez, J.L.; Tao, X.; Chu, T. Characterization of zinc stress response in Cyanobacterium Synechococcus sp. IU 625. *Aquat. Toxicol.* **2017**, *186*, 159–170. [CrossRef]
51. Mikalsen, B.; Boison, G.; Skulberg, O.M.; Fastner, J.; Davies, W.; Gabrielsen, T.M.; Rudi, K.; Jakobsen, K.S. Natural Variation in the Microcystin Synthetase Operon mcyABC and Impact on Microcystin Production in Microcystis Strains. *J. Bacteriol.* **2003**, *185*, 2774–2785. [CrossRef]
52. Pacheco, A.; Guedes, I.; Azevedo, S. Is qPCR a Reliable Indicator of Cyanotoxin Risk in Freshwater? *Toxins* **2016**, *8*, 172. [CrossRef]
53. Chiu, Y.-T.; Chen, Y.-H.; Wang, T.-S.; Yen, H.-K.; Lin, T.-F. A qPCR-Based Tool to Diagnose the Presence of Harmful Cyanobacteria and Cyanotoxins in Drinking Water Sources. *Ijerph* **2017**, *14*, 547. [CrossRef]
54. Wallinder, I.O.; Leygraf, C. A Critical Review on Corrosion and Runoff from Zinc and Zinc-Based Alloys in Atmospheric Environments. *Corrosion* **2017**, *73*, 1060–1077. [CrossRef]
55. Vera, R.; Guerrero, F.; Delgado, D.; Araya, R. Atmospheric Corrosion of Galvanized Steel and Precipitation Runoff from Zinc in a Marine Environment. *J. Braz. Chem. Soc.* **2013**, *,24*, 449–458. [CrossRef]
56. Li, X.; Liu, L.; Wang, Y.; Luo, G.; Chen, X.; Yang, X.; Gao, B.; He, X. Integrated Assessment of Heavy Metal Contamination in Sediments from a Coastal Industrial Basin, NE China. *PLoS ONE* **2012**, *7*, 10. [CrossRef]
57. Surosz, W.; Palinska, K.A. Effects of Heavy-Metal Stress on Cyanobacterium Anabaena flos-aquae. *Arch. Environ. Contam. Toxicol* **2004**, *48*, 40–48. [CrossRef]
58. Zhou, W.; Juneau, P.; Qiu, B. Growth and photosynthetic responses of the bloom-forming cyanobacterium Microcystis aeruginosa to elevated levels of cadmium. *Chemosphere* **2006**, *65*, 1738–1746. [CrossRef]
59. Zeng, J.; Yang, L.; Wang, W.X. Cadmium and zinc uptake and toxicity in two strains of Microcystis aeruginosa predicted by metal free ion activity and intracellular concentration. *Aquat. Toxicol.* **2009**, *91*, 212–220. [CrossRef]

60. Gouvêa, S.P.; Boyer, G.L.; Twiss, M.R. Influence of ultraviolet radiation, copper, and zinc on microcystin content in Microcystis aeruginosa (Cyanobacteria). *Harmful Algae* **2008**, *7*, 194–205. [CrossRef]
61. Rzymski, P.; Poniedzialek, B.; Niedzielski, P.; Tabaczewski, P.; Wiktorowicz, K. Cadmium and lead toxicity and bioaccumulation in Microcystis aeruginosa. *Front. Environ. Sci. Eng.* **2014**, *8*, 427–432. [CrossRef]
62. Wang, S.; Zhang, D.; Pan, X. Effects of arsenic on growth and photosystem II (PSII) activity of Microcystis aeruginosa. *Ecotoxicol. Environ. Saf.* **2012**, *84*, 104–111. [CrossRef]
63. Xiao, M.; Li, M.; Reynolds, C.S. Colony formation in the cyanobacterium Microcystis. *Biol. Rev. Camb. Philos. Soc.* **2018**, *93*, 1399–1420. [CrossRef]
64. Ybarra, G.R.; Webb, R. Effects of Divalent Metal Cations and Resistance Mechanisms of the Cyanobacterium Synechococcus SP. Strain PCC 7942. *J. Hazard. Subst. Res.* **1999**, *2*, 11. [CrossRef]
65. Tillett, D.; Dittmann, E.; Erhard, M.; von Döhren, H.; Börner, T.; Neilan, B.A. Structural organization of microcystin biosynthesis in Microcystis aeruginosa PCC7806: an integrated peptide–polyketide synthetase system. *Chem. Biol.* **2000**, *7*, 753–764. [CrossRef]
66. Kaebernick, M.; Neilan, B.A. Ecological and molecular investigations of cyanotoxin production. *Fems Microbiol. Ecol.* **2001**, *35*, 1–9. [CrossRef]
67. Yu, L.; Kong, F.; Zhang, M.; Yang, Z.; Shi, X.; Du, M. The Dynamics of Microcystis Genotypes and Microcystin Production and Associations with Environmental Factors during Blooms in Lake Chaohu, China. *Toxins* **2014**, *6*, 3238–3257. [CrossRef]
68. Wu, H.; Lin, L.; Shen, G.; Li, M. Heavy-metal pollution alters dissolved organic matter released by bloom-forming Microcystis aeruginosa. *Rsc. Adv.* **2017**, *7*, 18421–18427. [CrossRef]
69. Gan, N.; Xiao, Y.; Zhu, L.; Wu, Z.; Liu, J.; Hu, C.; Song, L. The role of microcystins in maintaining colonies of bloom-forming Microcystis spp.: Microcystis colony maintenance by microcystins. *Environ. Microbiol.* **2012**, *14*, 730–742. [CrossRef]
70. Nohomovich, B.; Nguyen, B.T.; Quintanilla, M.; Lee, L.H.; Murray, S.R.; Chu, T.-C. Physiological effects of nickel chloride on the freshwater cyanobacterium Synechococcus sp. IU 625. *ABB* **2013**, *04*, 10–14. [CrossRef]
71. Babu, N.G.; Sarma, P.A.; Attitalla, I.H.; Murthy, S.D.S. Effect of Selected Heavy Metal Ions on the Photosynthetic Electron Transport and Energy Transfer in the Thylakoid Membrane of the Cyanobacterium, Spirulina platensis. *Acad. J. Plant. Sci* **2010**, *3*, 46–49.
72. Khattar, J.I.S.; Sarma, T.A.; Sharma, A. Effect of Cr6+ Stress on Photosynthetic Pigments and Certain Physiological Processes in the Cyanobacterium Anacystis nidulans and Its Chromium Resistant Strain. *J. Microbiol Biotech.* **2004**, *14*, 1211–1216.
73. Spoof, L.; Vesterkvist, P.; Lindholm, T.; Meriluoto, J. Screening for cyanobacterial hepatotoxins, microcystins and nodularin in environmental water samples by reversed-phase liquid chromatography-electrospray ionisation mass spectrometry. *J. Chromatogr A* **2003**, *1020*, 105–119. [CrossRef]
74. Schmittgen, T.D.; Livak, K.J. Analyzing real-time PCR data by the comparative CT method. *Nat. Protoc* **2008**, *3*, 1101–1108. [CrossRef]

Article

Isolation and Characterization of [D-Leu¹]microcystin-LY from *Microcystis aeruginosa* CPCC-464

Patricia LeBlanc [1], Nadine Merkley [1], Krista Thomas [1], Nancy I. Lewis [1], Khalida Békri [1,†], Susan LeBlanc Renaud [2], Frances R. Pick [2], Pearse McCarron [1], Christopher O. Miles [1] and Michael A. Quilliam [1,*]

1 Biotoxin Metrology, National Research Council, 1411 Oxford Street, Halifax, NS B3H 3Z1, Canada; Patricia.LeBlanc@nrc-cnrc.gc.ca (P.L.); Nadine.Merkley@nrc-cnrc.gc.ca (N.M.); Krista.Thomas@nrc-cnrc.gc.ca (K.T.); Nancy.Lewis@nrc-cnrc.gc.ca (N.I.L.); khalidabekri@hotmail.com (K.B.); Pearse.McCarron@nrc-cnrc.gc.ca (P.M.); Christopher.Miles@nrc-cnrc.gc.ca (C.O.M.)

2 Department of Biology, University of Ottawa, Ottawa, ON K1N 6N5, Canada; serenaud@gmail.com (S.L.R.); frpick@uOttawa.ca (F.R.P.)

* Correspondence: michael.quilliam@nrc.ca

† Current address: Agriculture and Agri-Food Canada, Sherbrooke Research and Development Centre, Sherbrooke, QC J1M 0C8, Canada.

Received: 16 December 2019; Accepted: 20 January 2020; Published: 23 January 2020

Abstract: [D-Leu¹]MC-LY (**1**) ($[M + H]^+$ m/z 1044.5673, Δ 2.0 ppm), a new microcystin, was isolated from *Microcystis aeruginosa* strain CPCC-464. The compound was characterized by ^1H and ^{13}C NMR spectroscopy, liquid chromatography–high resolution tandem mass spectrometry (LC–HRMS/MS) and UV spectroscopy. A calibration reference material was produced after quantitation by ^1H NMR spectroscopy and LC with chemiluminescence nitrogen detection. The potency of **1** in a protein phosphatase 2A inhibition assay was essentially the same as for MC-LR (**2**). Related microcystins, [D-Leu¹]MC-LR (**3**) ($[M + H]^+$ m/z 1037.6041, Δ 1.0 ppm), [D-Leu¹]MC-M(O)R (**6**) ($[M + H]^+$ m/z 1071.5565, Δ 2.0 ppm) and [D-Leu¹]MC-MR (**7**) ($[M + H]^+$ m/z 1055.5617, Δ 2.2 ppm), were also identified in culture extracts, along with traces of [D-Leu¹]MC-M(O_2)R (**8**) ($[M + H]^+$ m/z 1087.5510, Δ 1.6 ppm), by a combination of chemical derivatization and LC–HRMS/MS experiments. The relative abundances of **1**, **3**, **6**, **7** and **8** in a freshly extracted culture in the positive ionization mode LC–HRMS were ca. 84, 100, 3.0, 11 and 0.05, respectively. These and other results indicate that [D-Leu¹]-containing MCs may be more common in cyanobacterial blooms than is generally appreciated but are easily overlooked with standard targeted LC–MS/MS screening methods.

Keywords: microcystin; cyanotoxin; structure; PP2A inhibition; liquid chromatography; mass spectrometry; cyanobacteria

Key Contribution: Structure determination and toxicity evaluation of a new microcystin, [D-Leu¹]MC-LY, and detection of related toxins in a culture of *Microcystis aeruginosa* (CPCC-464). This study and others indicate that [D-Leu¹]-containing MCs are more common in cyanobacterial blooms than is generally appreciated but are easily overlooked with standard targeted LC–MS/MS screening methods.

---

1. Introduction

Microcystins (MCs), such as MC-LR (**2**), are cyclic heptapeptide hepatotoxins (Figure 1) produced primarily in cyanobacterial genera such as *Microcystis*, *Dolichospermum (Anabaena)*, *Nostoc* and *Planktothrix (Oscillatoria)* [1]. The common feature of MCs is their cyclic structure and possession of several rare, highly conserved amino acids moieties [2]. One of the unusual structural features of MCs is the β-amino acid (2*S*,3*S*,8*S*,9*S*)-3-amino-9-methoxy-2,6,8-trimethyl-10-phenyldeca-4(*E*),6(*E*)-dienoic acid (Adda) at position 5 (Figure 1) [3]. Adda^5 and Glu^6 appear to be primarily responsible for the characteristic biological activity of MCs [2,3]. Protein phosphatase inhibition is directly related to the toxins' mechanism of action and animal studies have demonstrated that MCs are potent tumor promoters [1]. To date, the number of identified MCs continues to increase and more than 250 analogues have been characterized [4]. However, due to a lack of standards for these analogues, very few studies have adequately assessed their distribution in natural waters.

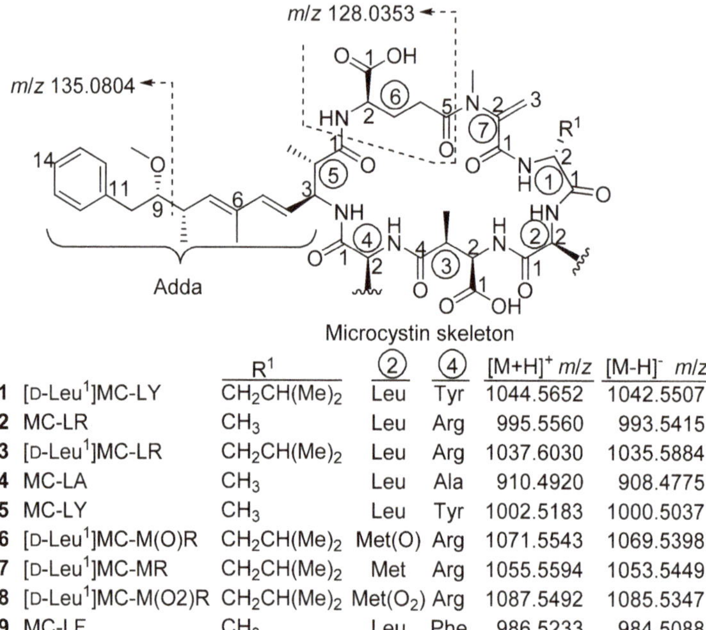

		R^1	②	④	$[M+H]^+$ *m/z*	$[M-H]^-$ *m/z*
1	[D-Leu^1]MC-LY	$CH_2CH(Me)_2$	Leu	Tyr	1044.5652	1042.5507
2	MC-LR	CH_3	Leu	Arg	995.5560	993.5415
3	[D-Leu^1]MC-LR	$CH_2CH(Me)_2$	Leu	Arg	1037.6030	1035.5884
4	MC-LA	CH_3	Leu	Ala	910.4920	908.4775
5	MC-LY	CH_3	Leu	Tyr	1002.5183	1000.5037
6	[D-Leu^1]MC-M(O)R	$CH_2CH(Me)_2$	Met(O)	Arg	1071.5543	1069.5398
7	[D-Leu^1]MC-MR	$CH_2CH(Me)_2$	Met	Arg	1055.5594	1053.5449
8	[D-Leu^1]MC-M(O2)R	$CH_2CH(Me)_2$	Met(O_2)	Arg	1087.5492	1085.5347
9	MC-LF	CH_3	Leu	Phe	986.5233	984.5088

Figure 1. Structures of the microcystins (MCs) mentioned in the text, showing their *m/z* values and characteristic mass spectral fragment ions in positive (*m/z* 135.0804) and negative (*m/z* 128.0353) ionization modes. Numbers in circles indicate the amino acid residue number, while atoms are numbered starting from the carboxyl carbon of each amino acid.

As part of feasibility studies for a cyanobacterial matrix reference material [5], a survey of cyanobacterial cultures from Canada was conducted using LC with UV and MS detection. Among the samples analyzed, two *Microcystis aeruginosa* cultures from Saskatchewan and Alberta, CPCC-464 and CPCC-299, showed the presence of a new microcystin tentatively identified as [D-Leu^1]MC-LY (**1**) [5] together with the previously reported [6,7] and well-characterized [8] [D-Leu^1]MC-LR (**3**). [D-Leu^1]MC-LY (**1**) was also tentatively identified recently by LC–HRMS/MS in a cyanobacterial bloom sample from southwestern Ontario, Canada [9], indicating that it may be a significant component of natural cyanobacterial blooms in this and other parts of the world. It is therefore necessary to verify the structure, and to evaluate its toxicity relative to other MCs because limited data are available on the toxicological consequences of varying the amino acid at position 1 and 2.

2. Results and Discussion

The analysis of *M. aeruginosa* culture CPCC-464 by LC–UV–MS/MS is shown in Figure 2. A very similar profile was observed with culture CPCC-299, with the only differences being in relative peak areas. The LC–UV chromatogram of CPCC-464 (Figure 2a) showed two major peaks due to **1** and **3**. In the same experiment, the MS was operated with a precursor scan using the *m*/*z* 135 product ion for Adda, which is characteristic of most MCs [10]. Examination of all peaks in the total ion current chromatogram (Figure 2b) revealed three major MCs of interest, **1**, **3** and **6**, with $[M+H]^+$ ions at *m*/*z* 1044.5, 1037.5 and 1071.5, respectively. These ions are plotted as extracted ion chromatograms in Figure 2c. LC–HRMS chromatograms with and without mercaptoethanol derivatization are shown in Figures S1 and S2. Compounds **1** and **3** were hypothesized to be [D-Leu[1]]MC-LY and [D-Leu[1]]MC-LR, respectively [5], while the MS characteristics of **6** are consistent with [D-Leu[1]]MC-M(O)R, tentatively identified recently by LC–HRMS/MS in cyanobacterial blooms in Canada and the USA [9,11].

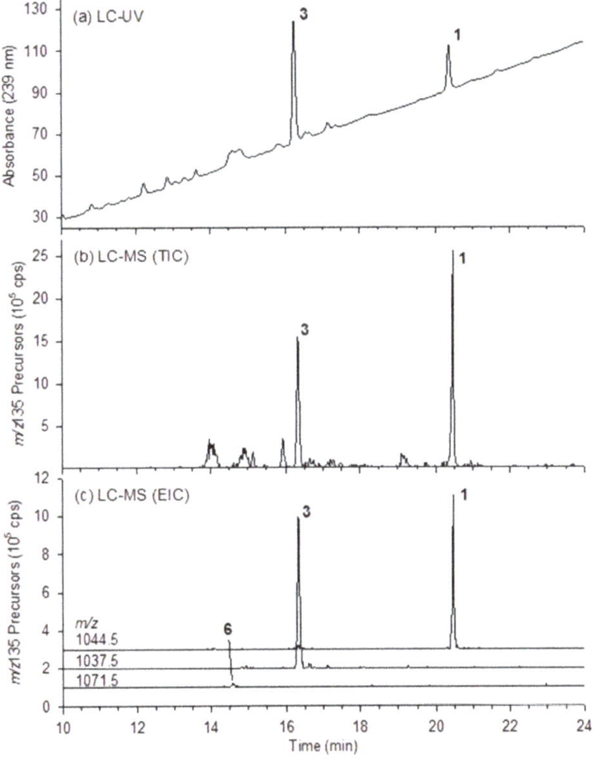

Figure 2. LC–UV–MS analysis of an extract of CPCC-464: (**a**) UV absorbance at 238 nm; (**b**) total ion current chromatogram from *m*/*z* 135 precursor scan; and (**c**) extracted ion chromatograms from *m*/*z* 135 precursor scan.

Large scale culturing of CPCC-464 followed by centrifugation provided 188 g of biomass for purification of **1**. This material was extracted with 70% MeOH–H_2O, then taken through a preparative isolation procedure consisting of a hexane partitioning, C18 LC, LH-20 gel permeation, C18-flash chromatography, and semi-preparative HPLC. The total yield of **1** was 28.7 mg containing a small amount of [D-Leu[1],D-Glu(OMe)[6]]MC-LY and a trace of what is believed to be [D-Leu[1],(6Z)-Adda[5]]MC-LY (observed by LC–MS selected reaction monitoring modes).

The structure of **1** was elucidated from NMR spectra acquired in CD_3OH in order to observe the exchangeable amide protons. The proton NMR spectrum had six resonances in the amide region with a profile similar to that of a peptide. Individual spin systems from each amide resonance were identified and assigned using 2D ^1H-^1H DIPSI-2 and ^1H-^1H COSY correlations (Table 1). Detailed spectra are provided in Figures S3–S10 and an overlay of chemical shifts on the proposed 2-dimensional chemical structure of **1** is shown in Figure S11.

Table 1. NMR spectroscopic data for [D-Leu[1]]MC-LY (**1**) in CD_3OH^a.

Unit	Position	δc, Type	δH, Multiplicity (J in Hz)	HMBC
D-Leu[1]	1	ND		
	2	52.0, CH	4.44, m	
	2-NH		7.89, d (7.7)	
	3a	39.2, CH_2	1.62, m	4
	3b		1.33, m	
	4	24.8, CH	1.57, m	
	4-Me	19.9, CH_3	0.83, d (6.3)	
	5	22.6, CH_3	0.83, d (6.3)	
Leu[2]	1	174.5, C		
	2	53.8, CH	4.23, ddd (10.4, 6.6, 4.0)	
	2-NH		8.28, d (6.5)	1
	3a	39.8, CH_2	1.92, ddd (13.7, 12.0, 4.0)	2,5/6
	3b		1.51, ddd (13.7, 10.4, 3.8)	
	4	24.5, CH	1.76, m	
	4-Me	20.0, CH_3	0.88, d (6.6)	
	5	22.8, CH_3	0.88, d (6.6)	
D-Masp[3]	1	175.6, C		
	2	54.6, CH	4.60, dd (9.0, 3.8)	1,4
	2-NH		7.72, d (9.0)	Leu[2]-1
	3	40.7, CH	3.04, dq (7.2, 3.8)	
	3-Me	13.8, CH_3	0.86, d (7.2)	2,4
	4	177.4, C		
Tyr[4]	1	170.6, C		
	2	54.1, CH	4.35, ddd (11.7, 9.3, 3.3)	1
	2-NH		8.89, d (9.3)	
	3a	36.3, CH_2	3.29, m	1
	3b		2.54, dd (14.1, 11.7)	2,4,5/9
	4	128.3, C		
	5/9	130.2, CH	6.99, d (8.5)	3,7,5/9
	6/8	115.0, CH	6.62, d (8.5)	4,7,6/8
	7	156.4, C		
Adda[5]	1	176.0, C		
	2	44.1, CH	2.76, m	
	2-Me	14.8, CH_3	1.08, d (6.9)	1,2,3
	3	55.5, CH	4.70, ~q (9.7)	1,2,4,5
	3-NH		7.36, d (9.2)	
	4	125.2, CH	5.46, dd (15.5, 9.0)	6
	5	138.0, CH	6.32, d (15.5)	3,7,6-Me
	6	132.8, C		
	6-Me	11.8, CH_3	1.63, s	5,6,7
	7	136.2, CH	5.49, d (10.1)	5,6,8,9,6-Me
	8	36.5, CH	2.61, m	

Table 1. *Cont.*

Unit	Position	δc, Type	δ_H, Multiplicity (*J* in Hz)	HMBC
	8-Me	15.5, CH_3	1.03, d (6.7)	7,8,9
	9	87.2, CH	3.27, m	
	9-OMe	57.6, CH_3	3.24, s	9
	10a	37.7, CH_2	2.83, dd (13.9, 4.7)	12/16
	10b		2.68, dd (13.9, 7.4)	9,11,12/16
	11	139.4, C		
	12/16	129.5, CH	7.19, m	10,14,12/16
	13/15	128.1, CH	7.25, t (7.6)	11,13/15
	14	125.9, CH	7.17, m	
D-Glu[6]	1	174.8, C		
	2	53.3, CH	4.31, ~q (7.2)	1,3
	2-NH		7.55, brs	
	3a	27.0, CH_2	2.10, m	
	3b		1.79, m	
	4a	32.0, CH_2	2.72, m	
	4b		2.59, m	
	5	175.2, C		
Mdha[7]	1	165.5, C		
	2	145.1, C		
	2-NMe	37.3, CH_3	3.36, s	2, D-Glu-5
	3*E*	113.3, CH_2	5.84, s	1
	3*Z*		5.43, s	1,2

[a] s, singlet; d, doublet; t, triplet; q, quartet; m, multiplet; br, broad; ND, not detected.

Carbon assignments were determined indirectly using ^{1}H-^{13}C HSQC and ^{1}H-^{13}C HMBC 2D NMR spectra. One carbon resonance was not assigned for Leu[1] (C1) due to spectral overlap. The Adda unit was assembled with the aid of the HMBC data, which determined the positions of the methyl groups. *Trans*-configuration of the 4,5-double bond is indicated by the large coupling constant between Adda-H4 and -H5 (15.5 Hz) and by the observation of a ROESY correlation between Adda-H4 and Adda-6-Me, and is consistent with the absence of a ROESY correlation between Adda-H4 to Adda-H5. The second double bond was also *trans* as a ROESY correlation was observed between Adda-H5 and Adda-H7 (Figure 3).

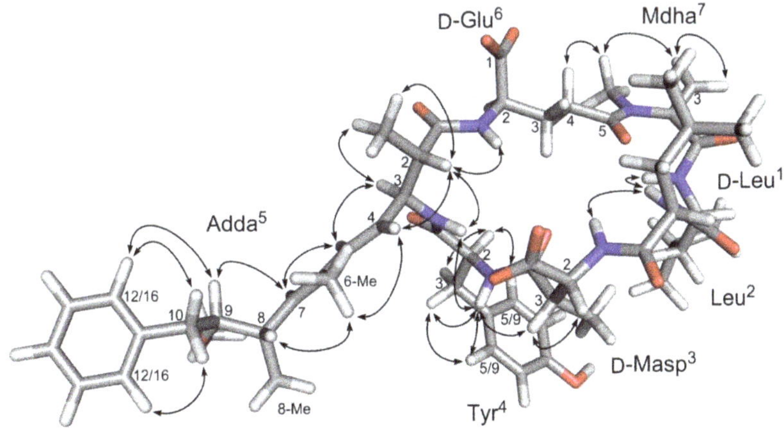

Figure 3. Three-dimensional structure of [D-Leu[1]]MC-LY (**1**), modeled from the solution structure of MC-LR (**2**) [12], showing NoE correlations observed in the ROESY NMR spectrum of **1**.

The relative stereochemistry of C2 and C3 of Adda was determined from the observation of a ROESY correlation between Adda-H3 and both Adda-H5 and Adda-2-Me, while Adda-H2 showed correlations to Adda-NH, Glu-NH, Adda-2-Me and Adda-H4, indicating that H2 and H3 are on the opposite faces of the Adda plane (Figure 3). This is consistent with the ca. 9.7 Hz coupling constant between Adda-H2 and -H3. The glutamic acid unit (Glu), *N*-methyldehydroalanine (Mdha) and *erthyo*-β-methylaspartic acid (Masp) were identified in a similar manner, and their proton and carbon resonances were very similar to those previously published for **3** [8]. Two leucine units were identified by the similarities of their ^{1}H and ^{13}C resonances to those in the BMRB database (http://www.bmrb.wisc.edu/ref_info/; accessed September 2011) and those published for **3** [10]. The relative stereochemistry for H2 and H3 of the Masp unit was determined from the absence of Masp-H2 to Masp-3-Me correlation and the presence of a ROESY correlation between Masp-H2 and -H3, which places H2 and H3 on the same side of the plane. The tyrosine unit (Tyr) was assigned from the presence of the two doublets at 6.99 and 6.62 ppm, characteristic of a para-substituted phenyl ring. The aromatic protons, Tyr-H5 and -H9 correlated to a carbon at 36.3 ppm, characteristic of an aromatic amino acid. The accurate mass from LC–HRMS/MS indicated that the substituent on the phenyl ring was a hydroxyl group (Table 2), establishing its identity as Tyr.

The amino acid subunits assigned in the ^{1}H-^{1}H DIPSI-2, ^{1}H-^{1}H COSY and ^{1}H-^{13}C HMBC spectra were linked through correlations observed in the ROESY, NOESY (Figure 3) and HMBC NMR spectra. In the HMBC spectra, a correlation between the *N*-methyl of Mdha and the carbonyl of Glu, and between the Masp-NH and the leucine carbonyl at 174.5 ppm, linked Glu^{6} to Mdha^{7} and Masp^{3} to Leu^{2}. Additionally, ROESY correlations were observed between Leu^{2}-NH and both Leu^{1}-NH and Masp^{3}-NH. Furthermore, the Tyr-NH showed ROESY or NOESY correlations to Masp^{3}-H3 and Adda^{5}-NH, Adda^{5}-NH showed correlations to Adda-H4 and Adda-H2, and Adda^{5}-H2 showed correlations to Adda-H4, Adda-2-Me and Glu^{6}-NH. These correlations show **1** to contain Leu-Leu-Masp-Tyr-Adda-Glu-Mdha, and the molecular formula established from LC–HRMS requires an amide linkage between Mdha^{7} and Leu^{1} moieties. That this linkage is present is demonstrated by the presence of numerous product ions in the HRMS/MS spectrum that are attributable to fragments containing both Leu^{1} and Mdha^{7}, such as those at *m*/*z* 169.1334, 197.1283 and 488.2745 (Table 2). The ROESY correlations observed for **1** (Figure 3), especially those between the amide protons, were consistent with those expected based on the established 3-dimensional solution structure for MC-LR [12], which is reported to be very similar to that of [D-Leu^{1}]MC-LR (**3**) [8]. Thus, **1** has the same relative stereochemistry as **2** and **3**. This is also supported by the close similarity of the ^{13}C NMR chemical shifts of **1** to those reported for [D-Leu^{1}]MC-LR (**3**) in the same solvent (Table S1). The fact both **1** and **3** are biosynthesized together by the MC synthetase of *M. aeruginosa* strain CPCC-464, and that **1** was subsequently found to have similar inhibitory potency to MC-LR (**2**) against protein phosphatase 2A (PP2A) (Figure 4), both indicate that **1** has the same absolute stereochemistry as **2** and **3** and that **1** is therefore [D-Leu^{1}]MC-LY (Figure 1).

Table 2. Reported product ions for MC-LA (**4**), with their exact *m/z* values and amino acid origins, and the corresponding product ions and their mass differences relative to the reported values for MC-LA, observed during LC–HRMS/MS analysis of MC-LA (**4**), MC-LY (**5**), and [D-Leu[1]]MC-LY (**1**) in the positive ionization mode[a].

AA Origin							Reported for MC-LA (4)		MC-LA (4)		MC-LY (5)		[D-Leu[1]]MC-LY (1)	
1	2	3	4	5	6	7	Calc. *m/z*	Formula	*m/z*	Diff.	*m/z*	Diff.	*m/z*	Diff.
							44.0495	$C_2H_6N^+$	ND	N/A	136.0757	92.0262	136.0756	92.0261
							56.0495	$C_3H_6N^+$	56.0497	0.0002	56.0497	0.0002	56.0497	0.0002
							84.0444	$C_4H_6NO^+$	84.0443	-0.0001	84.0444	0.0000	84.0444	0.0000
							86.0964	$C_5H_{12}N^+$	86.0964	0.0000	86.0964	0.0000	86.0964	0.0000
							103.0542	$C_8H_7{}^+$	103.0541	-0.0001	103.0542	0.0000	103.0542	0.0000
							107.0855	$C_8H_{11}{}^+$	107.0854	-0.0001	107.0855	0.0000	107.0855	0.0000
							127.0866	$C_6H_{11}N_2O^+$	127.0865	-0.0001	127.0866	0.0000	169.1334	42.0468
							135.0804	$C_9H_{11}O^+$	135.0803	-0.0001	135.0804	0.0000	135.0803	-0.0001
							135.1168	$C_{10}H_{15}{}^+$	135.1166	-0.0002	135.1167	-0.0001	135.1166	-0.0002
							155.0815	$C_7H_{11}N_2O_2{}^+$	155.0813	-0.0002	155.0813	-0.0002	197.1283	42.0468
							163.1117	$C_{11}H_{15}O^+$	163.1115	-0.0002	163.1117	0.0000	163.1115	-0.0002
							173.0921	$C_7H_{13}N_2O_3{}^+$	173.0917	-0.0004	265.1180	92.0259	265.1179	92.0258
							195.0764	$C_9H_{11}N_2O_3{}^+$	195.0763	-0.0001	195.0764	0.0000	195.0763	-0.0001
							213.0870	$C_9H_{13}N_2O_4{}^+$	213.0868	-0.0002	213.0869	-0.0001	213.0868	-0.0002
							218.1135	$C_8H_{16}N_3O_4{}^+$	218.1133	-0.0002	310.1393	92.0258	310.1392	92.0257
							265.1587	$C_{19}H_{21}O^+$	265.1581	-0.0006	265.1583	-0.0004	265.1581	-0.0006
							268.1656	$C_{13}H_{22}N_3O_3{}^+$	268.1650	-0.0006	268.1652	-0.0004	310.2119	42.0463
							292.1543	$C_{16}H_{22}NO_4{}^+$	292.1538	-0.0005	292.1541	-0.0002	292.1540	-0.0003
							314.1710	$C_{14}H_{24}N_3O_5{}^+$	314.1705	-0.0005	406.1967	92.0257	406.1965	92.0255
							331.1976	$C_{14}H_{27}N_4O_5{}^+$	331.1970	-0.0006	423.2230	92.0254	423.2229	92.0253
							375.1914	$C_{20}H_{27}N_2O_5{}^+$	375.1906	-0.0008	375.1912	-0.0002	375.1907	-0.0007
							385.2082	$C_{17}H_{29}N_4O_6{}^+$	385.2073	-0.0009	477.2359	92.0277	519.2807	134.0725
							402.2347	$C_{17}H_{32}N_5O_6{}^+$	402.2342	-0.0005	494.2602	92.0255	536.3072	134.0725
							446.2286	$C_{23}H_{32}N_3O_6{}^+$	446.2276	-0.0010	446.2283	-0.0003	488.2745	42.0459
							468.2453	$C_{21}H_{34}N_5O_7{}^+$	468.2443	-0.0010	560.2707	92.0254	602.3180	134.0727
							485.2718	$C_{21}H_{37}N_6O_7{}^+$	485.2708	-0.0010	577.2976	92.0258	619.3447	134.0729
							509.2646	$C_{29}H_{37}N_2O_6{}^+$	509.2637	-0.0009	509.2640	-0.0006	509.2640	-0.0006
							559.3126	$C_{29}H_{43}N_4O_7{}^+$	559.3117	-0.0009	559.3157	0.0031	601.3610	42.0484
							580.3017	$C_{32}H_{42}N_3O_7{}^+$	580.3008	-0.0009	580.3010	-0.0007	622.3479	42.0462
							597.2879	$C_{26}H_{41}N_6O_{10}{}^+$	597.2872	-0.0007	689.3134	92.0255	731.3605	134.0726
							693.3858	$C_{38}H_{53}O_8N_4{}^+$	693.3854	-0.0004	693.3854	-0.0004	735.4322	42.0464
							758.4083	$C_{37}H_{56}N_7O_{10}{}^+$	758.4076	-0.0007	850.4350	92.0267	892.4804	134.0721
							759.3923	$C_{37}H_{55}N_6O_{11}{}^+$	759.3917	-0.0006	851.4190	92.0267	893.4652	134.0729
							776.4189	$C_{37}H_{58}N_7O_{11}{}^+$	776.4187	-0.0002	868.4445	92.0256	910.4906	134.0717
							910.4920	$C_{46}H_{68}N_7O_{12}{}^+$	910.4908	-0.0012	1002.5175	92.0255	1044.5634	134.0714

[a] Reported fragments, their amino acid origins and calculated *m/z* for MC-LA based on data from Bortoli and Volmer 2014; Mayumi et al. 2006; Miles et al. 2013; Stewart et al. 2018 [13–16]; black squares indicate amino acids contributing to each product ion; gray squares indicate neutral loss of 134.0732 Da from Adda[5]; Diff., difference from calculated *m/z* for the corresponding product ion in MC-LA; ND, not detected.

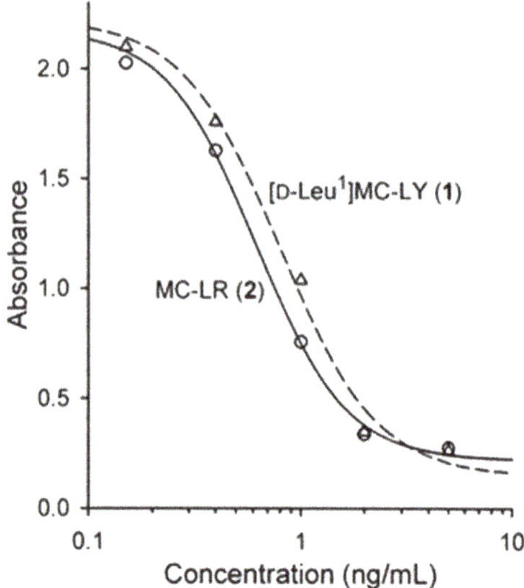

Figure 4. Protein phosphatase 2A (PP2A) inhibition curves for **2** and **1** fitted to a four-parameter logistic curve. The estimated IC_{50} values were 0.62 ng/mL (standard error 0.05) for **2**, and 0.80 ng/mL (SE 0.11) for **1** (0.62 and 0.76 nM, respectively).

The positive and negative LC–HRMS spectra of **1** were consistent with a molecular formula of $C_{55}H_{77}O_{13}N_7$ which, together with the presence of a prominent $[M + Na]^+$ adduct ion, a weak neutral loss of *m/z* 134.0727, and the late retention time, were consistent with a non-Arg-containing MC. This indicated that **1** contained an extra C_3H_6 (42.0470 Da) relative to MC-LY (**5**), and $C_9H_{10}O$ (134.0732 Da) relative to MC-LA (**4**), which differ from **1** only in their amino acids at position 1 (**5**), and at position 1 and 4 (**4**) (Table 2). The negative ion MS/MS spectrum obtained from the FS/DIA (full scan/data independent acquisition) LC–HRMS of **1** showed a prominent product ion at *m/z* 128.0355, consistent with the presence of a MC containing Glu at position 6, and a neutral loss of *m/z* 112.0190, consistent with the presence of Masp at position 3 [17]. Careful comparison of the positive ion targeted LC–HRMS/MS spectrum of **1** with those of standards of **4** and **5** (Table 2) showed that product ions in **1** that contained amino acid-1 were consistently heavier by 42.047 Da (Leu vs. Ala) than the corresponding product ions from **5**, and 42.047 (Leu^1 vs Ala^1), 92.026 (Tyr^4 vs Ala^4) or 134.073 Da (Leu^1 and Tyr^4 vs. Ala^1 and Ala^4) heavier than the corresponding product ions from **4** that contained amino acid-2, amino acid-4, or both amino acid-1 and -4, respectively (Table 2, Figures S12–S18). Furthermore, the UV spectrum of **1** obtained during LC–UV analysis was identical to that of **5**, and differed from that of **2** (Figure S19), suggesting the presence of Tyr in **1** and **5** in addition to the UV-absorbing chromophores also present in **2** (i.e., $Adda^5$ and $Mdha^7$). The LC–MS/MS and LC–UV results are therefore entirely consistent with **1** being [D-Leu^1]MC-LY.A portion of the purified **1** was used to prepare a stock solution. This was quantitated using qNMR [18] and LC with chemiluminescence nitrogen detection (CLND) [19], then accurately diluted with 1:1 MeOH–H_2O to prepare a reference material (RM) (~7.7 μM). LC–UV analysis of this RM showed the relative concentration of [D-Leu^1,D-Glu(OMe)^6]MC-LY to be 3.1%. The putative [D-Leu^1,(6Z)-$Adda^5$]MC-LY was below the limit of quantitation in LC–UV, but the relative concentration was estimated be to below 0.5% using HRMS/MS. Because MCs containing (6Z)-$Adda^5$ or D-Glu(OMe)^6 do not inhibit protein phosphatases [20], the RM of **1** was used for the PP2A inhibition assay without correcting for impurities. In the PP2A assay, the IC_{50} for a certified RM (CRM) of MC-LR (**2**) was 0.62 nM (0.62 ng/mL), while that for the RM of **1** was 0.76 nM (0.80 ng/mL) (Figure 4). Matthiensen et al. [7] reported that MC-LR (**2**) and [D-Leu^1]MC-LR (**3**) had similar toxicities to mice when injected intraperitoneally, and that the IC_{50} values of **2** and **3** in a PP1 assay were 3.1 and 4.4 nM, respectively. Similarly, Park et al. [6] independently found that **2** and **3** both had the same IC_{50} value of 0.3 nM in their PP1 assay. Ikehara et al. [21] found that the IC_{50} of MC-LF(**9**), which differs from MC-LY (**5**) only by the absence of a phenolic hydroxyl group on residue-2, was 3-fold higher than that of MC-LR (**2**) (0.096 vs. 0.032 nM) in their PP2A assay. Taken together with the data presented here, these results suggest that the replacement of D-Ala with D-Leu at position 1 in the MC structure has only a minor effect on the toxicity of MCs or on their inhibitory effects on PP1 and PP2A.

Authentic **3** in a cyanobacterial bloom extract from Poplar Island, MD, USA, whose structure has been verified as [D-Leu^1]MC-LR by purification and NMR analysis [11], had identical retention time and product ion spectra to the peak for **3** in CPCC-464 when analyzed by LC–HRMS/MS, thus verifying its identity as proposed by Hollingdale et al. [5]. LC–HRMS/MS also showed the presence of a minor microcystin with $[M + H]^+$ at *m/z* 1071.5556 ($C_{51}H_{79}O_{13}N_{10}S^+$, Δ 1.2 ppm) consistent with [D-Leu^1]MC-M(O)R (**6**) tentatively identified in bloom samples from south-western Ontario, Canada [9] and Maryland, USA [11], by untargeted LC–HRMS/MS methods and selective chemical oxidation. The identity of this compound in the culture was further verified by targeted LC–HRMS/MS, together with oxidation to its sulfone (**8**) using Oxone [22] (see Figures S20 and S21). In LC–HRMS/MS, **6** showed the neutral loss of methylsulfenic acid (63.9983 Da, CH_3SOH) characteristic of methyl sulfoxides, and was slowly oxidized to its sulfone by Oxone. The $[M + H]^+$ ion for both sulfone-**8**, and sulfoxide-**6** after neutral loss of CH_3SOH, showed the expected characteristic series of microcystin product ions. Further examination of the LC–HRMS/MS chromatograms revealed the presence of a peak corresponding to [D-Leu^1]MC-MR (**7**) ($[M + H]^+$ *m/z* 1055.5608, $C_{51}H_{79}O_{12}N_{10}S^+$, Δ 1.3 ppm) which showed the expected product ions and which was rapidly converted to **6** by oxidation with Oxone, thus verifying the structures of both **6** and **7**. Trace amounts of the corresponding sulfone (**8**) ($[M + H]^+$ *m/z* 1087.5491, $C_{51}H_{79}O_{14}N_{10}S^+$, Δ −0.1 ppm) were also detected in the culture extract, something that was recently also reported by Foss et al. [11] in a cyanobacterial bloom sample, together with **6** and **7**. Microcystins **6–8** from this sample and from *M. aeruginosa* CPCC-464 showed identical retention times and mass spectral characteristics. Methionine sulfoxide analogues of MCs appear to be formed by autoxidation [22], and it appears that the same process can also lead to formation of the corresponding sulfones. The stereochemistry of **6–8** cannot be verified by LC–MS methods. However, because **7** is presumably biosynthesized in the culture by the same synthetase that produces **1** and **3**, and that **6** and **8** are autoxidation products of **7**, **6–8** can therefore be assumed to have the same stereochemistry as **1** and **3** (Figure 1).

A careful non-targeted LC–MS analysis of a field sample by Foss et al. [11] recently reported more than 20 Leu^1-containing MCs in a cyanobacterial bloom, with [D-Leu^1]MC-LR (**3**) as the major component, but no **1** was detected. Including the present study, **1** now appears to have been detected in samples originating from three different locations in Canada [5,9], but so far, nowhere else in the world. Geographical differences in the distribution of microcystins are being reported [23,24]. Leu^1-containing MCs have been implicated in bird deaths in both Canada and the USA [6,11] and have been reported in samples from cyanobacterial blooms in Brazil and Argentina [7,25,26] as well as in lichens from Argentina, USA, China, Japan, Norway, Sweden, and Finland [25,27]. Leu^1 variants may be more common and widespread than these studies indicate, as many analyses for MCs are conducted using highly targeted LC–MS/MS methods, and the Leu^1-containing variants are heavier by 42 Da than the more common (and more commonly targeted) Ala^1-containing MCs. Both types of variants would be readily detected if they were targeted in the LC–MS/MS method, or if untargeted LC–MS methods were used. Protein phosphatase inhibition assays, or immunoassays with appropriate cross-reactivities [28,29], can also be expected to detect both D-Ala^1- and D-Leu^1-containing MCs although they cannot indicate which type of variant is present.

3. Conclusions

[D-Leu^1]MC-LR (**3**) has been reported previously and its structure confirmed by NMR spectroscopy [8]. [D-Leu^1]MC-LY (**1**) and [D-Leu^1]MC-M(O)R (**6**) have been tentatively identified by LC–HRMS/MS in a Canadian cyanobacterial bloom sample and in cultures [5,9]. The results presented here firmly establish the identity of **1** and show that it has similar inhibitory potency towards PP2A as MC-LR (**2**). A calibration reference material has been prepared that can be used to identify and quantitate **1** in field samples and cultures. Microcystins containing D-Leu at position 1 may be fairly common in the Americas, and the data presented here and elsewhere suggest these to be only slightly less toxic than their more common D-Ala^1-containing congeners. It is therefore important to consider the possible presence of a range of D-Leu^1-containing MCs when analyzing bloom samples.

4. Materials and Methods

4.1. General Experimental Procedures

Purified **1** (250 μg) was dissolved in 30 μL of CD_3OH for NMR spectroscopy. NMR spectra were acquired on a Bruker Avance III 600 MHz spectrometer (Bruker Biospin Ltd., Billerica, MA, USA) operating at a ^{1}H frequency of 600.28 MHz and ^{13}C frequency of 150.94 MHz using TOPSPIN 2.1 acquisition software with a 1.7 mm TXI gradient probe at 277 K. Standard Bruker pulse sequences were used for structure elucidation: one dimensional ^{1}H spectrum with composite pulse pre-saturation of water, double quantum filtered ^{1}H-^{1}H COSY, ^{1}H-^{1}H DIPSI-2 (mixing time 120 ms), ^{1}H-^{13}C HSQC, ^{1}H-^{13}C HMBC (60 and 90 ms delay for long range coupling evolution), ^{1}H-^{1}H ROESY (mixing time 400 ms) and ^{1}H-^{1}H NOESY (mixing time 200 ms). A second sample of **1** (1 mg in 210 μL) was prepared in a 3 mm NMR tube for acquisition of the ^{13}C spectrum on a Bruker Avance III 700 MHz spectrometer operating at ^{1}H frequency of 700.15 MHz and a ^{13}C frequency of 176.07 MHz equipped with a 5 mm cryogenically cooled probe at 278 K. The residual ^{1}H resonance of CD_3OH was referenced to 3.31 ppm and the carbon to 48.0 ppm. Data were processed with TOPSPIN 2.1 and analyzed using the NMR assignment software Sparky (T. D. Goddard and D. G. Kneller, SPARKY 3, University of California, San Francisco: https://www.cgl.ucsf.edu/home/sparky/).

The initial survey of cultures for the presence of MCs was performed by LC–UV–MS using an Agilent (Mississauga, ON, Canada) 1200 LC coupled with a SCIEX (Concord, ON, Canada) API 4000 Q-Trap mass spectrometer with UV monitoring at 238 nm and positive electrospray ionization MS, with full scans, *m/z* 135 precursor scans, product ion scans, and selected reaction monitoring. The LC column (50 × 2.1 mm; Agilent) was packed with 1.8 μm Zorbax SB-C18 and maintained at 40 °C. The flow rate was 0.3 mL/min, with a gradient of 10%–80% B over 30 min. Solvent A was water and B was 95% acetonitrile, each with 50 mM formic acid and 2 mM ammonium formate.

LC–UV spectra of **1**, **2** and **5** were acquired using an Agilent 1260 LC System with diode array detector (DAD) with UV monitoring at 238 and 210 nm. Separations were on an Acquity HSS T3 1.8 μm column (100 × 2.1 mm; Waters, Milford, MA, USA) held at 40 °C. Isocratic elution was performed with 50% MeOH-H_2O (0.1% *v/v* trifluoroacetic acid) at 0.25 mL min^{-1} with injection volumes of 5 μL.

LC–HRMS was conducted with a Q Exactive-HF Orbitrap mass spectrometer equipped with a HESI-II heated electrospray ionization interface (ThermoFisher Scientific, Waltham, MA, USA) with an Agilent 1200 G1312B binary pump, G1367C autosampler, and G1316B column oven. Analyses were performed with a 3.5 μm Symmetry Shield C18 column (100 × 2.1 mm; Waters) held at 40 °C with mobile phases A and B of H_2O and CH_3CN, respectively, each of which contained formic acid (0.1% *v*/*v*). A linear gradient (0.3 mL min^{-1}) was used from 20% to 90% B over 18 min, then to 100% B over 0.1 min, followed by a hold at 100% B (2.9 min), then returned to 20% B over 0.1 min with a hold at 20% B (3.9 min) to equilibrate the column. Injection volume was typically 1–5 μL. In positive ion mode the mass spectrometer was calibrated from *m/z* 74–1622, the spray voltage was 3.7 kV, the capillary temperature was 350 °C, and the sheath and auxiliary gas flow rates were 25 and 8 units, respectively, with MS data acquired from 2 to 20 min. Mass spectral data were collected using a combined FS/DIA method. FS data were collected from *m/z* 500–1400 using the 60000 resolution setting, an AGC target of 1×10^6 and a max IT of 100 ms. DIA data were collected using the 15000 resolution setting, an AGC target of 2×10^5, max IT set to 'auto' and a stepped collision energy of 30, 60 and 80 V. Precursor isolation windows were 62 *m/z* wide and centered at *m/z* 530, 590, 650, 710, 770, 830, 890, 950, 1010, 1070, 1130, 1190, 1250, 1310, and 1370. DIA chromatograms were extracted for product ions at *m/z* 121.1011, 121.0647, 135.0804, 135.1168, 375.1915, 389.2072, 361.1758, 213.0870, 426.2096, 440.2252, 454.2409, 412.1939, 393.2020, 379.1864, 585.3395, 599.3552, and 613.3709. Putative MCs detected using the above FS/DIA method were further probed in a targeted manner using the PRM scan mode with a 0.7 *m/z* precursor isolation window, typically using the 30,000 resolution setting, an AGC target of 5×10^5 and a max IT of 400 ms. Typical collision energies were: stepped CE at 30 and 35 eV for MCs with no Arg, and stepped CE at 60, 65 and 70 for MCs with one Arg. LC–MS/MS/MS spectra of *m/z* 1071.5 → 1007 for [D-Leu^1]MC-M(O)R (**6**) were obtained from an LC–HRMS/MS chromatogram run in PRM mode at *m/z* 1007.5 with in-source fragmentation energy set at 100 eV. In negative mode, the mass spectrometer was calibrated from *m/z* 69–1780 and the spray voltage was −3.7 kV, while the capillary temperature, sheath and auxiliary gas flow rates were the same as for positive mode. Mass spectral data were collected in FS/DIA scan mode as above using a scan range of *m/z* 750–1400, a resolution setting of 60000, AGC target of 1×10^6 and max IT of 100 ms. DIA data were collected using a resolution setting of 15,000, AGC target of 2×10^5, max IT set to 'auto', and stepped collision energy 65 and 100 V. Isolation windows were 45 *m/z* wide and centered at *m/z* 772, 815, 858, 902, 945, 988, 1032, 1075, 1118, 1162, 1205, 1248, 1294, 1335, and 1378. DIA chromatograms were extracted for product ions at *m/z* 128.0353.

The Oxone oxidations of **6** and **7**, based on Miles et al. [22] were performed by adding 1 μL of Oxone (10 mg/mL) in water to 19 μL of 1:1 MeOH–H_2O, then 20 μL of culture extract was added with vortex mixing. The reaction was monitored by LC–HRMS/MS after 15 min and then periodically thereafter.

4.2. Toxins and Other Materials

Distilled H_2O was further purified using a UV purification system (ThermoFisher Scientific) or a Milli-Q water purification system (Millipore Ltd., Oakville, ON, Canada). MeOH and CH_3CN (Optima LC–MS grade) were from ThermoFisher Scientific. Hexanes was from Caledon. Formic acid and trifluoroacetic acid were from Sigma-Aldrich (Oakville, ON, Canada). A certified reference material for **2** (CRM-MCLR (Lot # 20070131)) and in-house reference materials for **4** and **5** were from the National Research Council Canada (Biotoxin Metrology, Halifax, NS, Canada).

4.3. Biological Material

M. aeruginosa cultures CPCC-464 and CPCC-299 were obtained from the University of Toronto Culture Collection (now the Canadian Phytoplankton Culture Collection housed at the University of Waterloo, ON, Canada). CPCC-464 was isolated from Trampling Lake, Saskatchewan, Canada, July 1998 and deposited by D. Parker as UWOCC#E7. CPCC-299 was isolated from Pretzlaff Pond, Alberta, Canada, August 1990 and deposited by E. Prepas and A. Lam as sample #45-2A. Bulk cultures of CPCC-464 were prepared in two aerated Brite-boxes (250 and 300 L), which are self-contained fiberglass boxes that optimize temperature and light to maximize biomass production. All cultures were grown on BG11 medium [30,31] made using filtered (1 μM) lake water that had pasteurized for 6 h at 85 °C. Light was provided by internally mounted cool white fluorescent tubes shaded with nylon mesh for an approximate intensity of 75–100 μmol m^{-2} s^{-1} on a 14:10 h light:dark cycle. Temperature was maintained at 20 °C and pH was monitored and remained constant at 8.6. When cultures reached late exponential stage, 188 g of wet biomass was harvested using a tangential flow centrifuge (IEC Centra MP-4R CEPA Z41 with an 804S rotor (GMI, Ramsey, MN, USA)) with a flow rate of 2–3 L min^{-1}. The biomass was stored at −20 °C. An extract of lyophilized material from a cyanobacterial bloom at Poplar Island, MD, USA, which contained authentic 3 as well as the tentatively identified 6–8, was available from an earlier study [11].

4.4. Toxin Isolation from Culture Biomass

Wet cell biomass of CPCC-464 (104.8 g) was extracted four times with 70% MeOH–H_2O (400 mL). After centrifugation, the supernatants were pooled (1.7 L) and partitioned with hexanes (700 mL). The hexane portion was back-extracted with 85% MeOH–H_2O (300 mL) and combined with the first extract. The cleaned extract was adjusted to 85% MeOH and partitioned a second time with hexane (300 mL). The combined MeOH–H_2O extracts were partially evaporated, pre-adsorbed on ~14 g of Waters 55-105µm prep C18 and packed on top of a vacuum liquid chromatography column (4 cm × 11 cm) containing Waters 55–105 µm Prep C18-silica. Fractions were eluted with increasing percentages of MeOH and analyzed by LC–MS. [D-Leu[1]]MC-LY (**1**) was present in fractions containing 40–80% MeOH–H_2O and further purified on a Sephadex LH-20 column (1.6 cm × 68.0 cm), which was eluted isocratically with MeOH. Fractions containing **1** were combined and subjected to a flash chromatography column (Bakerbond 40 µm C18-silica, 1.5 cm × 22.0 cm) eluted with 55% MeOH–H_2O. After analyzing the fractions, those containing **1** were purified using a 3 µm Luna C18(2) column (250 × 10 mm; Phenomenex, Torrance, CA, USA) eluted isocratically with 48% CH_3CN–H_2O containing 0.05% TFA at 2 mL/min, with UV monitoring at 238 nm. To remove TFA the collected fractions were diluted to 14% CH_3CN–H_2O and loaded on Waters Oasis HLB (6 cc, 500 mg) cartridges, washing the acid out with H_2O and eluting **1** with MeOH.

Purity assessment of the final purified material by LC–MS was carried out on a SCIEX API 4000 mass spectrometer using full scan (*m*/*z* 700–1200) and selected reaction monitoring with positive electrospray ionization. Separations were performed by linear gradient elution on an Agilent 2.7 µm Poroshell 120 SB-C18 column (2.1 × 150 mm) held at 40 °C with mobile phase A: H_2O, and B: 95% CH_3CN, each containing 2 mM ammonium formate and 50 mM formic acid. The gradient was 25–75% B over 25 min, increased to 100% B over 2 min, and held for 11 min, at 0.2 mL/min. LC–UV analysis of the pure product was conducted on an Agilent 1290 Infinity LC System with diode array detector (DAD) with UV monitoring at 238 and 210 nm (same LC column and conditions as LC–MS monitoring except mobile phase A: H_2O, and B: CH_3CN, each containing 0.1% trifluoroacetic acid).

[D-Leu[1]]MC-LY (**1**): white solid; ^1H and ^{13}C NMR (Table 1); HRMS $[M + H]^+$ *m*/*z* 1044.5660 (calculated for $C_{55}H_{78}O_{13}N_7^+$ 1044.5652 (Δ 0.8 ppm)), LC–HRMS/MS (Table 2); HRMS $[M - H]^-$ *m*/*z* 1042.5515(calculated for $C_{55}H_{76}O_{13}N_7^-$ 1042.5507 (Δ 0.8 ppm)), LC–HRMS/MS (DIA) 1024.5418 (42%), 587.2759 (7), 325.2246 (28), 307.2141 (17), 128.0355 (100). λ_{max} (LC–UV) 233, 280 nm (Figure S19).

4.5. Preparation of Reference Material

An aliquot containing [D-Leu[1]]MC-LY (1) (4.3 mg) was evaporated under N_2 and dissolved in 3.0 mL 90% CD_3OH–H_2O. This stock solution was quantitated directly by ^1H NMR using high purity caffeine as the external calibrant as described previously [18]. A dilution of the stock solution was prepared with 50% MeOH–H_2O for analysis by LC–UV-CLND [19] using an Agilent 1100 HPLC system with a 1050 UV detector connected to a model 8060 CLND (Antek PAC, Houston, TX, USA). Separations were performed on an Agilent 3.5 µm Poroshell SB-C8 (2.1 × 150 mm) maintained at 40 °C. Isocratic elution was at 0.2 mL/min, using 65% MeOH–H_2O (0.2% HCOOH) for 1. The external calibrant was also caffeine, with serial dilutions prepared gravimetrically in deionized H_2O. Caffeine was eluted with 40% MeOH–H_2O (0.2% HCOOH). The concentration of contaminating [D-Leu[1],D-Glu(OMe)[6]]MC-LY was measured using the UV detector at 238 nm, with an accurate dilution of the RM of 1 as the calibrant.

After quantitation, the stock solution was quantitatively transferred using 50% high purity degassed MeOH–H_2O to a calibrated volumetric flask, then diluted to the mark with the same. The solution was packaged under argon in flame sealed ampoules using an automatic ampouling machine (Cozzolli, Model FPS1-SS-428, NJ, USA), then stored at −80 °C.

4.6. Protein Phosphatase Inhibition Assay

Ampoules of the RM of **1**, along with the CRM of **2** (CRM-MCLR), were sent to Abraxis LLC (Warminster, PA, USA) for evaluation of toxicity. PP2A assays were performed using the microcystin-PP2A plate kit according to the kit's standard procedures [32].

Supplementary Materials: The following are available online at http://www.mdpi.com/2072-6651/12/2/77/s1. Figure S1: LC–HRMS of CPCC-464 with thiol derivatization (positive mode), Figure S2: LC–HRMS of CPCC-464 (positive and negative modes), Figure S3: ^{1}H NMR Spectrum of [Leu^{1}]MC-LY (**1**) in CD_3OH, Figure S4: COSY NMR Spectrum of [Leu^{1}]MC-LY (**1**) in CD_3OH, Figure S5: DIPSI NMR Spectrum of [Leu^{1}]MC-LY (**1**) in CD3OH, Figure S6: HSQC NMR Spectrum of [Leu^{1}]MC-LY (**1**) in CD_3OH, Figure S7: HMBC NMR Spectrum of [Leu^{1}]MC-LY (**1**) in CD_3OH, Figure S8: ^{13}C NMR Spectrum of [Leu1]MC-LY (**1**) in CD_3OH, Figure S9: ROESY NMR Spectrum of [Leu1]MC-LY (**1**) in CD_3OH, Figure S10: NOESY NMR Spectrum of [Leu^{1}]MC-LY (**1**) in CD_3OH, Figure S11: ^{1}H and ^{13}C chemical shifts overlaid on the structure of [Leu^{1}]MC-LY (**1**), Figure S12: MS/MS spectra of MC-LA (**4**), MC-LY (**5**) and [Leu^{1}]MC-LY (**1**), Figures S13–S18: MS/MS spectra of MC-LA (**4**), MC-LY (**5**) and [Leu^{1}]MC-LY (**1**) (expanded), Figure S19: UV spectra of MC-LR (**2**), MC-LY (**5**) and [Leu^{1}]MC-LY (**1**), Figure S20: LC–HRMS of CPCC-464 with Oxone oxidation (positive mode), Figure S21: MS/MS spectra of [Leu1]MC-M(O)R (**6**) and [Leu^{1}]MC-M(O)R (**7**), Table S1: Comparison of the ^{13}C chemical shift assignments for 1 in CD_3OH with those reported for 3 in CD_3OD.

Author Contributions: Conceptualization: F.R.P. and M.A.Q.; culture: N.I.L.; toxin isolation: P.L.; measurements: P.L., N.M., K.T. K.B. and S.L.R.; reference material preparation: K.T.; data analysis: N.M., K.T., C.O.M., P.M. and M.A.Q.; writing: P.L., F.R.P., P.M., C.O.M. and M.A.Q.; funding acquisition: F.R.P. and M.A.Q. All authors have read and agreed to the published version of the manuscript.

Funding: This research was funded by the Ontario Ministry of the Environment (Best in Science projects 6729 and 78571). The helium cooled probe for the 700 MHz at the NRC (Halifax, NS) was provided by Dalhousie University through an Atlantic Canada Opportunities Agency Grant.

Acknowledgments: The authors thank Agriculture and Agri-Food Canada for access to their 600 MHz NMR spectrometer in Charlottetown, PE. The authors are grateful to R. Syvitski for the acquisition of the 600 MHz NMR spectra, F. Rubio (Abraxis LLC, Warminster, PA, USA) for performing the PP2A assays, and A. Foss (GreenWater Laboratories/CyanoLab, Palatka, FL, USA) for an extract of Poplar Island bloom material.

Conflicts of Interest: The authors declare no conflict of interest.

References

1. Catherine, A.; Bernard, C.; Spoof, L.; Bruno, M. Microcystins and nodularins. In *Handbook of Cyanobacterial Monitoring and Cyanobacterial Analysis*, 1st ed.; Meriluoto, J., Spoof, L., Codd, G.A., Eds.; John Wiley & Sons: Chichester, UK, 2017; p. 109.
2. Gulledge, B.M.; Aggen, J.B.; Eng, H.; Sweimeh, K.; Chamberlin, A.R. Microcystin analogues comprised only of adda and a single additional amino acid retain moderate activity as PP1/PP2A inhibitors. *Bioorg. Med. Chem. Lett.* **2003**, *13*, 2907–2911. [CrossRef]
3. Gulledge, B.M.; Aggen, J.B.; Chamberlin, A.R. Linearized and truncated microcystin analogues as inhibitors of protein phosphatases 1 and 2A. *Bioorg. Med. Chem. Lett.* **2003**, *13*, 2903–2906. [CrossRef]
4. Spoof, L.; Catherine, A. Appendix 3. Tables of microcystins and nodularins. In *Handbook of Cyanobacterial Monitoring and Cyanobacterial Analysis*; Meriluoto, J., Spoof, L., Codd, G.A., Eds.; John Wiley & Sons: Chichester, UK, 2017; pp. 526–537.
5. Hollingdale, C.; Thomas, K.; Lewis, N.; Bekri, K.; McCarron, P.; Quilliam, M.A. Feasibility study on production of a matrix reference material for cyanobacterial toxins. *Anal. Bioanal. Chem.* **2015**, *407*, 5353–5363. [CrossRef] [PubMed]
6. Park, H.; Namikoshi, M.; Brittain, S.M.; Carmichael, W.W.; Murphy, T. [D-Leu^{1}] microcystin-LR, a new microcystin isolated from waterbloom in a Canadian prairie lake. *Toxicon* **2001**, *39*, 855–862. [CrossRef]
7. Matthiensen, A.; Beattie, K.A.; Yunes, J.S.; Kaya, K.; Codd, G.A. [D-Leu^{1}] Microcystin-LR, from the cyanobacterium *Microcystis* RST 9501 and from a *Microcystis* bloom in the Patos Lagoon estuary, Brazil. *Phytochemistry* **2000**, *55*, 383–387. [CrossRef]
8. Schripsema, J.; Dagino, D. Complete assignment of the NMR spectra of [D-Leu^{1}]-microcystin-LR and analysis of its solution structure. *Mag. Res. Chem.* **2002**, *40*, 614–617. [CrossRef]

9. Ortiz, X.; Korenkova, E.; Jobst, K.J.; MacPherson, K.A.; Reiner, E.J. A high throughput targeted and non-targeted method for the analysis of microcystins and anatoxin-A using on-line solid phase extraction coupled to liquid chromatography-quadrupole time-of-flight high resolution mass spectrometry. *Anal. Bioanal. Chem.* **2017**, *409*, 4959–4969. [CrossRef]
10. Caixach, J.; Flores, C.; Spoof, L.; Meriluoto, J.; Schmidt, W.; Mazur-Marzec, H.; Hiskia, A.; Kaloudis, T.; Furey, A. Liquid chromatography–mass spectrometry. In *Handbook of Cyanobacterial Monitoring and Cyanobacterial Analysis*, 1st ed.; Meriluoto, J., Spoof, L., Codd, G.A., Eds.; John Wiley & Sons: Chichester, UK, 2017; pp. 218–257.
11. Foss, A.J.; Miles, C.O.; Samdal, I.A.; Løvberg, K.E.; Wilkins, A.L.; Rise, F.; Jaabæk, J.A.H.; McGowan, P.C.; Aubel, M.T. Analysis of free and metabolized microcystins in samples following a bird mortality event. *Harmful Algae* **2018**, *80*, 117–129. [CrossRef]
12. Trogen, G.; Edlund, U.; Larsson, G.; Sethson, I. The solution NMR structure of a blue-green algae hepatotoxin, microcystin-RR. A comparison with the structure of microcystin-LR. *Eur. J. Biochem.* **1998**, *258*, 301–312. [CrossRef]
13. Stewart, A.K.; Strangman, W.K.; Percy, A.; Wright, J.L.C. The biosynthesis of (15)N-labeled microcystins and the comparative MS/MS fragmentation of natural abundance and their (15)N-labeled congeners using LC-MS/MS. *Toxicon* **2018**, *144*, 91–102. [CrossRef]
14. Bortoli, S.; Volmer, D. Account: Characterization and identification of microcystins by mass spectrometry. *Eur. J. Mass Spectrom.* **2014**, *20*, 1–19. [CrossRef] [PubMed]
15. Miles, C.O.; Sandvik, M.; Nonga, H.E.; Rundberget, T.; Wilkins, A.L.; Rise, F.; Ballot, A. Identification of microcystins in a Lake Victoria cyanobacterial bloom using LC–MS with thiol derivatization. *Toxicon* **2013**, *70*, 21–31. [CrossRef]
16. Mayumi, T.; Kato, H.; Imanishi, S.; Kawasaki, Y.; Hasegawa, M.; Harada, K. Structural characterization of microcystins by LC/MS/MS under ion trap conditions. *J. Antibiot.* **2006**, *59*, 710. [CrossRef] [PubMed]
17. Dorr, F.A.; Oliveira-Silva, D.; Lopes, N.P.; Iglesias, J.; Volmer, D.A.; Pinto, E. Dissociation of deprotonated microcystin variants by collision-induced dissociation following electrospray ionization. *Rapid. Commun. Mass Spectrom.* **2011**, *25*, 1981–1992. [CrossRef] [PubMed]
18. Burton, I.W.; Quilliam, M.A.; Walter, J.A. Quantitative ^{1}H NMR with external standards: Use in preparation of calibration solutions for algal toxins and other natural products. *Anal. Chem.* **2005**, *77*, 3123–3131. [CrossRef] [PubMed]
19. Thomas, K.; Crain, S.; Quilliam, M.A.; Chen, Y.M.; Wechsler, D. Analysis of natural toxins by liquid chromatography-chemiluminescence nitrogen detection and application to the preparation of certified reference materials. *J. AOAC Int.* **2016**, *99*, 1173–1184. [CrossRef]
20. Gulledge, B.M.; Aggen, J.B.; Huang, H.; Nairn, A.C.; Chamberlin, A.R. The Microcystins and Nodularins: Cyclic polypeptide inhibitors of PP1 and PP2A. *Curr. Med. Chem.* **2002**, *9*, 1991–2003. [CrossRef]
21. Ikehara, T.; Imamura, S.; Sano, T.; Nakashima, J.; Kuniyoshi, K.; Oshiro, N.; Yoshimoto, M.; Yasumoto, T. The effect of structural variation in 21 microcystins on their inhibition of PP2A and the effect of replacing cys269 with glycine. *Toxicon* **2009**, *54*, 539–544. [CrossRef]
22. Miles, C.O.; Melanson, J.E.; Ballot, A. Sulfide oxidations for LC-MS analysis of methionine-containing microcystins in *Dolichospermum flos-aquae* NIVA-CYA 656. *Environ. Sci. Technol.* **2014**, *48*, 13307–13315. [CrossRef]
23. Pick, F.R. Blooming algae: A Canadian perspective on the rise of toxic cyanobacteria. *Can. J. Fish. Aquat. Sci.* **2016**, *73*, 1149–1158. [CrossRef]
24. Mantzouki, E.; Lürling, M.; Fastner, J.; De Senerpont Domis, L.; Wilk-Woźniak, E.; Koreivienė, J.; Seelen, L.; Teurlincx, S.; Verstijnen, Y.; Krztoń, W.; et al. Temperature effects explain continental scale distribution of cyanobacterial toxins. *Toxins* **2018**, *10*, 156. [CrossRef] [PubMed]
25. Shishido, T.; Kaasalainen, U.; Fewer, D.; Rouhiainen, L.; Jokela, J.; Wahlsten, M.; Fiore, M.; Yunes, J.; Rikkinen, J.; Sivonen, K. Convergent evolution of [D-Leucine[1]] microcystin-LR in taxonomically disparate cyanobacteria. *BMC Evol. Biol.* **2013**, *13*, 86. [CrossRef] [PubMed]
26. Qi, Y.; Rosso, L.; Sedan, D.; Giannuzzi, L.; Andrinolo, D.; Volmer, D.A. Seven new microcystin variants discovered from a native *Microcystis aeruginosa* strain—Unambiguous assignment of product ions by tandem mass spectrometry. *Rapid. Commun. Mass Spectrom.* **2015**, *29*, 220–224. [CrossRef] [PubMed]

27. Kaasalainen, U.; Fewer, D.P.; Jokela, J.; Wahlsten, M.; Sivonen, K.; Rikkinen, J. Cyanobacteria produce a high variety of hepatotoxic peptides in lichen symbiosis. *Proc. Natl. Acad. Sci. USA* **2012**, *109*, 5886–5891. [CrossRef]
28. Samdal, I.A.; Ballot, A.; Lovberg, K.E.; Miles, C.O. Multihapten approach leading to a sensitive ELISA with broad cross-reactivity to microcystins and nodularin. *Environ. Sci. Technol.* **2014**, *48*, 8035–8043. [CrossRef]
29. Fischer, W.J.; Garthwaite, I.; Miles, C.O.; Ross, K.M.; Aggen, J.B.; Chamberlin, A.R.; Towers, N.R.; Dietrich, D.R. Congener-independent immunoassay for microcystins and nodularins. *Environ. Sci. Technol.* **2001**, *35*, 4849–4856. [CrossRef]
30. Rippka, R.; Deruelles, J.; Waterbury, J.B.; Herdman, M.; Stanier, R.Y. Generic assignments, strain histories and properties of pure cultures of cyanobacteria. *Microbiology* **1979**, *1*, 111. [CrossRef]
31. Allen, M.M.; Stanier, R.Y. Growth and division of some unicellular blue-green algae. *J. Gen. Microbiol.* **1968**, *51*, 199–202. [CrossRef]
32. Abraxis Microcystins/Nodularins PP2A, Microtiter Plate: Test for the Detection of Microcystins and Nodularins in Water. Available online: https://www.abraxiskits.com/wp-content/uploads/2017/03/Microcystins-PP2A-Plate.pdf (accessed on 14 February 2019).

Article

Absence of Cyanotoxins in Llayta, Edible Nostocaceae Colonies from the Andes Highlands

Alexandra Galetović [1], Joana Azevedo [2], Raquel Castelo-Branco [2], Flavio Oliveira [2], Benito Gómez-Silva [1] and Vitor Vasconcelos [2,3,*]

[1] Laboratorio de Bioquímica, Departamento Biomédico, Facultad Ciencias de la Salud, and Centre for Biotechnology and Bioengineering, CeBiB, Universidad de Antofagasta, Antofagasta 1270300, Chile; alexandra.galetovic@uantof.cl (A.G.); benito.gomez@uantof.cl (B.G.-S.)

[2] Centro Interdisciplinar de Investigação Marinha e Ambiental- CIIMAR, 4450-208 Matosinhos, Portugal; joana.azevedo@ciimar.up.pt (J.A.); raquel.castelobranco.12@gmail.com (R.C.-B.); oliveira_flavio@outlook.pt (F.O.)

[3] Faculdade de Ciências da Universidade do Porto, 4169-007 Porto, Portugal

* Correspondence: vmvascon@fc.up.pt

Received: 8 April 2020; Accepted: 5 June 2020; Published: 9 June 2020

Abstract: Edible Llayta are cyanobacterial colonies consumed in the Andes highlands. Llayta and four isolated cyanobacteria strains were tested for cyanotoxins (microcystin, nodularin, cylindrospermopsin, saxitoxin and β-N-methylamino-L-alanine—BMAA) using molecular and chemical methods. All isolates were free of target genes involved in toxin biosynthesis. Only DNA from Llayta amplified the *mcy*E gene. Presence of microcystin-LR and BMAA in Llayta extracts was discarded by LC/MS analyses. The analysed Llayta colonies have an incomplete microcystin biosynthetic pathway and are a safe food ingredient.

Keywords: cyanobacteria; cyanotoxins; Llayta; microcystin; *Nostoc*

Key Contribution: No known cyanotoxins were found in naturally collected Llayta and isolated *Nostoc* strains using molecular and chemical methods.

---

1. Introduction

Members of the cyanobacteria genera *Microcystis, Anabaena, Oscillatoria, Planktothrix* and *Nostoc* are able to synthesize cyanotoxic secondary metabolites such as microcystin, nodularin, cylindrospermopsin, saxitoxin or β-N-methylamino-L-alanine (BMAA), a non-proteinaceous amino acid form [1–5]. Thus, any food or ingredient originating from cyanobacterial biomass and destined for human consumption must be seriously scanned for the presence of these toxins [6]. This is particularly important in the case of Llayta. Llayta is a foodstuff consumed by rural Andean communities since pre-Columbian times [7,8]. The vernacular name Llayta identifies the dry biomass of macro colonies of a filamentous cyanobacterium that grows at Andean wetlands over 3000 m of altitude. The cyanobacterium has been isolated from Llayta colonies (denominated *Nostoc* sp. strain LLA-15) and its taxonomy and biochemical composition has been previously reported [9,10]. After harvesting and sun drying, Llayta can be purchased today at food markets in Tacna (southern Peru) and Arica and Iquique (northern Chile) and used for the preparation of local dishes [11]. Studies on the biochemical composition of Llayta showed that 60% of total amino acids were essential amino acids for humans and 32% of total fatty acids were polyunsaturated fatty acids [10]. To date, no data on the potential for toxin producing in Llayta have been reported, although there are no epidemiological reports that show that consumption of Llayta may cause human illnesses. Based on this body of evidence, we hypothesize that Llayta available at southern Peruvian and northern Chilean food markets for human consumption is a biomass free of cyanotoxins. To support this hypothesis, we provide genetic and analytic evidence.

2. Results and Discussion

The target genes *mcyA*, *cyrJ*, *anaC*, *sxtA* and *sxtG* were not amplified in any sample. The *mcy*A gene can be used in the early warning of cyanotoxins with good certainty, but according to Jasser et al. (2017) [12], they preferentially amplify *Microcystis* and *Planktothrix* genera. On the other hand, the *mcy*E gene was only amplified from Llayta DNA. Since *mcy*E gene primer sequences are found in several microcystin producer genera, the detection of this gene in opposition to the absence of *mcy*A gene is reasonable. However, the gene *mcyE* was amplified from natural Llayta DNA only. Interestingly, DNA from strain LLA-15 did not amplify the *mcyE* gene, even though LLA-15 was isolated from the dry biomass of Llayta. The *mcyE* amplicon from Llayta DNA was recovered and sequenced, and the BLASTn at the NCBI data bank showed 96% identity with gen *mcyE* from *Nostoc* sp. 152, a known microcystin-producing strain [13]. Then, the amplified *mcy*E gene could belong to other microcystin-producing cyanobacteria present in Llayta colonies; in fact, the metagenomics of the closely associated microbiome of edible Llayta colonies show an abundance of bacteria from the phylum Proteobacteria and several cyanobacterial phyla dominated by *Nostoc* (38%), *Anabaena* (25%) and *Nodularia* (20%), with minor contributions of *Microcystis* and *Cylindrospermopsis* (unpublished data). In addition, it is important to highlight that molecular screening for cyanotoxins only gives us preliminary data for potential production, being crucial to confirm the data by analytical methodologies. For that reason, after molecular screening, the MC-LR evaluation by HPLC-PDA was performed in Llayta biomass extracts. MC-LR standard injection showed a peak at 8.6 min with an absorption maximum at 238 nm. The Llayta HPLC chromatogram showed a weak signal at 8.79 min with absorption maxima at 238 nm and 390 nm. Then, it was decided to conduct a MS for this weak signal, since this might be evidence for the presence of microcystin in the Llayta extract. As expected, the standard MC-LR microcystin LR showed an RT of 7.21 min and mass fragments of 995.37-977.36-866-599 m/z (Figure 1A). The Llayta sample in LC-MS did not show a signal at RT 7.21 min. The LC signal region with RT 6–7 min was analysed by MS, and did not show the fragment pattern expected for microcystin LR. Together, the evidence supports the notion that Llayta biomass does not contain a microcystin LR-like toxin (Figure 1B). Interestingly, the Llayta LC chromatogram showed a strong signal at approximately 14 min, which becomes a subject for future work, in order to discover which molecules are there.

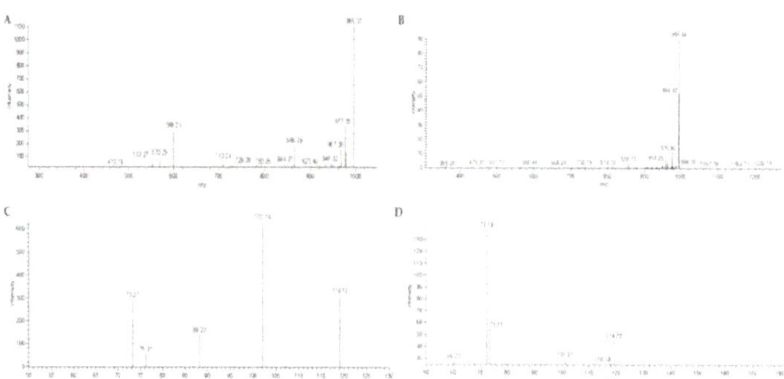

Figure 1. Mass spectra for standards MC-LR and BMAA and Llayta extracts. (**A**): Microcystin LR standard (SIGMA). (**B**): Llayta extract in 50% methanol. (**C**): BMAA standard (SIGMA). (**D**): Llayta hydrolyzate in 50% methanol.

Dominant microcystins produced by *Nostoc* strains are DMadda or ADMadda variants, like the toxic [ADMAdda^5] MC-LR (m/z 1009), the [DMAdda^5] MC-LR (m/z 981) and the [D-Asp^3, ADMAdda^5]

MC-LR (m/z 1023) [14]. The Llayta spectra (Supplementary Materials: Figures S1–S6) obtained did not contain any of the fragments reported for this MC-LR variants or for DMadda or ADMadda moieties due to the absence of the diagnostic ions at m/z 553 [Mdha-Ala-Leu-MeAspArg + H^+] and m/z 627 [Arg-ADMAdda-Glu + H^+]. DMAdda (m/z 121.06) and ADMAdda (m/z 163.08) specific fragment ions were also absent. The absence of all those reported [15,16] characteristic fragments confirms the safety of Llayta as a food ingredient with respect to cyanobacteria toxins.

MC-LR biosynthesis is dependent on the expression of a gene cluster that includes genes *mcy*A and *mcy*E and cyanobacteria are not able to produce the toxin if one or more of these genes are lost during evolutionary processes, as has been reported for *Microcystis aeruginosa*, *Microcystis viridis*, *Microcystis wesenbergii* and *Microcystis ichtyobable* [17–20]. A similar explanation can be inferred from our results, in which Llayta DNA amplified the gene *mcy*E but not the gene *mcy*A, and MC-LR was not present in Llayta extracts.

The LC chromatogram and MS spectrum for standard BMAA (Figure 1C) showed that BMAA migrated with an approximate RT of 5 min and yielded fragments of 119-102-88-76-73 m/z, in agreement with the literature [21–23]. In contrast, the Llayta hydrolysate showed a weak signal with an RT 5.5 min (Figure 1D), and its MS spectrum showed a different fragment pattern to that of standard BMAA.

As has been previously documented, cyanobacteria food supplements may be contaminated with cyanotoxins, either microcystins [24] or anatoxin-a [25], and the monitoring of these toxins is fundamental to prevent human intoxications. In our work, extracts of Llayta were analysed and did not show the presence of cyanotoxins. However, these results need further work to confirm the presence or absence of cyanotoxins in cyanobacterial samples used for human consumption, from different geographical regions in Peru and northern Chile, as well the seasonal effect on cyanotoxin production.

3. Conclusions

The results support the hypothesis that the Llayta biomass analysed does not contain MC-LR or BMAA toxins, and it can be considered a safe ingredient for human consumption.

4. Materials and Methods

To evaluate whether Llayta is a safe ingredient for human consumption, we evaluated the presence of target cyanotoxin genes in the genomes of Llayta (dry biomass) and four strains of cyanobacteria (LLC-10, LLA-15, CAQ-15, LCHI-10) isolated from various water bodies in northern Chile. Additionally, the presence of microcystin-LR (MC-LR) and BMAA was evaluated by HPLC-MS in the dried biomass of Llayta. Genomic DNA was extracted and purified from all isolates and Llayta (dry biomass) using the PureLink Genomic DNA Mini Kit (Invitrogen, USA), according to the manufacturer's instructions. Table 1 shows the target genes (*mcy*A, *mcy*E, *cyr*J, *ana*C, *sxt*A and *sxt*G) and the primers used for PCR. The cyanobacterial strains *Aphanizomenon gracile* LMECYA-040, *Anabaena* sp. LEGE X-002, *Cylindrospermopsis raciborskii* LEGE 97047 and *Microcystis aeruginosa* LEGE 91339 were used as positive control for saxitoxin, anatoxin, cylindrospermopsin and microcystin, respectively, and obtained from Blue Biotechnology and Ecotoxicology Culture Collection (LEGE Culture Collection) of CIIMAR/University of Porto.

Table 1. Primers used to show the presence/absence in Llayta samples of target genes required for toxin biosynthesis.

Target Gene	Primer Pair	Target Group Producers	Size (bps)	References
*mcy*A	*mcy*A-Cd1F/*mcy*A-Cd1R	Microcystin	297	[26]
*mcy*E	HEPF/HEPR	Microcystin and nodularin	472	[27]
*cyr*J	cynsulF; cylnamR	Cylindrospermopsin	586	[28]
*ana*C	*ana*C-genF/*ana*C-genR	Anatoxin	366	[28]
*sxt*A	*sxt*AF/*sxt*AR	Saxitoxin	683	[29]
*sxt*G	*sxt*GF/*sxt*GR	Saxitoxin	893	[29]

After molecular screening of the samples, the dry Llayta biomass was also examined by liquid chromatography coupled with mass spectrometry (LC-MS) in order to confirm the presence of MC-LR and BMAA. One gram of Llayta (dry biomass) was macerated with 8 mL of methanol 75% in liquid nitrogen. After two hours, 12 mL MeOH 75% was added and the methanolic suspension (20 mL) was sonicated (5 times, 60 s, 60 Hz) on ice to extract the intracellular toxins. After centrifugation (4000× *g* for 2 min, at 4 °C), the supernatant was recovered and concentrated in a rotary evaporator. This residue was suspended in 20 mL MeOH 75%, sonicated, centrifuged and evaporated as before. The final residue was dissolved in 50% (v/v) methanol and analysed by LC-MS to detect microcystin. To detect the presence of BMAA by LC-ESI-MS/MS, the resulting pellet from the maceration of Llayta was subjected to acid hydrolysis dissolving it in 1 mL 6M HCL at 110 °C for 24 h [21].

The LC-MS system used to identify and quantify MC-LR in Llayta was a Liquid Phase Chromatograph Finnigan Surveyor (Thermo Scientific, San Jose, CA, USA), coupled with a spectrometry detector (MS Mass LCQ FleetTM ion trap), with electrospray interface (ESI), including a Surveyor LC pump, a Surveyor auto sampler and a Surveyor photoelectric diode array detector (PDA). The program used for data acquisition and processing was XcaliburTM version 2 (Thermo Scientific, San Jose, CA, USA). The mass spectrometer was operated in full scan mode. The capillary voltage and tube lens were maintained at 22 and 120 kV, respectively; the spray voltage was 4.5 kV. Nitrogen was used as a sheath and auxiliary gas. The sheath gas flow rate was set at 80 (arbitrary units) and the auxiliary gas at 10. The capillary temperature was held at 350 °C. Helium was used as a collision gas in the ion trap at a pressure of 3 bar. Separation was achieved on C18 Hypersil Gold column (100 × 4.6 mm I.D., 5 μm, Thermo Scientific, Waltham, MA, USA) kept at 25 °C, with a flow rate of 0.7 mL/min. The injected volume was 10 μL in loop partial mode. Samples were injected in positive polarity mode, in Full scan (270–2000 m/z). The standards and samples were injected in duplicate and a blank and two standards of different concentration were introduced at each set of 6 samples. The standard solution of MC-LR was purchased from DHI LAB Products (Hørsholm, Denmark, Batch nº MCLR-110), with a concentration of 11.026 μg/mL. The system was calibrated using seven dilutions of the standard solution of MC-LR (between 8.5 and 180 μg/L) diluted in 50% acetonitrile (ACN). A gradient elution was used with mobile phase A (ACN) and B (water), both acidified with 0.1% formic acid (55% A and 45% B at 0 min, 90% A and 10% B at 12 min, 100% A at 12.5 min, 100% A at 15 min, 45% A and 55% B at 15.01 and 25 min). Under these conditions the MC-LR retention time (RT) was 7.21 min and the LOD and LOQ were 5.7 μg/L and 8.5 μg/L, respectively. Samples were analysed using the mass-to-charge ratio (m/z) transition of 995 > 599, at 35 eV collision energy. The MC-LR transition was monitored for 1 microscan time. The precursor ion (m/z 995) and MC-LR reference fragment ions with m/z values of 375, 553, 599, 866 and 977 were monitored in the MS/MS mode, in order to validate the presence of the toxin. To assess the presence/absence of other microcystin variants, precursor ions of [DMAdda^5] MC-LR (m/z 981), [ADMAdda^5] MC-LR (m/z 1009) and [D-Asp^3, ADMAdda^5] MC-LR (m/z 1023) were searched in the MS/MS obtained spectra. Diagnostic ions at m/z 553 [Mdha-Ala-Leu-MeAspArg + H^+] and m/z 627 [Arg-ADMAdda-Glu + H^+] were also scanned, as well as DMAdda (m/z 121.06) and ADMAdda (m/z 163.08) specific fragment ions. The qualitative analysis of BMAA in Llayta hydrolysates was done by LC-ESI-MS/MS as described above with the following modifications: the capillary voltage and tube lens were maintained at 33 and 115kV/a, respectively; the spray voltage was 5 kV; nitrogen was used as a sheath and auxiliary gas at a flow rate of 60 (arbitrary units) and the auxiliary gas at 20; the column used was a HILIC (100 × 4.6 mm I.D., 2.6 μm, Phenomenex, USA) kept at 40 °C, with a flow rate of 0.5 mL/min; samples were injected in positive polarity mode, in full scan (50–500 m/z); the standard solution was a mixture of BMAA and L-2, 4-Diaminobutyric acid dihydrochloride (DAB) from Fluca (Batch nº 0001418988), with a concentration of 0.145 μg/g; a gradient elution was used with mobile phases A (MeOH) and B (water), both acidified with 0.1% formic acid (90% A until 10 min, 60% A until 20 min, 50% A until 26 min), returning the start conditions and equilibrating 10 min and finally, samples were analysed using the mass-to-charge ratio (m/z) of 119 at 20 eV collision energy.

Supplementary Materials: The following are available online at http://www.mdpi.com/2072-6651/12/6/382/s1, Figure S1. Extracted Ion Chromatogram and spectrum of Blank solution (methanol 50% + 0.1% Formic acid), Figure S2. Extracted Ion Chromatogram and Full MS^2 spectrum of MC-LR Standard (20 ppb) at 3.01 min, Figure S3. Extracted Ion Chromatogram and Full MS^2 spectra of Llayta extract at *m/z* 995.50, Figure S4. Extracted Ion Chromatogram and Full MS^2 spectra of Llayta extract at *m/z* 981, Figure S5. Extracted Ion Chromatogram and Full MS^2 spectra of Llayta extract at *m/z* 1009, Figure S6. Extracted Ion Chromatogram and MS^2 spectra of Llayta extract at *m/z* 1023, Figure S7. Extracted Ion Chromatogram and MS^2 spectrum of Llayta extract at *m/z* 121.06, Figure S8. Extracted Ion Chromatogram and MS^2 spectrum of Llayta extract at *m/z* 163.07.

Author Contributions: Conceptualization, A.G. and V.V.; methodology, A.G., J.A., R.C.-B., V.V., formal analysis, A.G., J.A., R.C.B, F.O.; investigation, all authors.; resources, A.G., B.G.-S., V.V.; writing—original draft preparation, A.G. writing—review and editing, all authors; funding acquisition, A.G., B.G.-S., V.V. All authors have read and agreed to the published version of the manuscript.

Funding: This work was supported by MINEDUC-UA, Universidad de Antofagasta, Project [project code ANT 1755, 2018]; Proyecto Semillero de Investigación, Universidad de Antofagasta [SI-5305, 2018]; CeBiB, CONICYT-Chile [FB-0001, 2018]; and the CIIMAR [FCT Project UIDB/04423/2020 and UIDP/04423/2020 and by the Atlantic Interreg Project—EnhanceMicroAlgae—High added-value industrial opportunities for microalgae in the Atlantic Area (EAPA_338/2016)].

Acknowledgments: Special thanks to CIIMAR members from the Laboratory Blue Biotechnology and Ecotoxicology.

Conflicts of Interest: The authors declare no conflict of interest. The funders had no role in the design of the study; in the collection, analyses, or interpretation of data; in the writing of the manuscript, or in the decision to publish the results.

References

1. Welker, M.; von Dören, H. Cyanobacterial peptides—Nature's own combinatorial biosynthesis. *FEMS Microbiol. Rev.* **2006**, *30*, 530–563. [CrossRef] [PubMed]
2. Johnson, H.; King, S.R.; Banack, S.A.; Webster, C.; Callanaupa, W.J.; Cox, P.A. Cyanobacteria (*Nostoc commune*) used as a dietary item in Peruvian highlands produce the neurotoxic amino acid BMAA. *J. Ethnopharmacol.* **2008**, *118*, 159–165. [CrossRef] [PubMed]
3. Baptista, M.S.; Cianca, R.C.C.; Almeida, C.M.R.; Vasconcelos, V.M. Determination of the non protein amino acid β-N-methylamino-l-alanine in estuarine cyanobacteria by capillary electrophoresis. *Toxicon* **2011**, *58*, 410–414. [CrossRef] [PubMed]
4. Monteiro, M.; Costa, M.; Moreira, C.; Vasconcelos, V.M.; Baptista, M.S. Screening of BMAA-producing cyanobacteria in cultured isolates and in in situ blooms. *J. Appl. Phycol.* **2017**, *29*, 879–888. [CrossRef]
5. Cirés, S.; Caser, M.C.; Quesada, A. Toxicity at the edge of life: A review on cyanobacterial toxins from extreme environments. *Mar. Drugs* **2017**, *15*, 233. [CrossRef] [PubMed]
6. Buono, S.; Langellitti, A.L.; Martello, A.; Rinna, F.; Fogliano, V. Functional ingredients from microalgae. *Food Funct.* **2014**, *5*, 1669–1685. [CrossRef] [PubMed]
7. Bertonio, L. Vocabulario de la Lengua Aymara. Colección Biblioteca Nacional de Chile: Impreso en la Compañía de Jesús, Perú. 1612. Available online: http://www.memoriachilena.cl/602/w3-article-8656.html (accessed on 26 July 2019).
8. Aldave-Pajares, A. Algas andino peruanas como recurso hidrobiológico alimentario. *Bol. Lima* **1985**, *7*, 66–72.
9. Gómez-Silva, B.; Mendizabal, C.; Tapia, I.; Olivares, H. Microalgas del norte de Chile. IV. Composición química de *Nostoc* commune Llaita. *Rev. Invest. Cient. Tecnol. Cienc. Mar.* **1994**, *3*, 19–25.
10. Galetović, A.; Araya, J.; Gómez-Silva, B. Composición bioquímica y toxicidad de colonias comestibles de la cianobacteria andina *Nostoc* sp. Llayta. *Rev. Chil. Nutr.* **2017**, *44*, 360–370. [CrossRef]
11. Rivera, M.; Galetović, A.; Licuime, R.; Benito Gómez-Silva, B. A microethnographic and ethnobotanical approach to Llayta consumption among the Andes feeding practices. *Foods* **2018**, *7*, 202. [CrossRef]
12. Jasser, I.; Bukowska, A.; Humbert, J.F.; Haukka, K.; Fewer, D.P. Analysis of Toxigenic Cyanobacterial Communities through Denaturing Gradient Gel Electrophoresis. In *Molecular Tools for the Detection and Quantification of Toxigenic Cyanobacteria*; John Wiley & Sons, Ltd.: Chichester, UK, 2017; pp. 263–275.
13. Sivonen, K.; Carmichael, W.; Namikoshi, K.; Rinehart, I.; Dahlem, A.M.; Niemela, S.I. Isolation and Characterization of Hepatotoxic Microcystin Homologs from the Filamentous Freshwater Cyanobacterium *Nostoc* sp. Strain 152. *Appl. Environ. Microbiol.* **1990**, *56*, 2650–2657. [CrossRef] [PubMed]

14. Oksanen, I.; Jokela, J.; Fewer, D.P.; Wahlsten, M.; Rikkinen, J.; Sivonen, K. Discovery of Rare and Highly Toxic Microcystins from Lichen-Associated Cyanobacterium *Nostoc* sp. Strain IO-102-I. *Appl. Environ. Microbiol.* **2004**, *70*, 5756–5763. [CrossRef] [PubMed]
15. Roy-Lachapelle, A.; Solliec, M.; Sauvé, S.; Gagnon, C. A Data-Independent Methodology for the Structural Characterization of Microcystins and Anabaenopeptins Leading to the Identification of Four New Congeners. *Toxins* **2019**, *11*, 619. [CrossRef] [PubMed]
16. Bouaïcha, N.; Miles, C.O.; Beach, D.G.; Labidi, Z.; Djabri, A.; Benayache, N.Y.; Nguyen-Quang, T. Structural Diversity, Characterization and Toxicology of Microcystins. *Toxins* **2019**, *11*, 714. [CrossRef] [PubMed]
17. Mikalsen, B.; Boison, G.; Skulberg, O.M.; Fastner, J.; William Davies, W.; Gabrielsen, T.M.; Rudi, K.; Jakobsen, K.S. Natural Variation in the Microcystin Synthetase Operon mcyABC and Impact on Microcystin Production in *Microcystis* Strains. *J. Bacteriol.* **2003**, *185*, 2774–2785. [CrossRef]
18. Via-Ordorika, L.; Fastner, J.; Kurmayer, R.; Hisbergues, M.; Dittmann, E.; Komarek, J.; Erhard, M.; Chorus, I. Distribution of Microcystin-Producing and Non-Microcystin-Producing *Microcystis* sp. in European Freshwater Bodies: Detection of Microcystins and Microcystin Genes in Individual Colonies. *Syst. Appl. Microbiol.* **2004**, *27*, 592–602. [CrossRef]
19. Kurmayer, R.; Christiansen, G. The genetic basis of toxin production in Cyanobacteria. *Freshw. Rev.* **2009**, *2*, 31–50. [CrossRef]
20. Rantala, A.; Fewer, D.P.; Hisbergues, M.; Rouhiainen, L.; Vaitomaa, J.; Börner, T.; Sivonen, K. Phylogenetic evidence for the early evolution of microcystin synthesis. *Proc. Natl. Acad. Sci. USA* **2004**, *101*, 568–573. [CrossRef]
21. Li, A.; Fan, H.; Ma, F.; McCarron, P.; Thomas, K.; Tanga, X.; Quilliam, M.A. Elucidation of matrix effects and performance of solid-phase extraction for LC-MS/MS analysis of b-N-methylamino-L-alanine (BMAA) and 2,4-diaminobutyric acid (DAB) neurotoxins in cyanobacteria. *Analyst* **2012**, *137*, 1210–1219. [CrossRef]
22. McCarron, P.; Logan, A.C.; Giddings, S.D.; Quilliam, M.A. Analysis of β-N-methylamino-L-alanine (BMAA) in *Spirulina*-containing supplements by liquid chromatography-tandem mass spectrometry. *Aquat. Biosyst.* **2014**, *10*, 2–7. [CrossRef]
23. Beach, D.G.; Kerrin, E.S.; Giddings, S.D.; Quilliam, M.A.; McCarron, P. Differential Mobility-Mass Spectrometry Double Spike Isotope Dilution Study of Release of β-Methylaminoalanine and Proteinogenic Amino Acids during Biological Sample Hydrolysis. *Sci. Rep.* **2018**, *8*, 1–11. [CrossRef] [PubMed]
24. Saker, M.L.; Jungblut, A.-D.; Neilan, B.; Rawn, T.; Vasconcelos, V.M. Detection of microcystin synthethase genes in health food supplements containing the freshwater cyanobacterium *Aphanizomenon flos-aquae*. *Toxicon* **2005**, *46*, 555–562. [CrossRef]
25. Rellan, S.; Osswald, J.; Saker, M.; Gago, A.; Vasconcelos, V.M. First detection of anatoxin-a in human and animal dietary supplements containing cyanobacteria. *Food Chem. Toxicol.* **2009**, *47*, 2189–2195. [CrossRef]
26. Hisbergues, M.; Christiansen, G.; Rouhiainen, L.; Sivonen, K.; Borner, T. PCR-based identification of microcystin-producing genotypes of different cyanobacterial genera. *Arch. Microbiol.* **2003**, *180*, 402–410. [CrossRef]
27. Jungblut, A.D.; Neilan, B.A. Molecular identification and evolution of the cyclic peptide hepatotoxins, microcystin and nodularin, synthetase genes in three orders of cyanobacteria. *Arch. Microbiol.* **2006**, *185*, 107–114. [CrossRef]
28. Mihali, T.K.; Kellmann, R.; Muenchhoff, J.; Barrow, K.D.; Neilan, B.A. Characterization of the gene cluster responsible for cylindrospermopsin biosynthesis. *Appl. Environ. Microbiol.* **2008**, *74*, 716–722. [CrossRef] [PubMed]
29. Casero, M.C.; Ballot, A.; Agha, R.; Quesada, A.; Cirés, S. Characterization of saxitoxin production and release and phylogeny of sxt genes in paralytic shellfish poisoning toxin-producing *Aphanizomenon gracile*. *Harmful Algae* **2014**, *37*, 28–37. [CrossRef]

Article

Analysis of the Use of Cylindrospermopsin and/or Microcystin-Contaminated Water in the Growth, Mineral Content, and Contamination of *Spinacia oleracea* and *Lactuca sativa*

Maria Llana-Ruiz-Cabello [1], Angeles Jos [1], Ana Cameán [1], Flavio Oliveira [2], Aldo Barreiro [2], Joana Machado [2], Joana Azevedo [2], Edgar Pinto [3,4], Agostinho Almeida [3], Alexandre Campos [2,*], Vitor Vasconcelos [2,5] and Marisa Freitas [2,4,*]

1. Area of Toxicology, Faculty of Pharmacy, Universidad de Sevilla, Profesor García González n°2, 41012 Seville, Spain; mllana@us.es (M.L.-R.-C.); angelesjos@us.es (A.J.); camean@us.es (A.C.)
2. CIIMAR/CIMAR, Interdisciplinary Centre of Marine and Environmental Research, University of Porto, Terminal de Cruzeiros do Porto de Leixões, Av. General Norton de Matos, s/n, 4450-208 Porto, Portugal; up201510053@fc.up.pt (F.O.); aldo.barreiro@gmail.com (A.B.); joana.ffmachado@gmail.com (J.M.); joana_passo@hotmail.com (J.A.); vmvascon@fc.up.pt (V.V.)
3. LAQV/REQUIMTE, Departament of Chemical Sciences, Faculty of Pharmacy, University of Porto, Rua Jorge de Viterbo Ferreira 228, 4050-313 Porto, Portugal; edgarpinto7@gmail.com (E.P.); aalmeida@ff.up.pt (A.A.)
4. Polytechnic Institute of Porto, Department of Environmental Health, School of Health, CISA/Research Center in Environment and Health, Rua Dr. António Bernardino de Almeida, 400, 4200-072 Porto, Portugal
5. Biology Department, Faculty of Sciences, University of Porto, Rua do Campo Alegre, s/n, 4169-007 Porto, Portugal

* Correspondence: acampos@ciimar.up.pt (A.C.); maf@ess.ipp.pt (M.F.)

Received: 22 August 2019; Accepted: 25 October 2019; Published: 28 October 2019

Abstract: Cyanobacteria and cyanotoxins constitute a serious environmental and human health problem. Moreover, concerns are raised with the use of contaminated water in agriculture and vegetable production as this can lead to food contamination and human exposure to toxins as well as impairment in crop development and productivity. The objective of this work was to assess the susceptibility of two green vegetables, spinach and lettuce, to the cyanotoxins microcystin (MC) and cylindrospermopsin (CYN), individually and in mixture. The study consisted of growing both vegetables in hydroponics, under controlled conditions, for 21 days in nutrient medium doped with MC or CYN at 10 μg/L and 50 μg/L, or CYN/MC mixture at 5 + 5 μg/L and 25 + 25 μg/L. Extracts from *M. aeruginosa* and *C. ovalisporum* were used as sources of toxins. The study revealed growth inhibition of the aerial part (Leaves) in both species when treated with 50μg/L of MC, CYN and CYN/MC mixture. MC showed to be more harmful to plant growth than CYN. Moreover spinach leaves growth was inhibited by both 5 + 5 and 25 + 25 μg/L CYN/MC mixtures, whereas lettuce leaves growth was inhibited only by 25 + 25 μg/L CYN/MC mixture. Overall, growth data evidence increased sensitivity of spinach to cyanotoxins in comparison to lettuce. On the other hand, plants exposed to CYN/MC mixture showed differential accumulation of CYN and MC. In addition, CYN, but not MC, was translocated from the roots to the leaves. CYN and MC affected the levels of minerals particularly in plant roots. The elements most affected were Ca, K and Mg. However, in leaves K was the mineral that was affected by exposure to cyanotoxins.

Keywords: cyanobacteria; microcystin-LR; cylindrospermopsin; cyanotoxins mixture; plant growth; toxin bioaccumulation

Key Contribution: This work provides new data concerning the toxicological potential of cyanotoxins in edible vegetables and also on the differential susceptibility of vegetables to these toxins. In particular, this work may contribute to the clarification of the maximum cyanotoxin concentrations acceptable for irrigation waters. Moreover, reliable ultra-pressure liquid chromatography-tandem mass spectrometry (UPLC-MS/MS) data concerning the accumulation and distribution of MC and CYN in vegetables is reported, being a valuable source of data to undertake risk assessment related with food safety and human exposure.

---

1. Introduction

Cyanobacteria are anaerobic photoautotrophic group of primitive microorganisms widely distributed in freshwater [1]. The overgrowth of cyanobacteria is favored by several factors related with the eutrophication of aquatic ecosystems, which has increased in the last decades due to the intensification of agricultural and industrial activities, and also with climate change [1,2]. The concerns raised about the cyanobacteria are related with the cyanotoxins that many of these species and strains produce. Among the most prevalent cyanotoxins found in freshwaters are microcystins (MCs). MCs are cyclic heptapeptides and more than 200 chemical variants have been described so far [3]. Moreover they are potent tumor promoters and carcinogenic compounds [4]. Microcystin-LR (MC-LR, herein referred as MC), with leucine and arginine respectively in the positions X and Z of the molecule common structure is the most toxic variant known. The main mechanism of toxicity of these compounds is the specific binding and inhibition of protein phosphatases [5–7]. In fresh waters MCs concentrations can reach values from 108 μg/L up to 10,000 μg/L [8–10], however, the most common concentration range of MCs in surface and irrigation waters vary from 4–50 μg/L [11].

Cylindrospermopsin (CYN) is a tricyclic alkaloid with the following chemical formula: $C_{15}H_{21}N_5O_7S$. CYN has been associated with kidney toxicity and liver failure. CYN is also referred to display genotoxic activity [12–15]. Furthermore the toxin has been recognized as a nationwide threat due to the invasive nature of its main producer, *Cylindrospermopsis raciborskii* [16]. At the molecular level CYN may interact with the ribosomes and inhibit the synthesis of proteins in the cells [17]. CYN has been reported in surface waters at concentrations up to 173 μg/L [1] and also to co-occur with MCs [18].

The main route of human exposure of these cyanotoxins is the consumption of contaminated water or fish. However, the consumption of contaminated vegetables and food supplements may also contribute as an important source of human exposure [19]. Indeed vegetables are potential repositories of cyanotoxins when irrigated with contaminated freshwater [20]. This agricultural practice is potentially detrimental to plant development, plant yield and quality. MCs have been shown to affect root and shoot development, to inhibit seed germination and to affect plant metabolism [21–24]. Changes in photosynthesis, alterations in activity of oxidative stress defense enzymes, increase in lipid peroxidation and accumulation of reactive oxygen species (ROS) have also been attributed to exposure to MCs [2,25–27].

CYN also has shown to be adverse to plants, especially at high concentrations in the water (above 100 μg/L) [28–30]. CYN can cause oxidative stress [31]; inhibit germination [28] and growth [32,33] and trigger a range of biochemical effects [34]. Recently, programed cell death symptoms were reported in two model vascular plants after exposure to CYN 100 μg/L [35]. Contrarily, exposure to low concentrations of CYN (below 100 μg/L) might not be detrimental to plants. For instance, absence of effects, or even stimulation of growth, were reported by Freitas et al. (2015) [30] in hydroponic cultures of lettuce exposed 1 or 10 μg/l MC, CYN or MC+CYN [30], and by Machado et al. (2017) [36] in soil grown carrot plants irrigated for 1 month with 50 μg/L MC. In addition, Guzman-Guillén et al. (2017) [37] reported increased growth of carrots exposed to 50 μg/L CYN, and aromatic plants like parsley (*Petroselinum crispum* L.) and coriander (*Coriandrum sativum* L.) also showed tolerance to MC and CYN [38].

Alongside with the impairment of growth, MCs have shown to alter other plant traits of agronomic relevance and to affect nutritional parameters in cultivated plants. Freitas et al. (2015) [30] showed that cyanotoxins such as MC and CYN interfere with mineral accumulation in lettuce. Moreover, changes in the metabolism of plants exposed to cyanotoxins were also suggested to lead to the accumulation of potential allergenic proteins [39]. The exposure to MC could also be a plausible cause of the decrease in the accumulation of ascorbic acid (vitamin c) in this root vegetable [36].

El Khalloufi et al. (2016) [40] and Lahrouni et al. (2016) [41] showed that toxic cyanobacterial extracts containing MC affect plant-bacterial symbioses and decrease the bacterial growth rate of rhizospheric microbiota. Leguminous crops that develop symbiotic interactions with *Rhizobia* to uptake nitrogen can be particularly affected in such growth conditions. For instance, the chronic exposure of fava beans to contaminated water (50 and 100 μg/L MC) was shown to reduce significantly plant-Rhizobia nodule number and nitrogen assimilation (measured as dry matter) in the plants, as well as plant growth and photosynthetic activity [42].

Although a provisional upper limit in drinking water of 1 μg/L for MC has been proposed by the World Health Organization (WHO); and the United States Environmental Protection Agency (US-EPA) has set a health advisory for CYN of 3 μg/L, in general, there are very few countries with legislation concerning cyanotoxin levels in food. The legislation is addressed mainly to fish and shellfish products but does not cover vegetables or other food products and supplements [43]. The lack of legislation is probably a consequence of the lack of knowledge concerning the contamination and the impact of cyanotoxins in the quality and safety of vegetables. For instance, more than one type of cyanotoxin is frequently detected in the environment, nevertheless the bioactivity and toxicology of such combinations in plants has been overlooked. Moreover, the genetics may play a critical role in the sensitivity of plants to cyanotoxins. This factor has also been poorly considered in research, and there is a general lack of understanding of the susceptibility of plant crops to cyanotoxins and the genetic factors that determine the susceptible phenotype. This work aim to cover some of the knowledge gaps in this field of research, namely the susceptibility of green-vegetables to cyanotoxins at environmentally relevant concentrations, and cyanotoxins accumulation in plants. In the present work *Spinacea oleracea* and *Lactuca sativa*, two common vegetables used in the human diet, were exposed to environmentally relevant concentrations (10 and 50 μg/L) of MC and CYN, or their mixtures, through crude extracts of *M. aeruginosa* and *C. ovalisporum*, respectively. The effects of these cyanotoxins on growth, photosynthesis, mineral content and bioaccumulation on spinach and lettuce are here reported and discussed in relation to susceptibility to cyanotoxins, nutritional value, food safety, and human exposure.

2. Results and Discussion

2.1. Effects of CYN, MC and CYN/MC Mixture on Spinach and Lettuce Growth

The effects of CYN, MC and CYN/MC mixture on spinach and lettuce growth were studied by comparing the fresh weight of the control and treated plants at different environmental concentrations (CYN and MC: 10 and 50 μg/L; CYN/MC mixture: 5 + 5 μg/L and 25 + 25 μg/L, respectively) (Figure 1).

In general, for all exposure conditions, the response at the physiological level was concentration-dependent and more noticeable decreased growth was observed for spinach plants (roots and leaves). The results also show that spinach plants were more vulnerable to MC than CYN and this was also verified at morphological level, once for the treatment applied with MC 50 μg/L spinach plants were negatively affected resulting in mortality and deleterious effects in leaves (necrosis) (Figure 2). Decreased growth caused by MC on terrestrial plants was already reported in previous studies [21,23,26,44–48]. However, most of them were conducted under unrealistic conditions, where plants were exposed to quantities 100–1000-fold above the environmentally realistic concentrations. Nevertheless, recent studies in which exposure experiments were carried out with environmentally realistic concentrations (5 μg/L and 10 μg/L) have shown that depending on the plant species, MC can negatively affect plant productivity and quality, being the edible fraction reduced by the MC exposure [49]. Some examples can be enumerated: (1) lettuce plants treated with 5 and 10 μg/L MC were shorter, had fewer leaves per head and weighed less than the control group [49]; (2) carrot plants exposed to 1, 5, 10 μg/L MC decreased the total mass of the head and diameter of the roots relative to the control group [49]; (3) green beans treated with 1, 5, 10 μg/L MC had lower total mass and fewer beans per plant than the control [49]; and (4) the growth of cucumber at different growth stages was inhibited with exposure to 10 μg/L MC, and the order of growth inhibition was seedling stage > early flowering stage > fruiting stage [50].

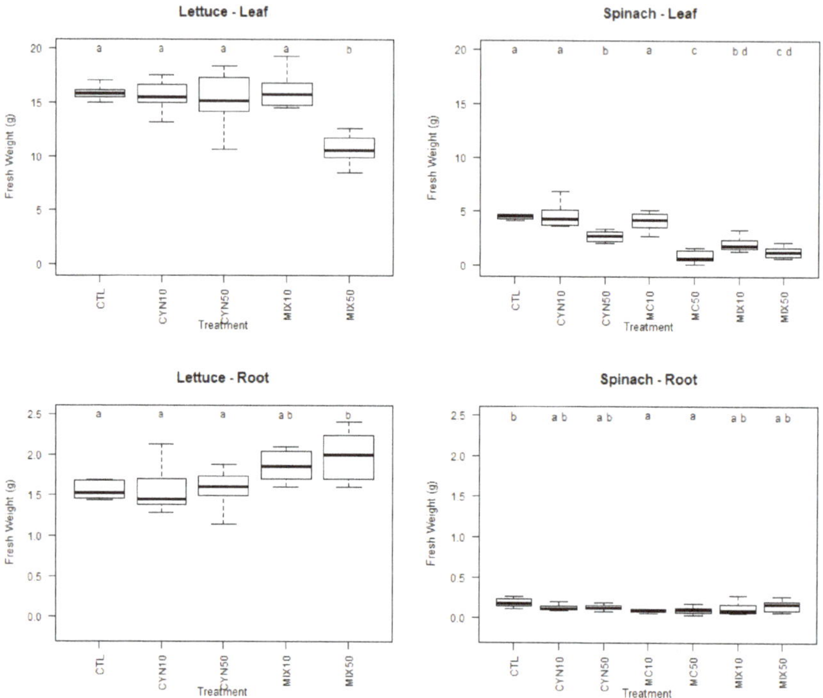

Figure 1. Box-whiskers plots of fresh weight of spinach and lettuce plants (leaves and roots) exposed to cylindrospermopsin (CYN) and cyanotoxins microcystin (MC) at the concentrations of 10 μg/L (CYN10 and MC10) and 50 μg/L (CYN50 and MC50) and CYN/MC mixture at the concentrations of 5 + 5 μg/L (MIX10) and 25 + 25 μg/L (MIX50), respectively, for 21 days. Different letters (**a**–**d**) mean significant differences ($p < 0.05$). Control plants (CTL). Number of sample replicates (n = 8).

Figure 2. Representative images of spinach plants grown for 21 days with CYN and MC at the concentrations of 10 μg/L (CYN10 and MC10) and 50 μg/L (CYN50 and MC50) and CYN/MC mixture at the concentrations of 5 + 5 μg/L (MIX10) and 25 + 25 μg/L (MIX50), respectively, for 21 days. Control plants (CTL).

However, contrarily, some studies showed no effects in crop plants due to MC exposure at environmentally realistic levels. Cao et al. (2018) [51] reported that irrigation with natural concentrations of MC-contaminated water did not affect the growth and yield of lettuce and rice. Liang et al. (2016) [52] also reported that low concentration of MCs (1 μg/L) did not affect the growth and photosynthesis of rice. Furthermore, interestingly, the productivity and nutritional quality of some agricultural plants may even be enhanced when exposed to ecologically relevant concentrations of MCs, with a weight of leaves increasing, as it was shown in tomato and lettuce plants, respectively [30,53]. In this study, the results obtained for spinach exposed to MC corroborate the findings of Pflugmacher et al. (2007) [26], in which the exposure of different variants of spinach to cyanobacterial crude extract containing 0.5 μg/L MC, resulted for some variants, in the reduction of growth due to the lower leaf size and the lower number of leaves than in control groups. In line with the study of Pflugmacher et al. (2007) [26], here we report differences in the sensitivity of the two vegetables (spinach and lettuce) to CYN, which shows that sensitivity to cyanotoxins is, in part, determined by genetic factors. Despite the differences in the experimental designs and although the toxicity mechanisms involved remain unknown, together, these results raise attention for species that seem to be most vulnerable to adverse effects of MC exposure, which should be considered by the regulatory agencies and policy-makers when outlining legislation for irrigation water.

Less is known about the potential effects of CYN and CYN/MC mixture on productivity and quality of crop plants. In this work, spinach plants showed to be more sensitive to environmentally realistic concentrations of cyanotoxins than lettuce plants. The exposure of spinach plants to CYN led to a significant decrease in the fresh weight of leaves at the highest concentration tested (50 μg/L) ($p < 0.05$) (although lower than for MC at the same concentration). For lettuce, plants it is important to note that CYN at 10 and 50 μg/L did not result in significant differences in root and leaf weight compared to control group. Previous studies have shown that although low concentrations of CYN (1–10 μg/L) can stimulate the root and leaf growth, the exposure of lettuce plants to 100 μg/L CYN negatively affected the leaf yield by the reduction of its fresh weight [30]. So, as the exposure conditions were similar in both studies, this can be indicative that lettuce plants are able to tolerate CYN at least till concentrations of 50 μg/L without deleterious effects on their productivity. Data also exists for rice plants which were exposed to 2.5 μg/L CYN, resulting in a significant increase in root fresh weight upon 48 h of exposure and a significant decrease in leaf fresh weight of leaves after nine days of exposure [31].

Regarding the cyanotoxins mixture, interestingly, only the fresh weight of spinach leaves was significantly lower than the control group ($p < 0.05$), being the effect more prominent for the highest concentration. For lettuce, a significant increase in fresh weight of roots and a significant decrease in fresh weight of leaves was only observed for plants exposed to CYN/MC mixture at the highest concentration (25 + 25 μg/L CYN/MC) ($p < 0.05$). Our results are in agreement with those reported by Freitas et al. (2015) [30], which have suggested that lettuce plants are able to cope with low concentrations (1 and 10 μg/L) of CYN and CYN/MC mixture by ensuring the maintenance of and even increasing their fresh weight. Prieto et al. (2011) [31] have studied the interaction effects of CYN (0.13 μg/L) and MC (50 μg/L), on rice plants and the mixture did not produce significant changes in fresh and dry weight of roots and leaves after 48 h of exposure. In our study, it can be hypothesized that the increased metabolic activity in roots required to maintain or improve the growth could compromise the leaf growth of the spinach and lettuce plants, respectively. It is also important to point out that, in this study, when plants were exposed to CYN/MC mixtures, even in concentrations that individually plants seem to be susceptible of homeostatic compensation (e.g., 25 μg/L), the effects observed were exacerbated.

Collectively, these results demonstrate that MC can severely inhibit the growth and performance of spinach plants, and that the combined effect of CYN and MC was more intense than their individual effects for lettuce plants. This suggests that the presence of multiple cyanotoxins in irrigation water, even at environmentally realistic concentrations may influence the productivity of crop plants in a species-dependent manner. As the simultaneous presence of multiple cyanotoxins in aquatic ecosystems is frequent, being MC the most prevalent cyanotoxin and CYN has been increasingly recurrent [11], the potential interaction between the mechanisms of toxicity of CYN (inhibition of protein synthesis) [54,55] and MC (inhibition of serine/threonine PP) [56,57] should be further studied at cellular level in plants.

2.2. Effects of CYN, MC and CYN/MC Mixture on Lettuce and Spinach Photosynthetic Capacity

The treatment with 50 μg/L MC or with the toxin mixture, as reported in Figures 1 and 2, was highly toxic to spinach, leading to extended leaf necrosis and even to the death of some plants. Thereby measurements of chlorophyll fluorescence were carried out only in plants that survived those treatments. In non-stressed plants Fv/Fm values are known to vary between 0.75 and 0.85 [58]. As reported in Figure 3 all Fv/Fm values for the spinach plants analyzed were above 0.76 meaning that chlorophyll fluorescence and the photosynthetic capacity of spinach plants was not affected by treatments with MC and CYN toxins. Of notice is, nevertheless, the increased variability of Fv/Fm values in the 50 μg/L MC and 25 + 25 μg/L CYN/MC treatments, which might evidence some adverse effects of the cyanobacterial extracts in particular spinach plants. Indeed, some of spinach plants in 50 μg/L MC and 25 + 25 μg/L CYN/MC groups evidenced necrotic symptoms in leaves and reduced growth.

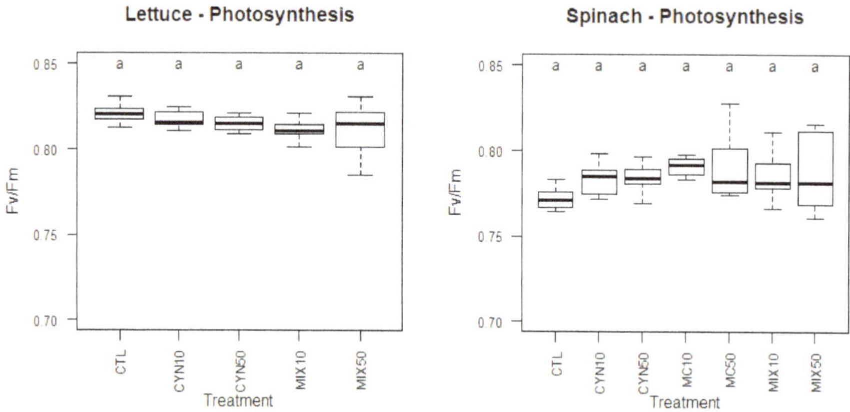

Figure 3. Box-whiskers plots of maximum fluorescence (Fv/Fm) of spinach and lettuce plants exposed CYN and MC at the concentrations of 10 μg/L (CYN10 and MC10) and 50 μg/L (CYN50 and MC50) and CYN/MC mixture at the concentrations of 5 + 5 μg/L (MIX10) and 25 + 25 μg/L (MIX50), respectively, for 21 days. Different letters (a, b, c and d) mean significant differences ($p < 0.05$). Control plants (CTL). Number of sample replicates (n = 8).

Maximum fluorescence of lettuce plants was not affected by any of the treatments and all values reported for lettuce were above 0.76. Maximum fluorescence yield is directly related to reaction center II (PSII) function. Moreover the decrease of this parameter has been attributed to PSII malfunction, or inhibition of electron transport linked with excitation of reaction centers [59]. A variety of effects of cyanotoxins on plant photosynthesis have been reported. Inhibition of photosynthesis has been consistently reported in plants exposed to high concentrations (usually above 100 μg/L) of MC. Freitas et al. (2015) [39], in a proteomics study in lettuce provided insights regarding the putative molecular events that could be causing the inhibition of photosynthesis by CYN and MC. The authors reported alterations in the expression of key proteins associated with primary photosynthesis reactions (light reactions) such as quinone oxidoreductase, oxygen-evolving enhancer proteins, chlorophyll a-b-binding proteins, chloroplast PsbO4, cytochrome b6/f heme-binding protein 2 and ferredoxin-NADP reductase, but also proteins involved in the Calvin cycle (carbon fixation reactions), ribulose-1,5-bisphosphate carboxylase/oxygenase activase (RuBPactivase), ribulose bisphosphate carboxylase/oxygenase activase 1 (RuBisCO activase 1), phosphoribulokinase (PRK), and sedoheptulose-1,7-bisphosphatase (SBPase), photorespiration (gamma carbonic anhydrase-like 2).

On the other hand, contrasting effects have been reported in studies with toxin concentrations below 100 µg/L. While in tomato and *Vicia faba* the exposure to respectively 100 µg/L MC and 50–100 µg/L MC resulted in a decrease of Fv/Fm [42,60], in carrots for instance, the exposure to water contaminated with 10–50 µg/L MC and 50 µg/L CYN [36] resulted in an increase in the maximum fluorescence yield in this root vegetable. The increase in Fv/Fm after exposure to cyanotoxins was interpreted by Machado et al. (2017) [36] as a plant mechanism of defense related with the enhancement of the physiological condition through improving the photosynthetic capacity. Also Bittencourt-Oliveira et al. (2016) [61] related the increase in the net photosynthetic rate in lettuce plants exposed to *M. aeruginosa* toxic extracts (0.65 to 13 µg/L total MC) as a mechanism of lettuce plants to synthesize additional substrates to supply energy and to be utilized in the biosynthesis of important antioxidant molecules and enzymes that protect from oxidative stress induced by MC. The presented results are thus consistent with previous studies and show that MC or CYN, in the range of 10–50 µg/L, is not detrimental to plant photosynthesis.

2.3. *Effects of CYN, MC and CYN/MC Mixture on Lettuce and Spinach Mineral Content*

Essential mineral elements are usually classified as macronutrients or micronutrients, according to their relative concentration in plant tissue [62]. In this study, the content of macronutrients (Ca, Mg and K) and micronutrients (Mn, Na, Cu, Zn and Fe) was determined in roots and leaves of spinach and lettuce plants exposed to CYN, MC and CYN/MC mixture, in order to assess the physiological condition and nutritional quality of the edible parts of these two vegetables.

2.3.1. Macronutrients

The effects of CYN, MC and CYN/MC mixture on the macronutrients content in spinach and lettuce plants (roots and leaves) are presented in Figure 4.

Overall, the detrimental effects of exposure to CYN, MC and CYN/MC mixture were more pronounced in roots than in leaves of spinach and lettuce plants. The results show that there were no significant differences in Ca and Mg content in the edible portion (leaves) of spinach and lettuce plants exposed to CYN, MC and CYN/MC mixture in comparison to the respective control groups. The exception was for K, which content was significantly decreased in spinach leaves ($p < 0.05$) in several exposure conditions, being MC at both exposure concentrations (10 µg/L and 50 µg/L) the most deleterious toxin. Indeed, some studies have demonstrated that MC at ecologically realistic concentrations promotes negative effects in mineral content of leaf plants, generally higher than CYN and MC/CYN mixture. Regarding lettuce, it was reported that exposure to different concentrations of MC (1, 10 and 100 µg/L) resulted in a decrease of the leaf mineral content, and the effects were more pronounced at the highest time and concentration of exposure [30]. In addition, the content of K and Ca in the shoots of *V. faba* have been reported to decrease after two months of exposure to *M. aeruginosa* extract containing 50 and 100 µg/L of MCs [42].

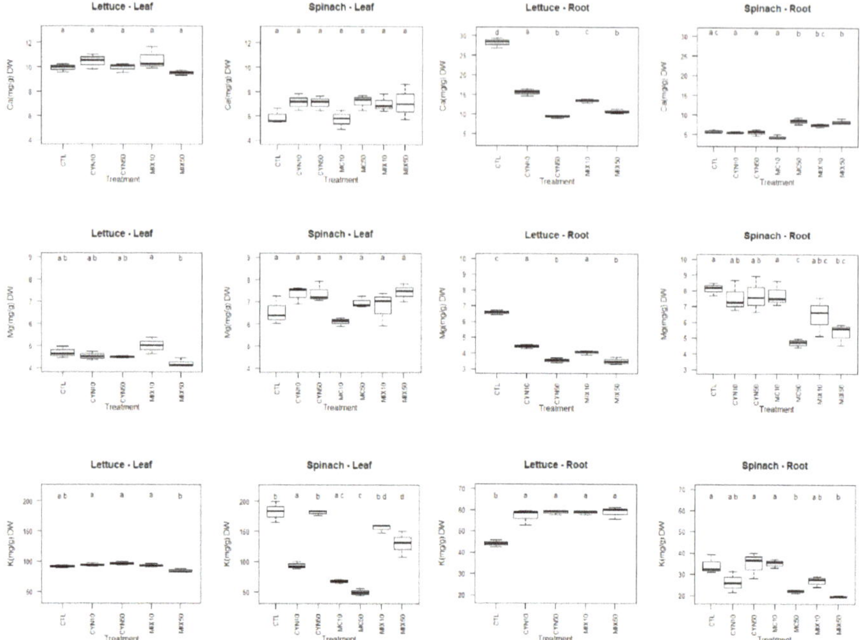

Figure 4. Box-whiskers plots of macronutrients (Ca, Mg and K) content of spinach and lettuce plants (leaves and roots) exposed to CYN and MC at the concentrations of 10 µg/L (CYN10 and MC10) and 50 µg/L (CYN50 and MC50) and CYN/MC mixture at the concentrations of 5 + 5 µg/L (MIX10) and 25 + 25 µg/L (MIX50), respectively, for 21 days. Different letters (a, b, c and d) mean significant differences ($p < 0.05$). Control plants (CTL). Number of sample replicates (n = 3).

The effects of CYN and CYN/MC exposure on the mineral content in plants have been studied to a much lesser extent than MC. According to our knowledge, this is the first study reporting the effects of CYN and CYN/MC mixture on the mineral content in spinach plants. For lettuce leaves, Freitas et al. (2015) [30] stated similar results to those obtained in this study, with no significant effects on Ca, Mg and K content after 10 days of exposure to CYN and CYN/MC mixture. However, it should be highlighted the tendency for the decreasing in Mg and K content in the lettuce leaves at the highest concentration of CYN/MC mixture. For spinach, interestingly, despite the significant decrease in K content after plant exposure to CYN 10 µg/L and to the mixture at both concentrations ($p < 0.05$), the effects were less deleterious than for MC alone (at both exposure concentrations).

More complex responses were obtained for the mineral content in roots of spinach and lettuce plants. The roots of spinach plants were significantly affected by 50 µg/L MC and CYN/MC mainly at the highest concentration of exposure, resulting in a significant increase in Ca, and significant decreases in Mg and K contents in roots, compared to control group ($p < 0.05$). Similarly to the range of concentrations used in this study, the K in the roots of *V. faba* were decreased, after two months of exposure to *M. aeruginosa* extract containing 50 µg/L MC [42]. However, the responses of spinach roots were not parallel with those obtained for carrot roots at the same concentrations of exposure (10 and 50 µg/L MC) [36], being even contrary for Mg and K (significant increase in carrots and decrease in spinach). In lettuce, in this study, it was also obtained a significant increase in K and a significant decrease in Ca and Mg content in roots in all exposure conditions (CYN and CYN/MC), being the effects more pronounced at the highest concentrations ($p < 0.05$). Similar results for K content in roots were also obtained for *Lycopersicon esculentum* [24].

These results show that the physiological response and consequently the changes in roots mineral content may vary according to the plant species. This was already suggested by Saqrane et al. (2009) [46], which found a concentration-dependent increase in the macro mineral content (Na, K, Ca, P, and N) in roots of *Triticum durum, Zea mays, Pisum sativum, Lens esculenta* after plants exposure to MC. Guzmán-Guillén et al. (2017) [37] also suggest that carrots, a root vegetable, seem to be more vulnerable to CYN (in terms of mineral content) than leafy vegetables, such as lettuce plants. The results obtained for spinach (roots) are in agreement with that suggestion, once for the concentrations tested (10 and 50 μg CYN/L) there were no changes in Ca, Mg and K content in comparison to the control group. However, in this study, the content of Ca and Mg in roots of lettuce plants followed a pattern similar to the one reported in carrots [36]. Lahrouni et al. (2013) [42] proposed that the changes in minerals might result from changes of plant membrane permeability caused by cyanotoxins such as MC. Furthermore, the antioxidant response to stress promoted by cyanotoxins is usually more pronounced in roots than in leaves of exposed plants [30,31]. In fact, since roots are the part of the plant in direct contact with cyanotoxins, it is understood that the uptake of nutrients in spinach and lettuce plants may have been considerably affected by oxidative stress, cellular damage and changes in membrane permeability caused by the toxins. Minerals are essential to plant development since they are intrinsic components of all cellular structures and molecules, and key-players in the metabolism and function of plants [62]. On the other hand, imbalances in a specific element may induce deficiency or excessive accumulation of another element [62]. Although each essential element participates in many different metabolic reactions, it is possible to anticipate which general functions of plant metabolism will be affected due to the disturbance of mineral uptake and translocation promoted by extracts of CYN, MC and CYN/MC.

Correspondingly, it is well known that Mg and K content in plants can significantly affect photosynthesis and K, which is required as a cofactor for more than 40 enzymes, is the principal cation in establishing cell turgor and maintaining cell electroneutrality of plant cells [62]. Mg, besides being part of the ring structure of the chlorophyll molecule, also has a specific role in the activation of enzymes involved in respiration, photosynthesis and the synthesis of DNA and RNA [63]; and Ca is required for the normal functioning of plant membranes and has been implicated as a second messenger in plant responses to both environmental and hormonal stimuli (calcium acts as a signal to regulate key enzymes in the cytosol) [62]. It is also used in the synthesis of new cell walls and in the mitotic spindle during cell division [62].

The ability of crop plants to cope with abiotic stress, maintaining or even increasing the nutritional value is of utmost importance for food security. In this study, the mineral content (Ca, Mg and K) of the edible parts of lettuce and spinach plants were almost not changed in comparison to the control group (except for K in spinach plants), which enables to conclude that, at least at this level, lettuce plants are able to maintain its mineral content if exposed till 50 μg/L of CYN and MC. Nevertheless, the effects of the CYN/MC mixtures should still be studied in greater detail.

2.3.2. Micronutrients

The impact of cyanotoxins on micronutrients content in crop plants has been studied to a less extent than macronutrients [24,42,63]. The effects of CYN, MC and CYN/MC mixture on the micronutrients content in spinach and lettuce plants (roots and leaves) are presented in Figure 5.

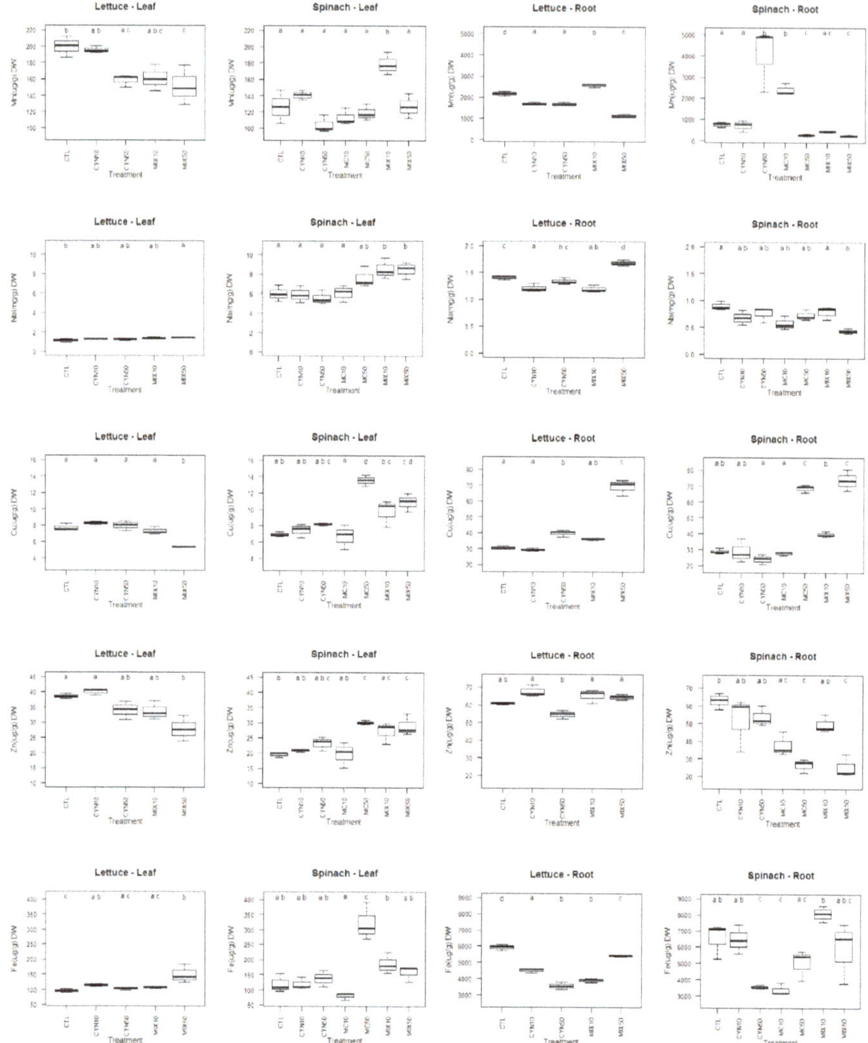

Figure 5. Box-whiskers plots of micronutrients (Mn, Na, Cu, Zn and Fe) content of spinach and lettuce plants (leaves and roots) exposed to CYN and MC at the concentrations of 10 µg/L (CYN10 and MC10) and 50 µg/L (CYN50 and MC50) and CYN/MC mixture at the concentrations of 5 + 5 µg/L (MIX10) and 25 + 25 µg/L (MIX50), respectively, for 21 days. Different letters (**a**–**d**) mean significant differences ($p < 0.05$). Control plants (CTL). Number of sample replicates (n = 3).

In general, the changes in micronutrients content were more pronounced in roots than in leaves of lettuce and spinach plants and the effects of CYN/MC mixture were greater than the single cyanotoxins. In lettuce leaves, comparatively to the control group, the content in Mn, Cu and Zn was significantly decreased and the content in Na and Fe was significantly increased, mainly due to the exposure to the MC/CYN mixture at the highest concentration (25 + 25 µg/L). The results obtained for Mn and Zn are in agreement with those reported by Freitas et al. (2015) [30], in which the mixture of MC/CYN (100 + 100 µg/L) led to a significant decrease of their content in lettuce leaves. However, being the same plant species (*Lactuca sativa*) and the experimental conditions similar, the content of Cu in leaves, followed a contrary tendency when compared to the results of Freitas et al. (2015) [30]. Although speculative, a possible explanation can be related with differences of the use of purified cyanotoxins vs. cyanobacterial crude extracts. Thus, it can be hypothesized that the crude extracts may have some substances that function as Cu chelators. Furthermore, in the study developed by Freitas et al. (2015) [30], the mineral content significantly increased in lettuce leaves due to the CYN exposure, and apparently these effects were time and concentration-dependent. However, in this study, in general, there were no significant differences between the content in micronutrients in lettuce leaves exposed to CYN (10 and 50 μg/L) and the control group.

Regarding spinach leaves, overall, the content of all micronutrients analyzed (Mn, Na, Cu, Zn, Fe) was significantly increased; and most of the changes occurred for the conditions MC 50 μg/L and MC/CYN (25 + 25 μg/L). The result achieved for Na content in spinach leaves is corroborated with that obtained for shoots of *V. faba*, in which plants treated with 50 and 100 μg/L MC significantly increased its content [42].

Concerning the micronutrient content in roots, once again, the results suggest that although lettuce and spinach plants were affected by cyanotoxins, the susceptibility to the different conditions seems to be dependent of the plant species and genetic factors. Lettuce plants were especially affected by CYN/MC mixture at the highest concentration of exposure, while spinach plants were affected by MC 50 μg/L at the similar extent of CYN/MC mixture at the highest concentration of exposure. Overall, due to 25 + 25 μg/L CYN/MC exposure, the lettuce roots significantly increased its content in Na and Cu, significantly decreased its content in Mn and Fe and no changes were registered for Zn content in comparison to the control group. Furthermore, the exposure of lettuce plants to CYN resulted in a significant decrease in Mn and Cu content in roots. Contrasting to our results, Guzmán-Guillén et al. (2017) [37] reported that only Mn and Cu were significantly increased in roots of *Daucus carota* after exposure (30 days) to CYN (10 and 50 μg/L).

The spinach roots showed a significant increase in Cu and a significant decrease in Mn, Na and Zn content when compared to the control group due the exposure to MC 50μg/L and 25 + 25 μg/L CYN/MC. Again, contrary to our results, a significantly higher content of Na in roots of *Triticum durum, Zea mays, Pisum sativum, Lens esculenta* [46], *L. esculentum* [24], *V. faba* [42] and *Daucus carota* [36] was found after a prolonged exposure to MC. The uptake of minerals by the plant, especially micronutrients, can be affected by various factors. Thus, the contradictory results can be attributable to different plant species, different time and concentrations of exposure, use of purified toxins or crude extracts, different mechanisms of enzymatic and non-enzymatic defense system and the systems where plants grow (hydroponic vs. soil-grown systems).

Inadequate (deficit or excessive) uptake of micronutrients by crop plants can result in physiological disorders, which may have implications in its antioxidant defense system, yield and nutritional quality. The biological functions which could be affected by micronutrients imbalances comprises antioxidant activity and cell growth. Fe and Cu are associated with enzymes involved in redox reactions; and Na is involved in the synthesis of new cell walls, cell division and cell expansion [62]. Mn is required for activity of some dehydrogenases, decarboxylases, kinases, oxidases, and peroxidases, playing an important role in the structure of photosynthetic proteins as well as ATP synthesis [62]. In addition, Zn is required for many enzymes and also for chlorophyll biosynthesis [62].

Finally, although the enhancement of mineral content in the edible parts of plants is of utmost importance for its nutritional quality, it can result in unintended consequences for public health due to the potential accumulation of cyanotoxins in edible tissues. Furthermore, it is important to point out that the assays were performed in controlled environmental condition, but the changes in mineral nutrition can predispose more the plants to diseases, which can be of concern due to the actual threatens of climate change.

2.4. Accumulation of CYN and MC in Plant Leaves

The mean CYN concentration found in the leaves of lettuce exposed to the toxin, at 10 and 50 µg/L CYN, were 2.4 and 9.4 µg CYN/kg fw (Table 1), respectively, determined with the quantification method described by Prieto et al. (2018) [64]. This means that the total amount of CYN taken-up, in relation to the amount present in the culture medium, varied between 23.7% and 18.8%, depending on the CYN concentration tested, and globally it could be considered around 20%. For spinach, in the same exposure conditions, higher concentrations of CYN (9.52 and 36.97 µg/kg fw, respectively) were detected (Table 1). MC was not detected at any concentration tested. For CYN-exposed spinach, the uptake of plant leaves was 95% and 74%, for the exposure to 10 and 50 µg/L CYN, respectively. These results evidence that spinach leaves accumulate more CYN than lettuce, and this might be related with the mechanisms of transport of molecules in the plant tissues which might require specific molecular transporters. The increased CYN accumulation coincided with the inhibition of growth of spinach, being thereby one of the plausible causes due to the toxic effects of CYN in spinach tissues.

Table 1. Levels of MC and CYN in tissues of lettuce and spinach exposed to the 2 cyanotoxins, measured by UPLC-MS/MS as described previously [18]. These results were first published in Díez-Quijada et al. (2018) [18]. < LOD—below the detection limit of the method (0.06 ng MC/g fw and 0.07 ng CYN/g fw); nd—not determined. Number of sample replicates (n = 3). Concentrations of exposure were: 10 µg/L (CYN10 and MC10) and 50 µg/L (CYN50 and MC50) and CYN/MC mixture at the concentrations of 5 + 5 µg/L (MIX10) and 25 + 25 µg/L (MIX50), respectively, for 21 days.

Levels of MC and CYN in Tissues of Lettuce and Spinach		CYN10	CYN50	MC10	MC50	MIX10	MIX50	MIX10	MIX50
		CYN	CYN	MC	MC	MC	MC	CYN	CYN
		(µg/kg fw)	(µg/kg fw)	(µg/kg fw)	(µg/kg fw)	(µg/kg fw)	(µg/kg fw)	(µg/kg fw)	(µg/kg fw)
Lettuce	leaves	2.4 ± 0.89	9.4 ± 2.38	nd	nd	< LOD	< LOD	10.00 ± 2.40	41.92 ± 6.37
	roots	nd	nd	nd	nd	0.22 ± 0.08	1.10 ± 0.25	34.23 ± 11.58	110.00 ± 33.29
Spinach	leaves	9.52 ± 3.56	36.97 ± 10.25	< LOD	< LOD	< LOD	< LOD	12.57 ± 4.22	119.69 ± 43.93
	roots	nd	nd	nd	nd	0.53 ± 0.32	1.31 ± 0.14	39.16 ± 31.30	24.00 ± 3.76

Regarding the exposure to toxins mixture, data of CYN and MC in lettuce and spinach plants were already reported previously, in a work aiming to validate a new method for their simultaneous determination in vegetable matrices [18]. In the present work, results from the uptake of MC and CYN are again presented, aiming at a more detailed discussion of results in the context of the plant response and CYN and MC phytotoxicity (Table 1). The two toxins were analyzed in roots and leaves of plants exposed to CYN/MC mixtures at the concentrations of 5 + 5 µg/L and 25 + 25 µg/L. The accumulation of MC in roots of both vegetables increased concomitantly with the increase of concentration of exposure. No significant differences in the accumulation of MC were observed between the two vegetables (Table 1). MC content varied between 0.22 and 0.53 µg/kg fw in 5 + 5 µg/L CYN/MC group and between 1.10 and 1.31 µg/kg fw in 25 + 25 µg/L CYN/MC group. In contrast no MC was detected in leaves of both vegetables.

CYN accumulated at much higher levels than MC in both plant species. Moreover, CYN was detected in leaves as well as in roots of both plant species (Table 1). The accumulation of CYN increased with the increase in the concentration of exposure, in most of the tissues analyzed except in the roots of spinach. Moreover, in lettuce CYN accumulated more in roots than in leaves. These differences were not so evident in spinach and the highest accumulation was observed in the leaves of plants exposed to 25 + 25 µg/L CYN/MC. Mean CYN content estimated in leaves of lettuce and spinach exposed to 5 + 5 µg/L CYN/MC and 25 + 25 µg/L CYN/MC were respectively 10.0–41.9 µg/kg fw and 12.6–119.7 µg/kg fw. In lettuce roots, mean CYN content varied between 34.2 and 110.0 µg/kg fw, whilst in spinach between 39.2 and 24.0 µg/kg fw. These results evidence that CYN is more easily taken-up by the roots than MC. In contrast to MC, CYN was translocated from the roots to the leaves meaning that this toxin can be transported through the plant vascular system. The patterns of accumulation of CYN and MC in plant tissues might be related to differences in their chemical structures and properties and the mechanisms of transport. MC is rather hydrophobic and also possess polar functions [65], therefore the transport of the molecule into the cells requires specific membrane transporters. In animal cells the molecule is known to be transported specifically by OATP type membrane transporters [66] but in plant cells the mechanism of transport is unknown. CYN has approximately half of molecular mass of MC and is highly soluble in water, thus being more easily transported than MC.

Overall the results show that the detrimental effects of cyanotoxins in lettuce and spinach are related with the accumulation of these compounds in the plant tissues. Despite the absence of accumulation in leaves, MC proved to affect severely leaf development, but causes no effect or may even contribute to the stimulation of root development. CYN, on the other hand, is accumulated at higher levels in leaves and roots of spinach and lettuce; however, it seems to be less detrimental to plant development than MC, having lower effects on fresh weight and Fv/Fm. Nevertheless the increased accumulation of CYN in spinach leaves, in comparison to lettuce leaves, could in part explain the increased toxicity and inhibition of growth observed in spinach exposed to CYN50, MIX10 and MIX50.

Studies of MC accumulation in plants suggest that shoot translocation of the toxin may be dependent on plant species and exposure conditions (toxin concentration, time of exposure). Crush et al. (2008) [67] did not detect toxin in leaves of clover, lettuce, ryegrass, or rape, irrigated with water containing 2.1 mg/L MCs. However Mohamed and Al Shehri (2009) [68] reported MCs in leaves of different vegetable crops irrigated with contaminated groundwater. Peuthert et al. (2007) [69] also reported low accumulation of MCs in leaves of 11 agricultural crops and higher accumulation in roots. Another study described a positive correlation between the accumulation of MCs in lettuce and the different exposure concentrations of MCs (MC and MC-RR) [61], which is in agreement with the pattern found in this work for MC.

Regarding CYN accumulation in vegetables, there are fewer data available. Kittler et al. (2012) [70] carried out a study of CYN uptake and accumulation in kale (*Brassica oleracea* var. sabellica) and mustard (*Brassica juncea*). The authors reported significant levels of toxin in the leaves (2.71 ± 0.65 and 3.78 ± 0.47 µg/kg fw in kale and in vegetable mustard respectively) in aeroponic cultures irrigated with toxic cyanobacterial extracts (18.2 µg/L CYN). Cordeiro-Araújo et al. (2017) [19] also reported a concentration-dependent CYN bioaccumulation in lettuce, and the mean value reported after exposure to 10 µg/L CYN for 7 days was 3.78 ± 0.25 µg/kg fw. Relatively higher values were observed in the present work in lettuce and spinach, which might be due to the increased time of exposure to the toxin.

Assuming a consumption of 40 g of vegetable per day for an adult of 60 kg (equivalent to 0.67 g/kg bw day) [19,36], the dietary intake of CYN would vary between 0.00067–0.028 µg/kg bw per day with the ingestion of contaminated lettuce leaves, and between 0.008–0.08 µg/kg bw day with the ingestion of contaminated spinach leaves. Taking into account the proposed TDI for CYN of 0.03 µg/kg per day, the consumption of spinach with the highest level of contamination would be potentially adverse to human health.

3. Conclusions

The contamination of vegetables and plant products raises growing concerns as a result of the increasing degradation of water resources used in irrigation and the presence of toxic cyanobacteria in irrigation waters. Despite the advances in the understanding of the effects of cyanotoxins on plants, for example, plant development and productivity, the multiple environmental factors that influence the action of cyanotoxins and the response of plants to cyanotoxins makes risk analysis extremely difficult, needing more scientific knowledge. This work provides new evidences concerning the effects of cyanotoxins at environmental concentrations. Overall MCs and CYN, separately or in combination, at a concentration of 50 µg/L in plant's growth medium (hydroponic conditions) were detrimental to the development of lettuce and spinach. When the same amount of toxin was present, the CYN+MC mixture showed to be more toxic then CYN alone. From this result, it may be speculated that cyanotoxins can act synergistically increasing the toxic potential of the water. Moreover, for purposes of toxicity assessment of environmental and irrigation waters, the presence of different cyanotoxins in the water should be monitored and their potential synergistic effects taken into account. On the one hand, MC concentrations in irrigation waters may raise more concerns due to the detrimental effects on plant growth. On the other hand, CYN is assimilated by the plant in greater amount than MC, leading to the conclusion that the use of water contaminated with this toxin is particularly concerning with regard to food safety and human exposure. Moreover some crops could be more sensitive to this toxin than others, and accumulate more of this toxin in the tissues. Given the risks identified with the exposure to CYN and/or MC at 50 µg/L, we suggest that concentrations lower than 50 µg/L should be considered for establishing the regulatory limits of cyanotoxins in irrigation waters.

Additionally, the diverse effects found in the mineral content of cyanotoxins exposed plants should be further studied because might have significant implications on plant development as well as in plant nutritional value.

4. Materials and Methods

4.1. Biological Material

Microcystis aeruginosa (LEGE 91094) and *Chrysosporum ovalisporum* (LEGE X-001) were grown as previously described by Campos et al. (2013) [71] in the Interdisciplinary Centre of Marine and Environmental Research, CIIMAR (Porto, Portugal). The lyophilized material was stored at room temperature in the dark for toxin extraction.

Spinacia oleracea (spinach) and *Lactuca sativa* (lettuce) were purchased as sprouts in a local market (Porto, Portugal). Plants were washed with deionized water in order to remove all remaining soil present in roots. Then, plants were cultivated in a hydroponic system according to Freitas et al. (2015) [30]. Briefly, plants were placed into 100 mL opaque glass jars, randomly distributed in groups and acclimated in Jensen culture medium [72] during a week (14–10 h, light-dark period and 21 ± 1 °C). After that, the plants were utilized for the exposure experiments.

4.2. Cyanobacterial Crude Extracts and Quantification of MC and CYN

MC and CYN extractions from cultures of *M. aeruginosa* and *C. ovalisporum*, respectively, were performed following the procedures described by Pinheiro et al. (2013) [73] and Welker et al. (2002) [74]. Analysis by high-performance liquid chromatography photodiode array detection (HPLC-PDA) showed a mean content of 0.2 mg MC/g and 2.9 mg CYN/g of lyophilized material, with retention times of 9.75 min (MC) and 6.305 min (CYN).

4.3. Exposure Experiments

After acclimation, five experimental groups for lettuce and seven for spinach were outlined comprehending the plants irrigated with non-contaminated water (control group), plants irrigated either with *M. aeruginosa* or *C. ovalisporum* extracts containing environmentally realistic toxin concentrations (10 and 50 µg/L MC or CYN), and plants irrigated with a mixture of both cyanobacterial extracts (toxin concentration of 5 µg/L MC + 5 µg/L CYN and 25 µg/L MC + 25 µg/L CYN). Eight replicates of each treatment were performed. In order to simplify the description, the water with *M. aeruginosa* extracts will be referred to as "MC-contaminated water" and the water with *C. ovalisporum* extracts will be referred to as "CYN-contaminated water" throughout the manuscript. The toxin concentration in the cyanobacterial extract was always quantified before preparing the artificially contaminated water to certify that the correct concentrations of MC or CYN in the irrigation water were present.

The plant culture media and toxins were replaced three times a week for 21 days. At the end of the experiments, plants were washed with deionized water and underwent different pre-treatment depending on the analysis.

4.4. Physiological Parameters: Plant Fresh Weight and Photosynthetic Capacity

Immediately after harvesting the plants, leaves and roots were separated and weighted (fresh weight, fw), and stored at −80 °C for further analysis. Plant growth was expressed as the mean fresh weight (fw) ± SD from eight plants (n = 8) per treatment.

Photosynthetic capacity was determined through pulse amplitude modulation (PAM) fluorometry, with PAM 2000 (Walz, Effeltrich, Germany) instrument according to Machado et al. (2017) [36]. Plants were first kept in dark conditions for at least 30 min and leaves subsequently illuminated with a pulse of saturating light. The fluorescence emitted was recorded by the instrument. This procedure allows measuring the maximum fluorescence yield of photosystem II (PSII)(Fv/Fm) that is directly related with the functional state of the PSII protein complex and the photosynthetic efficiency of the plants [59].

4.5. Mineral Content

Determination of mineral content was developed following Freitas et al. (2015) [30]. Briefly, freeze-dried samples (leaves and roots from spinach and lettuce) were digested by microwave-assisted acid digestion using a MLS 1200 Mega system (Milestone, Sorisole, Italy). Sample solutions were then analyzed by flame atomic absorption spectroscopy (FAAS) using a 200 Analyst equipment (Perkin Elmer, Überlingen, Germany) and by inductively coupled plasma-mass spectrometry (ICP-MS) using an iCAPTM Q (Thermo Fisher Scientific, Bremen, Germany) for total metal content. Results were expressed on a dry weight (dw) basis.

4.6. MC Extraction and Quantification from Spinach

MC were extracted from leaves of spinach exposed to 0, 10, and 50 µg/L MC following the method described by Prieto et al. (2011) [31] with slight modifications. Lyophilized spinach leaves (0.05 g) were mixed with 10 mL of 0.1 M acetic acid and 20 mL of a 1:1 (*v*/*v*) mixture of methanol–chloroform. Then, the mixture was stirred (15 min.) and sonicated (15 min.) three times at 4 °C and then centrifuged at 3400× *g*. When the supernatant was collected, the extracts were purified according to Guzmán-Guillén et al. (2011) [75], using Oasis HLB cartridges (500 mg/6 mL, Waters, Mildford, MA, USA). Chromatographic separation was performed using a UPLC Acquity (Waters, Mildford, MA, USA) coupled to a Xevo TQ-S micro (Waters, Mildford, MA, USA) consisting of a triple quadrupole mass spectrometer equipped with an electrospray ion source operated in positive mode. UPLC analyses were performed as described by Díez-Quijada et al. (2018) [18]. The transitions employed for MC are 996.5/135.0, 996.5/213.1 and 996.5/996.5.

4.7. CYN Extraction and Quantification from Spinach and Lettuce

The extraction and purification of CYN from plant materials was performed according to Prieto et al. (2018) [64]. Briefly, lyophilized biomass was extracted with 6 mL of 10% acetic acid, homogenized by ultraturrax, sonicated (15 min) and stirred (15 min). Then, the mixture was centrifuged for 15 min (12,000× *g*) and the supernatant was collected and purified. The purified CYN fractions were concentrated (solvent evaporation) and resuspended in 1 mL Milli-Q water prior to its UPLC/MS-MS analysis, following the method described by Prieto et al. (2018) [64].

4.8. CYN/MC MIX Simultaneous Extraction and Quantification from Spinach and Lettuce

The determination of toxins in CYN/MC mixtures was performed following the method described by Diez-Quijada et al. (2018) [18]. Lyophilized leaves and roots were extracted with 6 mL of 80% methanol, ultraturrax (1 min), sonicated (15 min) and stirred (15 min). Then, the mixture was centrifugated (3400× *g*, 15 min) and the supernatant collected for clean-up. An assembly of a C_{18} Bakerbond cartridge (500 mg, 6 mL, Dicsa, Andalucia, Spain) and a BOND ELUT Carbon cartridge (Agilent Technologies, Amstelveen, The Netherlands) was employed. After adjusting the supernatant to pH 11, the following reagents were passed through the assembled cartridges: 6 mL DCM, 6 mL

100% MeOH, 6 mL H_2O (pH 11) and sample (pH 11): then, cartridges were dried for 5 min and eluted with 10 mL DCM/MeOH (40/60) + 0.5 formic acid. Then, the extracts were evaporated to dryness and resuspended in 1 mL 20% MeOH for its analysis by UPLC-MS/MS, according to Diez-Quijada et al. (2018) [18].

4.9. Statistical Analysis

Leaf weight, root weight and photosynthetic rate data from the two plant species were analyzed with General Linear Models, in which plant dry weight, root dry weight and photosynthetic rate were used as response variables and the cyanobacterial treatment as factor. Assumptions of residual normality and homoscedasticity were tested with Shapiro-Wilks and Levene tests, respectively. If any of the assumptions was not met, data were transformed using the Box-Cox transformation. Tukey post-hoc tests between factor levels were performed when due. All these analyses were performed with R software, with functions from base, stats, MASS, car and multcomp packages.

Spinach and lettuce ionome (the abundances of chemical elements on individual plants) is a kind of quantitative data from a series of elements that constitute parts of a whole, usually given in proportions. This kind of data are considered in Statistics as Compositional data. These data usually involve redundancy and multiple correlations of spurious nature that make inefficient the use of most standard statistical techniques, and hence, need specific procedures for its analysis, that we describe here below.

A sample space of compositional data, such as the data of plants ionome, is defined by S^D, a positive vector of D components adding up to a constant k, such as 100% or some other constant sum. The "closure" operator C normalizes the contained vector as follows:

$$S^D = C(c_1, c_2, \ldots, c_D) = \left[\frac{c_1 K}{\sum_{i=1}^{D} c_i}, \frac{c_2 K}{\sum_{i=1}^{D} c_i}, \ldots, \frac{c_D K}{\sum_{i=1}^{D} c_i}\right]$$

where k is the unit of measurement and c_i is the ith part of a composition containing D-parts [76]. Due to the nature of the sample space, independence hypotheses must clearly take different forms from those associated with S_D. Thus, the sample space should be divided into non-overlapping subcompositions where each subcomposition can be interpreted independently. In order to do this, the ionome data were recalculated to sum up to a constant value (1 in our case) per individual plant.

For the elimination of redundancy and spurious correlations, as well as getting data distribution closer to normality, it is necessary to apply the isometric log-ratio (*ilr*) transformation. First, new orthogonal variables were created, through the combination of those already existing, thus eliminating linear dependency. This task was performed by constructing a sequential binary partition (SBP) from the whole ionome data. This technique consists in combining the elements in balanced ratios that are orthogonal. From n elements in the ionome matrix, n—1 orthogonal balances would be obtained. The orthogonality is determined by assigning orthogonal coefficients to *ilr* transformed balances in all possible subcompositions of the data set (see Parent et al. 2013) [77]. Those orthogonal balances are non-redundant and scale-invariant. This task was performed with the functions *acomp*, *gsi.merge2signary* and *gsi.buildilrBase*, from the *R* package *compositions*. The *ilr* transformation has the following formula:

$$ilr_j = \sqrt{\frac{r_j\, s_j}{r_j + s_j}} \ln \frac{g\left(c+\,_j\right)}{g\left(c-\,_j\right)}$$

where *ilrj* is each of the isometric log-ratios, r_j and s_j are those elements with positive and negative orthogonal coefficients, respectively, that integrate the jth balance. $g\ (c +\,_j)$ and $g\ (c -\,_j)$ are the geometric means of those elements with positive and negative coefficients, respectively.

Author Contributions: Conceptualization, M.L.-R.-C., A.J., A.C. (Ana Cameán) and A.C. (Alexandre Campos); Formal analysis, F.O., A.B., J.M., J.A., A.A. and E.P.; Funding acquisition, A.J. and A.C. (Ana Cameán); Investigation,

M.L.-R.-C.; Methodology, F.O., A.B., J.M., J.A., A.A. and E.P.; Supervision, A.J., A.C. (Ana Cameán), A.C. (Alexandre Campos) and V.V.; Writing—original draft, M.L.-R.-C., A.J., A.C. (Ana Cameán), F.O., A.B., J.A., E.P., A.C. (Alexandre Campos). and V.V.

Funding: Project AGL2015-64558-R, MINECO/FEDER, UE. M. Llana-Ruiz-Cabello was supported by 'Junta de Andalucía' grant associated to AGR-7252 project. This work has received funding from the European Union's Horizon 2020 research and innovation programme under the Marie Skłodowska-Curie grant agreement No 823860 and from the project UID/Multi/04423/2019 funded by Foundation for Science and Technology (FCT).

Conflicts of Interest: The authors declare no conflict of interest.

References

1. Buratti, F.M.; Manganelli, M.; Vichi, S.; Stefanelli, M.; Scardala, S.; Testai, E.; Funari, E. Cyanotoxins: Producing organisms, occurrence, toxicity, mechanism of action and human health toxicological risk evaluation. *Arch. Toxicol.* **2017**, *91*, 1049–1130. [CrossRef] [PubMed]
2. Nguyen, T.T.N.; Némery, J.; Gratiot, N.; Strady, E.; Tran, V.Q.; Nguyen, A.T.; Aimé, J.; Peyne, A. Nutrient dynamics and eutrophication assessment in the tropical river system of Saigon—Dongnai (southern Vietnam). *Sci. Total Environ.* **2019**, *653*, 370–383. [CrossRef] [PubMed]
3. Catherine, A.; Bernard, C.; Spoof, L.; Bruno, M. Microcystins and Nodularins. In *Handbook of Cyanobacterial Monitoring and Cyanotoxin Analysis*, 1st ed.; Meriluoto, J., Spoof, L., Codd, G.A., Eds.; John Wiley & Sons: Hoboken, NJ, USA, 2017; Volume 1, pp. 107–126.
4. International Agency for Research on Cancer. Ingested Nitrate and Nitrite, and Cyanobacterial Peptide Toxins. In *Monographs on the Evaluation of Carcinogenic Risks to Humans*; IARC Publications: Lyon, France, 2010; Volume 94, pp. 1–412.
5. Craig, M.; Luu, H.A.; McCready, T.L.; Williams, D.; Andersen, R.J.; Holmes, C.F.B. Molecular mechanisms underlying the interaction of motuporin and microycystins with type-1 and type-2A protein phosphatases. *Biochem. Cell Biol. Biol. Cell.* **1996**, *74*, 569–578. [CrossRef] [PubMed]
6. Liang, J.; Li, T.; Zhang, Y.-L.; Guo, Z.-L.; Xu, L.-H. Effect of microcystin-LR on protein phosphatase 2A and its function in human amniotic epithelial cells. *J. Zhejiang Univ. Sci. B* **2011**, *12*, 951–960. [CrossRef]
7. Christen, V.; Meili, N.; Fent, K. Microcystin-LR induces endoplasmatic reticulum stress and leads to induction of NFκB, interferon-alpha, and tumor necrosis factor-alpha. *Environ. Sci. Technol.* **2013**, *47*, 3378–3385. [CrossRef]
8. Singh, S.; Asthana, R.K. Assessment of microcystin concentration in carp and catfish: A case study from Lakshmikund Pond, Varanasi, India. *Bull. Environ. Contam. Toxicol.* **2014**, *92*, 687–692. [CrossRef]
9. Gkelis, S.; Zaoutsos, N. Cyanotoxin occurrence and potentially toxin producing cyanobacteria in freshwaters of Greece: A multi-disciplinary approach. *Toxicon* **2014**, *78*, 1–9. [CrossRef]
10. Trainer, V.L.; Hardy, F.J. Integrative monitoring of marine and freshwater harmful algae in Washington State for public health protection. *Toxins* **2015**, *7*, 1206–1234. [CrossRef]
11. Corbel, S.; Mougin, C.; Bouaïcha, N. Cyanobacterial toxins: Modes of actions, fate in aquatic and soil ecosystems, phytotoxicity and bioaccumulation in agricultural crops. *Chemosphere* **2014**, *96*, 1–15. [CrossRef]
12. Žegura, B.; Štraser, A.; Filipič, M. Genotoxicity and potential carcinogenicity of cyanobacterial toxins—A review. *Mutat. Res. Rev. Mutat. Res.* **2011**, *727*, 16–41. [CrossRef]
13. Štraser, A.; Filipič, M.; Žegura, B. Genotoxic effects of the cyanobacterial hepatotoxin cylindrospermopsin in the HepG2 cell line. *Arch. Toxicol.* **2011**, *85*, 1617–1626. [CrossRef] [PubMed]
14. Puerto, M.; Prieto, A.I.; Maisanaba, S.; Gutiérrez-Praena, D.; Mellado-García, P.; Jos, Á.; Cameán, A.M. Mutagenic and genotoxic potential of pure Cylindrospermopsin by a battery of in vitro tests. *Food Chem. Toxicol.* **2018**, *121*, 413–422. [CrossRef] [PubMed]
15. Pichardo, S.; Cameán, A.M.; Jos, A. In vitro toxicological assessment of cylindrospermopsin: A review. *Toxins* **2017**, *9*, 402. [CrossRef] [PubMed]
16. Svirčev, Z.; Lalić, D.; Savić, G.B.; Tokodi, N.; Backović, D.D.; Chen, L.; Meriluoto, J.; Codd, G.A. Global geographical and historical overview of cyanotoxin distribution and cyanobacterial poisonings. *Arch. Toxicol.* **2019**, *93*, 2429–2481. [CrossRef]
17. Froscio, S.M.; Humpage, A.R.; Burcham, P.C.; Falconer, I.R. Cylindrospermopsin-induced protein synthesis inhibition and its dissociation from acute toxicity in mouse hepatocytes. *Environ. Toxicol.* **2003**, *18*, 243–251. [CrossRef]

18. Díez-Quijada, L.; Guzmán-Guillén, R.; Ortega, A.P.; Llana-Ruíz-Cabello, M.; Campos, A.; Vasconcelos, V.; Jos, Á.; Cameán, A.M. New Method for Simultaneous Determination of Microcystins and Cylindrospermopsin in Vegetable Matrices by SPE-UPLC-MS/MS. *Toxins* **2018**, *10*, 406. [CrossRef]
19. Cordeiro-Araújo, M.K.; Chia, M.A.; do Carmo Bittencourt-Oliveira, M. Potential human health risk assessment of cylindrospermopsin accumulation and depuration in lettuce and arugula. *Harmful Algae* **2017**, *68*, 217–223. [CrossRef]
20. Bihn, E.A.; Smart, C.D.; Hoepting, C.A.; Worobo, R.W. Use of Surface Water in the Production of Fresh Fruits and Vegetables: A Survey of Fresh Produce Growers and Their Water Management Practices. *Food Prot. Trends* **2013**, *33*, 307–314.
21. Chen, J.; Song, L.; Dai, J.; Gan, N.; Liu, Z. Effects of microcystins on the growth and the activity of superoxide dismutase and peroxidase of rape (*Brassica napus* L.) and rice (*Oryza sativa* L.). *Toxicon* **2004**, *43*, 393–400. [CrossRef]
22. Pereira, S.; Saker, M.L.; Vale, M.; Vasconcelos, V.M. Comparison of sensitivity of grasses (*Lolium perenne* L. and *Festuca rubra* L.) and lettuce (*Lactuca sativa* L.) exposed to water contaminated with microcystins. *Bull. Environ. Contam. Toxicol.* **2009**, *83*, 81–84. [CrossRef]
23. Gehringer, M.M.; Kewada, V.; Coates, N.; Downing, T.G. The use of Lepidium sativum in a plant bioassay system for the detection of microcystin-LR. *Toxicon* **2003**, *41*, 871–876. [CrossRef]
24. El Khalloufi, F.; El Ghazali, I.; Saqrane, S.; Oufdou, K.; Vasconcelos, V.; Oudra, B. Phytotoxic effects of a natural bloom extract containing microcystins on *Lycopersicon esculentum*. *Ecotoxicol. Environ. Saf.* **2012**, *79*, 199–205. [CrossRef] [PubMed]
25. Pflugmacher, S.; Hofmann, J.; Hübner, B. Effects on growth and physiological parameters in wheat (*Triticum aestivum* L.) grown in soil and irrigated with cyanobacterial toxin contaminated water. *Environ. Toxicol. Chem.* **2007**, *26*, 2710–2716. [CrossRef] [PubMed]
26. Pflugmacher, S.; Aulhorn, M.; Grimm, B. Influence of a cyanobacterial crude extract containing microcystin-LR on the physiology and antioxidative defence systems of different spinach variants. *New Phytol.* **2007**, *175*, 482–489. [CrossRef]
27. Pflugmacher, S.; Jung, K.; Lundvall, L.; Neumann, S.; Peuthert, A. Effects of cyanobacterial toxins and cyanobacterial cell-free crude extract on germination of alfalfa (*Medicago sativa*) and induction of oxidative stress. *Environ. Toxicol. Chem.* **2006**, *25*, 2381–2387. [CrossRef]
28. Metcalf, J.S.; Barakate, A.; Codd, G.A. Inhibition of plant protein synthesis by the cyanobacterial hepatotoxin, cylindrospermopsin. *FEMS Microbiol. Lett.* **2004**, *235*, 125–129. [CrossRef]
29. Máthé, C.; Vasas, G. Microcystin-LR and cylindrospermopsin induced alterations in chromatin organization of plant cells. *Mar. Drugs* **2013**, *11*, 3689–3717. [CrossRef]
30. Freitas, M.; Azevedo, J.; Pinto, E.; Neves, J.; Campos, A.; Vasconcelos, V. Effects of microcystin-LR, cylindrospermopsin and a microcystin-LR/cylindrospermopsin mixture on growth, oxidative stress and mineral content in lettuce plants (*Lactuca sativa* L.). *Ecotoxicol. Environ. Saf.* **2015**, *116*, 59–67. [CrossRef]
31. Prieto, A.; Campos, A.; Cameán, A.; Vasconcelos, V. Effects on growth and oxidative stress status of rice plants (*Oryza sativa*) exposed to two extracts of toxin-producing cyanobacteria (*Aphanizomenon ovalisporum* and *Microcystis aeruginosa*). *Ecotoxicol. Environ. Saf.* **2011**, *74*, 1973–1980. [CrossRef]
32. Vasas, G.; Gáspár, A.; Surányi, G.; Batta, G.; Gyémánt, G.; M-Hamvas, M.; Máthé, C.; Grigorszky, I.; Molnár, E.; Borbély, G. Capillary electrophoretic assay and purification of cylindrospermopsin, a cyanobacterial toxin from *Aphanizomenon ovalisporum*, by plant test (Blue-Green Sinapis Test). *Anal. Biochem.* **2002**, *302*, 95–103. [CrossRef]
33. Beyer, D.; Surányi, G.; Vasas, G.; Roszik, J.; Erdodi, F.; M-Hamvas, M.; Bácsi, I.; Bátori, R.; Serfozo, Z.; Szigeti, Z.M.; et al. Cylindrospermopsin induces alterations of root histology and microtubule organization in common reed (*Phragmites australis*) plantlets cultured in vitro. *Toxicon* **2009**, *54*, 440–449. [CrossRef] [PubMed]
34. Mathe, C.; Garda, T.; Beyer, D.; Vasas, G. Cellular Effects of Cylindrospermopsin (Cyanobacterial Alkaloid Toxin) and its Potential Medical Consequences. *Curr. Med. Chem.* **2017**, *24*, 91–109. [CrossRef] [PubMed]
35. M-Hamvas, M.; Ajtay, K.; Beyer, D.; Jámbrik, K.; Vasas, G.; Surányi, G.; Máthé, C. Cylindrospermopsin induces biochemical changes leading to programmed cell death in plants. *Apoptosis* **2017**, *22*, 254–264. [CrossRef] [PubMed]

36. Machado, J.; Azevedo, J.; Freitas, M.; Pinto, E.; Almeida, A.; Vasconcelos, V.; Campos, A. Analysis of the use of microcystin-contaminated water in the growth and nutritional quality of the root-vegetable, *Daucus carota*. *Environ. Sci. Pollut. Res.* **2017**, *24*, 752–764. [CrossRef]
37. Guzmán-Guillén, R.; Campos, A.; Machado, J.; Freitas, M.; Azevedo, J.; Pinto, E.; Almeida, A.; Cameán, A.M.; Vasconcelos, V. Effects of Chrysosporum (Aphanizomenon) ovalisporum extracts containing cylindrospermopsin on growth, photosynthetic capacity, and mineral content of carrots (*Daucus carota*). *Ecotoxicology* **2017**, *26*, 22–31. [CrossRef]
38. Pereira, A.L.; Azevedo, J.; Vasconcelos, V. Assessment of uptake and phytotoxicity of cyanobacterial extracts containing microcystins or cylindrospermopsin on parsley (*Petroselinum crispum* L.) and coriander (*Coriandrum sativum* L.). *Environ. Sci. Pollut. Res.* **2017**, *24*, 1999–2009. [CrossRef]
39. Freitas, M.; Campos, A.; Azevedo, J.; Barreiro, A.; Planchon, S.; Renaut, J.; Vasconcelos, V. Lettuce (*Lactuca sativa* L.) leaf-proteome profiles after exposure to cylindrospermopsin and a microcystin-LR/cylindrospermopsin mixture: A concentration-dependent response. *Phytochemistry* **2015**, *110*, 91–103. [CrossRef]
40. El Khalloufi, F.; Oufdou, K.; Bertrand, M.; Lahrouni, M.; Oudra, B.; Ortet, P.; Barakat, M.; Heulin, T.; Achouak, W. Microbiote shift in the Medicago sativa rhizosphere in response to cyanotoxins extract exposure. *Sci. Total Environ.* **2016**, *539*, 135–142. [CrossRef]
41. Lahrouni, M.; Oufdou, K.; El Khalloufi, F.; Benidire, L.; Albert, S.; Göttfert, M.; Caviedes, M.A.; Rodriguez-Llorente, I.D.; Oudra, B.; Pajuelo, E. Microcystin-tolerant Rhizobium protects plants and improves nitrogen assimilation in Vicia faba irrigated with microcystin-containing waters. *Environ. Sci. Pollut. Res.* **2016**, *23*, 10037–10049. [CrossRef]
42. Lahrouni, M.; Oufdou, K.; El Khalloufi, F.; Baz, M.; Lafuente, A.; Dary, M.; Pajuelo, E.; Oudra, B. Physiological and biochemical defense reactions of *Vicia faba* L.-Rhizobium symbiosis face to chronic exposure to cyanobacterial bloom extract containing microcystins. *Environ. Sci. Pollut. Res.* **2013**, *20*, 5405–5415. [CrossRef]
43. Testai, E.; Buratti, F.M.; Funari, E.; Manganelli, M.; Vichi, S.; Arnich, N.; Biré, R.; Fessard, V.; Sialehaamoa, A. Review and analysis of occurrence, exposure and toxicity of cyanobacteria toxins in food. *EFSA Support. Publ.* **2017**, *13*, 998E. [CrossRef]
44. El Khalloufi, F.; Oufdou, K.; Lahrouni, M.; El Ghazali, I.; Saqrane, S.; Vasconcelos, V.; Oudra, B. Allelopatic effects of cyanobacteria extracts containing microcystins on Medicago sativa-Rhizobia symbiosis. *Ecotoxicol. Environ. Saf.* **2011**, *74*, 431–438. [CrossRef] [PubMed]
45. McElhiney, J.; Lawton, L.A.; Leifert, C. Investigations into the inhibitory effects of microcystins on plant growth, and the toxicity of plant tissues following exposure. *Toxicon* **2001**, *39*, 1411–1420. [CrossRef]
46. Saqrane, S.; Ouahid, Y.; El Ghazali, I.; Oudra, B.; Bouarab, L.; del Campo, F.F. Physiological changes in *Triticum durum*, *Zea mays*, *Pisum sativum* and *Lens esculenta* cultivars, caused by irrigation with water contaminated with microcystins: A laboratory experimental approach. *Toxicon* **2009**, *53*, 786–796. [CrossRef]
47. Lahrouni, M.; Oufdou, K.; Faghire, M.; Peix, A.; El Khalloufi, F.; Vasconcelos, V.; Oudra, B. Cyanobacterial extracts containing microcystins affect the growth, nodulation process and nitrogen uptake of faba bean (*Vicia faba* L., Fabaceae). *Ecotoxicology* **2012**, *21*, 681–687. [CrossRef]
48. Chen, J.; Zhang, H.Q.; Hu, L.B.; Shi, Z.Q. Microcystin-LR-induced phytotoxicity in rice crown root is associated with the cross-talk between auxin and nitric oxide. *Chemosphere* **2013**, *93*, 283–293. [CrossRef]
49. Lee, S.; Jiang, X.; Manubolu, M.; Riedl, K.; Ludsin, S.A.; Martin, J.F.; Lee, J. Fresh produce and their soils accumulate cyanotoxins from irrigation water: Implications for public health and food security. *Food Res. Int.* **2017**, *102*, 234–245. [CrossRef]
50. Zhu, J.; Ren, X.; Liu, H.; Liang, C. Effect of irrigation with microcystins-contaminated water on growth and fruit quality of Cucumis sativus L. and the health risk. *Agric. Water Manag.* **2018**, *204*, 91–99. [CrossRef]
51. Cao, Q.; Steinman, A.D.; Wan, X.; Xie, L. Bioaccumulation of microcystin congeners in soil-plant system and human health risk assessment: A field study from Lake Taihu region of China. *Environ. Pollut.* **2018**, *240*, 44–50. [CrossRef]
52. Liang, C.; Wang, W.; Wang, Y. Effect of irrigation with microcystins-contaminated water on growth, yield and grain quality of rice (*Oryza sativa*). *Environ. Earth Sci.* **2016**, *75*, 505. [CrossRef]
53. Corbel, S.; Bouaïcha, N.; Nélieu, S.; Mougin, C. Soil irrigation with water and toxic cyanobacterial microcystins accelerates tomato development. *Environ. Chem. Lett.* **2015**, *13*, 447–452. [CrossRef]

54. Terao, K.; Ohmori, S.; Igarashi, K.; Ohtani, I.; Watanabe, M.F.; Harada, K.I.; Ito, E.; Watanabe, M. Electron microscopic studies on experimental poisoning in mice induced by cylindrospermopsin isolated from blue-green alga *Umezakia natans*. *Toxicon* **1994**, *32*, 833–843. [CrossRef]
55. Runnegar, M.; Berndt, N.; Kaplowitz, N. Microcystin uptake and inhibition of protein phosphatases: Effects of chemoprotectants and self-inhibition in relation to known hepatic transporters. *Toxicol. Appl. Pharmacol.* **1995**, *134*, 264–272. [CrossRef] [PubMed]
56. MacKintosh, C.; Beattie, K.A.; Klumpp, S.; Cohen, P.; Codd, G.A. Cyanobacterial microcystin-LR is a potent and specific inhibitor of protein phosphatases 1 and 2A from both mammals and higher plants. *FEBS Lett.* **1990**, *264*, 187–192. [CrossRef]
57. Dawson, R.M. The toxicology of microcystins. *Toxicon* **1998**, *36*, 953–962. [CrossRef]
58. Bolhar-Nordenkampf, H.R.; Hofer, M.; Lechner, E.G. Analysis of light-induced reduction of the photochemical capacity in field-grown plants. Evidence for photoinhibition? *Photosynth. Res.* **1991**, *27*, 31–39. [CrossRef]
59. Maxwell, K.; Johnson, G.N. Chlorophyll fluorescence—A practical guide. *J. Exp. Bot.* **2000**, *51*, 659–668. [CrossRef]
60. Gutiérrez-Praena, D.; Campos, A.; Azevedo, J.; Neves, J.; Freitas, M.; Guzmán-Guillén, R.; Cameán, A.M.; Renaut, J.; Vasconcelos, V. Exposure of Lycopersicon Esculentum to microcystin-LR: Effects in the leaf proteome and toxin translocation from water to leaves and fruits. *Toxins* **2014**, *6*, 1837–1854. [CrossRef]
61. do Carmo Bittencourt-Oliveira, M.; Cordeiro-Araújo, M.K.; Chia, M.A.; de Toledo Arruda-Neto, J.D.; de Oliveira, Ê.T.; dos Santos, F. Lettuce irrigated with contaminated water: Photosynthetic effects, antioxidative response and bioaccumulation of microcystin congeners. *Ecotoxicol. Environ. Saf.* **2016**, *128*, 83–90. [CrossRef]
62. Taiz, L.; Zeiger, E. *Plant Physiology*, 3rd ed.; Sinauer Associates: Cary, NC, USA, 2003; ISBN 0878938230.
63. Runnegar, M.T.; Xie, C.; Snider, B.B.; Wallace, G.A.; Weinreb, S.M.; Kuhlenkamp, J. In vitro hepatotoxicity of the cyanobacterial alkaloid cyclindrospermopsin and related synthetic analogues. *Toxicol. Sci.* **2002**, *67*, 81–87. [CrossRef]
64. Prieto, A.I.; Guzmán-Guillén, R.; Díez-Quijada, L.; Campos, A.; Vasconcelos, V.; Jos, Á.; Cameán, A.M. Validation of a method for cylindrospermopsin determination in vegetables: Application to real samples such as lettuce (*Lactuca sativa* L.). *Toxins* **2018**, *10*, 63. [CrossRef] [PubMed]
65. Rivasseau, C.; Martins, S.; Hennion, M.C. Determination of some physicochemical parameters of microcystins (cyanobacterial toxins) and trace level analysis in environmental samples using liquid chromatography. *J. Chromatogr. A* **1998**, *799*, 155–169. [CrossRef]
66. Valério, E.; Vasconcelos, V.; Campos, A. New Insights on the Mode of Action of Microcystins in Animal Cells—A Review. *Mini-Reviews Med. Chem.* **2016**, *16*, 1032–1041. [CrossRef] [PubMed]
67. Crush, J.R.; Briggs, L.R.; Sprosen, J.M.; Nichols, S.N. Effect of irrigation with lake water containing microcystins on microcystin content and growth of ryegrass, clover, rape, and lettuce. *Environ. Toxicol.* **2008**, *23*, 246–252. [CrossRef]
68. Mohamed, Z.A.; Al Shehri, A.M. Microcystins in groundwater wells and their accumulation in vegetable plants irrigated with contaminated waters in Saudi Arabia. *J. Hazard. Mater.* **2009**, *172*, 310–315. [CrossRef] [PubMed]
69. Peuthert, A.; Chakrabarti, S.; Pflugmacher, S. Uptake of microcystins-LR and -LF (cyanobacterial toxins) in seedlings of several important agricultural plant species and the correlation with cellular damage (lipid peroxidation). *Environ. Toxicol.* **2007**, *22*, 436–442. [CrossRef]
70. Kittler, K.; Schreiner, M.; Krumbein, A.; Manzei, S.; Koch, M.; Rohn, S.; Maul, R. Uptake of the cyanobacterial toxin cylindrospermopsin in Brassica vegetables. *Food Chem.* **2012**, *133*, 875–879. [CrossRef]
71. Campos, A.; Araújo, P.; Pinheiro, C.; Azevedo, J.; Osório, H.; Vasconcelos, V. Effects on growth, antioxidant enzyme activity and levels of extracellular proteins in the green alga Chlorella vulgaris exposed to crude cyanobacterial extracts and pure microcystin and cylindrospermopsin. *Ecotoxicol. Environ. Saf.* **2013**, *94*, 45–53. [CrossRef]
72. Jensen, M.H.; Malter, A.J. *Protected Agriculture: A Global Review*; World Bank Tech. Pap.; The World Bank: Washington, DC, USA, 1995; Volume 253, p. 157.
73. Pinheiro, C.; Azevedo, J.; Campos, A.; Loureiro, S.; Vasconcelos, V. Absence of negative allelopathic effects of cylindrospermopsin and microcystin-LR on selected marine and freshwater phytoplankton species. *Hydrobiologia* **2013**, *705*, 27–42. [CrossRef]

74. Welker, M.; Bickel, H.; Fastner, J. HPLC-PDA detection of cylindrospermopsin—Opportunities and limits. *Water Res.* **2002**, *36*, 4659–4663. [CrossRef]
75. Guzmán-Guillén, R.; Prieto, A.I.; Moreno, I.; Soria, M.E.; Cameán, A.M. Effects of thermal treatments during cooking, microwave oven and boiling, on the unconjugated microcystin concentration in muscle of fish (*Oreochromis niloticus*). *Food Chem. Toxicol.* **2011**, *49*, 2060–2067. [CrossRef] [PubMed]
76. Egozcue, J.J.; Barceló-Vidal, C.; Martín-Fernández, J.A.; Jarauta-Bragulat, E.; Díaz-Barrero, J.L.; Mateu-Figueras, G. Elements of Simplicial Linear Algebra and Geometry. In *Compositional Data Analysis: Theory and Applications*; John Wiley and Sons: Hoboken, NJ, USA, 2011; pp. 139–157.
77. Parent, S.É.; Parent, L.E.; Egozcue, J.J.; Rozane, D.E.; Hernandes, A.; Lapointe, L.; Hébert-Gentile, V.; Naess, K.; Marchand, S.; Lafond, J.; et al. The plant ionome revisited by the nutrient balance concept. *Front. Plant Sci.* **2013**, *4*, 39. [CrossRef] [PubMed]

Article

Diagnosing Microcystin Intoxication of Canines: Clinicopathological Indications, Pathological Characteristics, and Analytical Detection in Postmortem and Antemortem Samples

Amanda J. Foss [1,*], Mark T. Aubel [1], Brandi Gallagher [2], Nancy Mettee [3], Amanda Miller [3] and Susan B. Fogelson [4]

1 GreenWater Laboratories, Palatka, FL 32177, USA
2 Animal Care Cancer Clinic, Stuart, FL 34994, USA
3 Pet Emergency of Martin County, Stuart, FL 34994, USA
4 Fishhead Labs LLC, Stuart, FL 34997, USA
* Correspondence: amandafoss@greenwaterlab.com; Tel.: +1-386-328-0882

Received: 13 June 2019; Accepted: 1 August 2019; Published: 3 August 2019

Abstract: In the summer of 2018, six dogs exposed to a harmful algal bloom (HAB) of *Microcystis* in Martin County Florida (USA) developed clinicopathological signs of microcystin (MC) intoxication (i.e., acute vomiting, diarrhea, severe thrombocytopenia, elevated alanine aminotransferase, hemorrhage). Successful supportive veterinary care was provided and led to survival of all but one patient. Confirmation of MC intoxication was made through interpretation of clinicopathological abnormalities, pathological examination of tissues, microscopy (vomitus), and analytical MC testing of antemortem/postmortem samples (vomitus, blood, urine, bile, liver, kidney, hair). Gross and microscopic examination of the deceased patient confirmed massive hepatic necrosis, mild multifocal renal tubular necrosis, and hemorrhage within multiple organ systems. Microscopy of a vomitus sample confirmed the presence of *Microcystis*. Three analytical MC testing approaches were used, including the MMPB (2-methyl-3-methoxy-4-phenylbutyric acid) technique, targeted congener analysis (e.g., liquid chromatography tandem-mass spectrometry of MC-LR), and enzyme-linked immunosorbent assay (ELISA). Total Adda MCs (as MMPB) were confirmed in the liver, bile, kidney, urine, and blood of the deceased dog. Urinalysis (MMPB) of one surviving dog showed a high level of MCs (32,000 ng mL^{-1}) 1-day post exposure, with MCs detectable >2 months post exposure. Furthermore, hair from a surviving dog was positive for MMPB, illustrating another testable route of MC elimination in canines. The described cases represent the first use of urine as an antemortem, non-invasive specimen to diagnose microcystin toxicosis. Antemortem diagnostic testing to confirm MC intoxication cases, whether acute or chronic, is crucial for providing optimal supportive care and mitigating MC exposure.

Keywords: HAB; microcystin; Adda; canine intoxication; MMPB; urinalysis; hair; ELISA; LC-MS/MS

Key Contribution: Multi-modal (clinal; pathological; analytical) investigation of an acute microcystin exposure event led to the discovery of ideal specimens for postmortem (kidney)and antemortem (urine) testing. The recommended method for detection; the MMPB technique; provided low level detection of total Adda MCs/NODs in urine and hair; which has not been previously described.

---

1. Introduction

Harmful algal blooms (HABs) resulting in animal intoxications are a worldwide occurrence, with reports of mortality becoming more prevalent [1,2]. Shoreline cyanobacteria blooms are one type of HAB that can lead to exposure of those living near lakes and streams, such as domestic dogs. Dogs represent a sentinel species due to their shared environment with humans [3], supporting the framework for a One Health approach when investigating cyanotoxins. Cyanobacteria poisoning of canines has been well documented, but comprehensive reporting on toxin levels detected in specimens is sparse [2,4–7]. Reports of canine intoxication by cyanobacterial neurotoxins are more prevalent, but intoxication to hepatotoxins such as microcystin (MC) have increased in frequency [2]. Over a four-year period (2007–2011), Departments of Health and/or Environment from 13 states reported 43 dogs suspected of poisoning by MCs with a moderate to high probability, based on clinical and diagnostic pathology [2]. Awareness of these events is spreading; however, MC intoxication events likely go under reported. This could be due to a multitude of contributing factors, such as; insufficient exposure history, lack of supportive environmental data, lack of standard HAB protocols or policies leading to inadequate sample acquisition and/or handling, improper analytical test selection, or misdiagnosis due to commonality of symptoms to other hepatotoxins. Furthermore, monetary restrictions may hinder testing beyond preliminary veterinary intervention. Therefore, providing veterinarians and analytical laboratories information on proper specimen collection protocols is of the utmost importance. Protocol dissemination will help to minimize costs, provide clinically relevant information, and compile data to inform the community of local environmental threats.

Typical MC canine toxicosis cases are the result of cyanobacteria ingestion. Post ingestion, it has been shown that MCs make their way through the gastrointestinal tract and into the liver, presumably through first pass effects through the bile-acid transport system [8–10]. MCs do not passively enter cells, but require active transport [11], mediated by organic anion transporters (OATs) [12]. These transporters are not only significantly expressed in the liver (Oatp1b2, OATP1A2, OATP1B1, OATP1B3) [12], but are also expressed in the brain and kidney (OATP1A2) [12,13]. The OAT facilitated uptake of MCs is one of the key steps in the pathogenesis of the reported hepatocellular damage and may account for the neurological effects observed in animals and humans following exposure [14,15]. The predominant OATs in the liver and kidney of canines (Oatp1b4 > Oatp2b1 > Oatp1a2) appear to exhibit similar substrate specificity to that of the human OATP1B3 [16]. Human OATP1B1 is abundant in lobular hepatocytes, while OATP1B3 is predominantly expressed in hepatocytes near the central vein [17], indicating interspecies differences in OAT location may play a role in clinical presentation of toxicosis. Once in the cytoplasm, MCs can affect a variety of cellular pathways, including regulation of DNA repair, regulation of protein activity, cell signaling, cell cycle, gene expression, apoptosis, and metabolism of endogenous or cytotoxic compounds [18]. The most studied of these pathways is the inhibition of the essential members of the protein phosphatase (PP) family. The reversible phosphorylation of proteins is an integral part of metabolism, which MCs inhibit by binding to serine/threonine PP1, PP2A [19] and PP3 [20]. PP inhibition can result in hyperphosphorylation [21], increases in reactive oxygen species (ROS) [22] and/or inflammation. These effects result in the disruption of cytoskeletal components, rearrangement of actin filaments within hepatocytes, and ultimately cellular death, which elucidates the observed morphological changes post-mortem [23].

The main pathway of hepatic elimination of MCs is Phase II biotransformation through conjugation with glutathione (via glutathione-S-transferase or non-enzymatically) [21,24–26] and through an elimination conjugation reaction with cysteine [24]. In mammals, MCs are primarily eliminated by both biliary and renal routes, with conjugated forms excreting mainly through the kidneys [27–29]. Since MC conjugates retain some of their toxic potential [30], metabolites may lead to continued insult to vital organ systems. The extent of intoxication and ability to recover from MC exposure is dependent on dose and the animal's capability to metabolize MCs. Since the antioxidant glutathione is integral to the detoxification and elimination of MCs, depletion after a high dose or in the presence of concomitant contaminants has been observed [31]. The loss of active glutathione coupled with continued hyperphosphorylation and resultant ROS formation likely contributes to the necrosis and apoptosis of hepatocytes, as well as the breakdown of the hepatocyte cytoskeleton. As the primary site of MC detoxification, the hepatic parenchyma exhibits the most striking damage to intoxication;

however, the renal nephron can also be negatively affected [5,32,33]. Although the cause of renal parenchymal damage has not been identified, tubular ischemia has been proposed as one mechanism [5]. Other probable mechanisms include hepatic shock and direct toxic action to the renal tubules of conjugated and free MCs.

The proper analytical approach is key to confirmation of MC exposure. Advances in MC research have elucidated numerous structurally related congeners, which have increased from 60+ known in the 1990's [34] to over 250 described to date [35]. Variations along the structure occur mainly in two amino acid positions (X^2, Y^4; Figure S1), but modifications, such as desmethylation, may happen along other parts of the structure. This structural variation coupled with protein binding and conjugate formation present significant analytical challenges. Therefore, widely available commercial enzyme-linked immunosorbent assays (ELISAs) are frequently used to test MCs due to their broad specificity to the various congeners (and potentially conjugates). However, MC ELISAs have a narrow range of applicability to complex matrix testing (e.g., urine, tissue, blood) from different animal species, with false positive results reported when analyzing mammalian livers [36]. While their availability and ease of use make them convenient, they also require an alternate method of confirmation, such as liquid chromatography tandem-mass spectrometry (LC-MS/MS). The specificity achieved when targeting MCs via LC-MS/MS provides quantitative accuracy, but also results in the under-reporting of total MCs [37,38]. This is due to a lack of commercially available reference materials for method calibration coupled with the extensive variability in MC forms. Other techniques utilized to address this include non-targeted high- and low-resolution LC-MS, but this requires an intimate knowledge of MC chemistry. An alternate approach involves the oxidative cleavage of the unique Adda (3-amino-9-methoxy-2,6,8-trimethyl-10-phenyl-4,6-decadienoic acid) side chain and subsequent quantitative analysis of MMPB (2-methyl-3-methoxy-4-phenylbutyric acid). The MMPB technique provides a relatively straightforward protocol accounting for total Adda containing MCs or nodularins [38–40]. This test allows for the quantification of free MCs and those modified during metabolism as long as there is conservation of the Adda side chain.

At present, the diagnosis of cyanobacteria poisoning requires a thorough history of the exposed patient in relation to the contaminated source. A two-tier analysis should include identification of the dominant cyanobacteria genera present and toxin analyses. In the absence of an algal grab sample, analyses can be conducted on the vomitus/stomach contents of a recently exposed individual, which is representative of unmetabolized cyanotoxins. However, due to the low pH of gastric contents, degradation of the organisms may impede identification of cyanobacteria genera. Thus, broad screening of multiple cyanobacterial toxins or targeted analysis for specific toxins based on clinicopathological data may be required. Once toxins are confirmed in the source, additional targeted analyses should be conducted on other specimens (e.g., liver, kidneys, feces, and urine) to confirm exposure and metabolism. Data achieved from the source of exposure, coupled with clinical/pathological observations and analytical data have been utilized to confirm MC intoxication in previous studies [4–6]. This process is time consuming, cost prohibitive, requires significant knowledge of cyanobacteria and the toxins they produce, and many specimens are not ideal for antemortem testing. Therefore, a standard protocol for diagnosing MC toxicosis should be developed for non-invasive specimen collection and sensitive accurate testing.

The present study illustrates the most comprehensive report on the pertinent clinicopathological data, pathological characteristics, supportive care, and novel diagnostic testing performed during and after an exposure event involving dogs. Results from this investigation provide support of viable antemortem testing methods for detection of MCs in canines during and after suspected exposure to microcystin producing cyanobacteria.

2. Results

2.1. Presentation, Clinical Data and Treatment

Between 26 August to 8 September 2018, six dogs were admitted for medical care at Pet Emergency of Martin County, Florida, USA. Information pertaining to the exposed animals and negative controls used in this study are presented in Table 1. The patient history for the six hospitalized cases included access to the Indian River and potential ingestion of decaying fish, organic debris, or water from the waterway. The onset of clinical signs varied from 2–48 h post exposure, with the most common signs being vomiting and depression. Weakness, collapse, tachycardia, petechia/ecchymosis and melena were also noted in a subset of patients. Clinicopathological findings included but were not limited to: elevated ALT (Alanine aminotransferase), thrombocytopenia, prolonged partial thromboplastin time (PTT) and prothrombin time (PT), peritoneal and/or pleural effusion, hypoglycemia and hyperbilirubinemia. For a full list of abnormalities and values, refer to Table 2.

Table 1. Subjects examined in this study shown with weights, age, sex, date of exposure and status. UE = Unexposed Individual. N = neutered, S = spayed.

ID	Breed	Age (years)	Sex	Weight (kg)	Date of Exposure	Status
C-SP	Standard Poodle	9	Male/N	23	4 September 2018	Deceased
C-GR #1	Golden Retriever	6	Female/S	32	8 September 2018	Living
C-GR #2	Golden Retriever	2	Female/S	30	8 September 2018	Living
C-GR #3	Golden Retriever	4	Female/S	31	1 September 2018	Living
C-Pom	Pomeranian	2	Female/S	2.3	26 August 2018	Living
C-Chih	Chihuahua	6	Male/N	5.5	26 August 2018	Living
C-LR	Labrador Retriever	<1	Male	36	UE	Living
C-GD	Goldendoodle	10	Female/S	34	UE	Deceased
C-CBR	Chesapeake Bay Retriever	11	Female	34	UE	Living

Table 2. Clinicopathological abnormalities noted during hospitalization of six dogs exposed to the St. Lucie River HAB event. ID = Identification, APTT/PT = activated partial thromboplastin time/prothrombin time, ALT = Alanine aminotransferase, >DL = greater than detection limit.

ID:	C-SP	C-GR #1	C-GR #2	[1]C-GR #3	C-Pom	[1]C-Chih
Vomiting:	Yes	Yes	Yes	Yes	Yes	Yes
Melena:	Yes	No	No	Yes	Yes	No
Tachycardia:	Yes	No	No	Yes	No	Yes
Body cavity effusion:	Yes	Yes	No	Yes	Yes	Unknown
APTT/PT:	>DL	>DL	Normal	>DL	>DL	128/17
Thrombocytopenia:	12K	15K	60K	69K	24K	77K
Bilirubin (mg/dL):	2.1	15	0.1	3.1	2.2	7.2
ALT (U/L):	>DL	10K	1889	3294	5287	>DL
Blood Glucose (mg/dL):	26	74	97	66	27	120

After presentation to the emergency clinic, a complete blood count, blood chemistry, and clotting profiles were performed. Decontamination through bathing was initiated in a subset of patients prior to arrival. One patient presented with productive emesis and vomitus was saved for cyanotoxin evaluation. The Animal Poison Control Center was contacted for advice on the cases but ultimately treatment was tailored for each dog by the attending veterinarian with the emphasis on acute liver injury. Therapy included intravenous fluids (with dextrose supplementation as indicated), gastroprotectants, antibiotics, antiemetics, analgesics, fresh frozen plasma (FFP), cholestyramine, vitamin K, N-acetylcysteine and various oral liver protectants. For the dogs that required intensive overnight care, continuous monitoring of electrocardiogram and blood pressure were performed. Furthermore, other essential parameters were monitored at varying intervals such as blood glucose, electrolytes, activated partial thromboplastin time/prothrombin time (APPT/PT), complete blood count (CBC), and blood chemistry. Patient hospitalization ranged from one day to nine days.

2.2. Pathology

One of the six dogs succumbed to fulminant liver failure, coagulopathy, and shock. A full postmortem examination was performed (Figure S2). On preliminary gross examination, the dog was in good body condition with a body condition score of 5/9 and mild post-mortem autolysis [41]. The skin in areas without hair appeared slightly yellow with multifocal areas of ecchymosis and petechiation. Diffuse icterus of the mucous membranes and subcutaneous adipose tissues was noted. Abundant dark red to black fluid drained from the nares and oral cavity upon manipulation of the head. Entry into the abdominal cavity showed up to 1L of serous red tinged fluid. The length of the intestines was dark pink to red with red streaking down the serosa, which had a granular appearance. Abundant edema and coagulated blood expanded the mesentery and omentum adjacent to the spleen as well as around the pancreas. Inspection of the esophagus, stomach, and intestines revealed abundant dark red to black fluid that filled the entire gastrointestinal tract. The gastric wall at the pylorus was diffusely expanded by submucosal hemorrhage and edema. The liver was diffusely dark red with sharp margins, a reticular pattern, and had a normal consistency. The gall bladder was filled with dark green bile and the wall of the bladder was thickened by edema. Abundant bright yellow, granular thick fluid was present in the urinary bladder. However, the kidneys and ureters were intact. The spleen was diffusely pale red and had multiple <3 mm fibrotic nodules on the serosal surface. The right lung lobes appeared pink except for a 3–4 cm red focus in the cranial lobe and mild dark red mottling in the middle lobe. The left lobes were diffusely red, wet, and oozed abundant red fluid upon transection. The adrenal glands had bilateral hyperplasia of the cortical layers. Multifocal petechia and ecchymosis were observed in the wall of the great vessels of the heart and in the endocardium of the left ventricle. There was mild multifocal, nodular thickening of the mitral valve leaflets.

Compared to normal canine microscopic anatomy (Figure 1A,C,E), tissues of the deceased dog showed several significant microscopic changes including; acute, severe, massive hepatocellular necrosis (Figure 1B), acute, moderate, multifocal, tubular necrosis with granular casts and intracellular iron (Figure 1D), acute, severe, multifocal to coalescing, hemorrhage in the thymus, gastrointestinal tract, mesentery, omentum, lymph nodes, pancreas, lungs, pulmonary artery, endocardium (Figure 1F), acute, diffuse, gall bladder edema, and acute, diffuse, splenic contraction with multifocal siderofibrotic plaques. Mild mitral valve endocardiosis was also noted in this patient as an incidental finding.

Ancillary testing included aerobic culture of the liver and leptospirosis PCR. Results of the culture revealed bacterial organisms *Enterococcus faecalis* and *Erysipelothrix rhusiopathiae*. Leptospirosis PCR was negative.

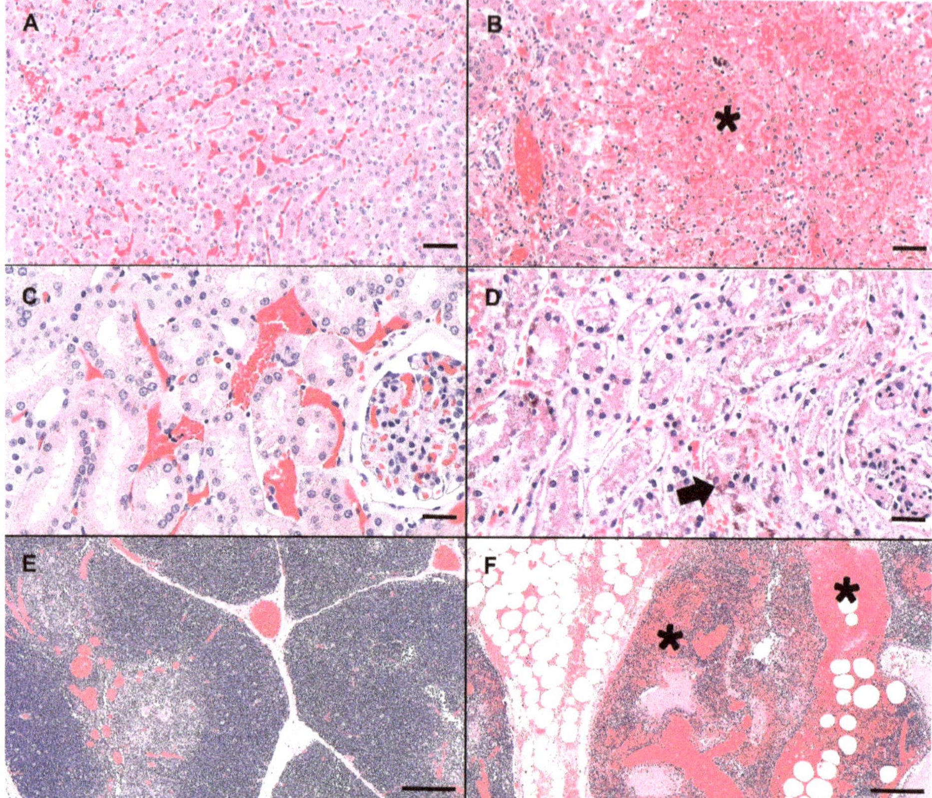

Figure 1. Photomicrographs (canine) of hematoxylin and eosin stained (H&E) normal liver, renal cortex, and thymus (**A**,**C**,**E**) as compared to the MC exposed dog (C-SP) (**B**,**D**,**F**). (**B**) Severely disrupted hepatic cords characterized by massive hepatocellular necrosis and hemorrhage (asterisk). Low numbers of hepatocytes adjacent to a central vein are spared. (**D**) Renal cortex with a locally extensive area of acute tubular necrosis. Note accumulation of brown granular pigment within tubular epithelial cytoplasm or sloughed cellular debris within the tubular lumina (arrow). (**F**) Mediastinal adipose and thymus expanded by hemorrhage, fibrin and edema (asterisks). Scale bars are 100, 50, and 500 μm for liver, renal cortex and thymus, respectively.

2.3. Phycology

Intact colonies of *Microcystis* were observed in the canine vomitus sample, confirming exposure to cyanobacteria (Figure 2). Other cyanobacteria were not observed. While a water sample was not submitted in conjunction with this exposure event, reports from the Florida Department of Environmental Protection (FDEP) support the vomitus phycological observations match the dominant genera (*Microcystis*) present in the St. Lucie River HAB at the time of exposure [42].

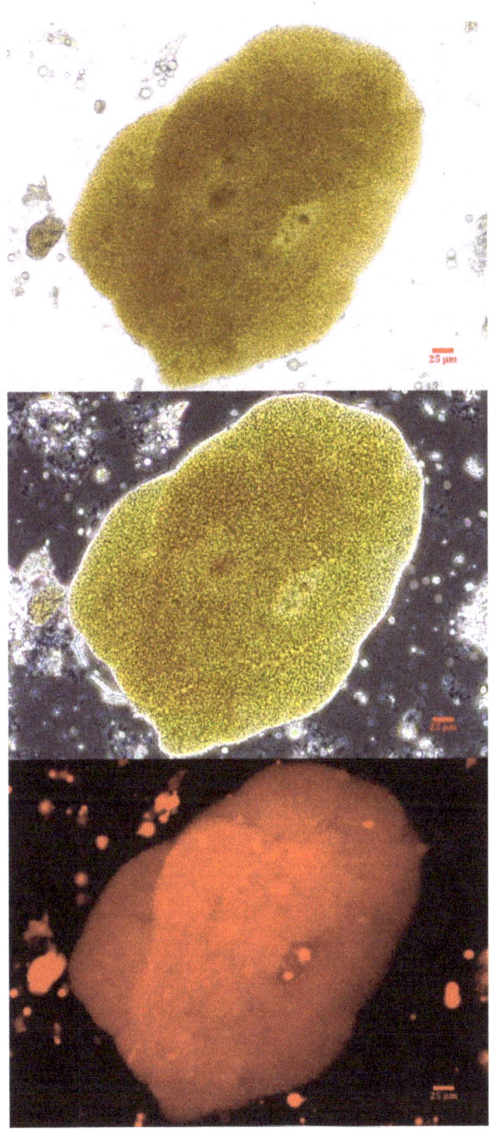

Figure 2. A *Microcystis* colony observed in the canine vomitus sample acquired within 6 h of exposure. The micrographs are at 400× with brightfield (top), phase-contrast (middle) and epi-fluorescence (bottom). The scale bar represents 25 μm.

2.4. Adda Microcystin/Nodularin (MC/NOD) Levels

Results from the three microcystin/nodularin (MC/NOD) analytical tests on all specimens are reported in Table 3. Enzyme-linked immunosorbent assay (ELISA) values (representing freely extractable Adda MCs/NODs) were, in general, supported by the MMPB (2-methyl-3-methoxy -4-phenylbutyric acid) data (representing total Adda MCs/NODs). This indicates that the Adda ELISA is able to react to conjugated forms of microcystin, as those excreted in urine. However, total Adda MCs (measured as MMPB) were higher in organ specimens, supporting that the MMPB method accounts for some fraction of protein bound MCs. The only non-metabolized specimen, the vomitus, had total Adda

MCs (MMPB) measured at 46,000 ± 8000 ng g^{-1}, confirming a high dose of exposure. The vomitus sample was significantly diluted with meal items (>30 grams submitted), so the dose was in excess of 1,380,000 ng total MCs. Analysis of the vomit using ELISA resulted in a lower level of MCs (25,000 ± 1800 ng g^{-1}), which may be due to MC losses to the matrix (high protein dietary items), partial MC degradation or due to differences in analytical technique used. Targeted LC-MS/MS of 19 MC variants and NOD-R confirmed the presence of 7 MC variants (Figure 3) and the absence of NOD-R. The dominant variant present was MC-LR (14,000 ± 100 ng g^{-1}), followed by [Dha^7]MC-LR (170 ± 21 ng g^{-1}), MC-HilR (140 ± 28 ng g^{-1}), [Asp^3]MC-LR (82 ± 0 ng g^{-1}), MC-LY (23 ± 16 ng g^{-1}), MC-LW (18 ± 3 ng g^{-1}), and MC-LF (14 ± 2 ng g^{-1}). Other variants were detected in a non-targeted MS scan, but due to a lack of standards available to verify identities or levels, they are not reported here (work ongoing). Targeted MCs in the vomit accounted for 56% of the ELISA measurement and 30% of the total Adda MCs by MMPB. The results of the targeted MC analysis were key to the decision to target MC-LR in the remaining collected specimens.

Table 3. Results of the Adda MC/NOD analyses conducted on all matrices. Data is reported in parts per billion, with organ and hair samples reported by weight (ng g^{-1}) and liquid samples by volume (ng mL^{-1}). Data is reported ± the standard deviation for samples with duplicate extractions. MMPB analysis represents total Adda MCs, ELISA represents freely extracted Adda MCs and MC-LR is LC-MS/MS analysis of the freely extracted variant. PE = Post exposure. UE = Unexposed individual.

ID	# Days PE	Specimen	Total Adda MCs (MMPB)	Spike Return	Adda ELISA	Spike Return	MC-LR	Spike Return
C-SP	2	Liver	530	21%	260 ± 78	193%	7.2 ± 1.5	97%
	2	Kidney	2800	12%	1100 ± 460	106%	20 ± 7.2	63%
	2	Heart Blood	73	9%	85 ± 28	146%	1.7 ± 0.0	68%
	2	Bile	5400 ± 750	72%	7500 ± 3200	211%	65 ± 26	78%
	2	Urine	41,000	79%	42,000 ± 4000	143%	670	57%
C-GR #1	1	Blood	50	27%	—	—	—	—
	1	Urine	32,000 ± 1600	42%	22,000 ± 4500	197%	110 ± 26	—
	11	Urine	4.4	65%	—	—	—	—
	24	Urine	2.6 ± 1.0	35%	—	—	—	—
	68	Urine	0.6 ± 0.0	22%	—	—	—	—
	72	Hair	180 ± 19	4%	< 30	0%	< 30	111%
C-GR #2	1	Blood	70	74%	33	—	< 1.0	—
	0	Vomit	46,000 ± 8000	114%	25,000 ± 1800		14,000 ± 1700	
	10	Urine	0.6 ± 0.3	72%	—	—	—	—
	24	Urine	< 0.2	36%	—	—	—	—
C-Pom	25	Blood	< 0.2	8%	—	—	—	—
	23	Urine	2.6	37%	—	—	—	—
	37	Urine	0.2[1]	26%	—	—	—	—
	85	Urine	< 0.2	11%	—	—	—	—
C-CBR	UE	Hair	< 20	4%	< 30	133%	< 30	108%
C-GD	UE	Liver	< 4.0	7%	52[2]	33%	< 5	108%
C-LR	UE	Urine	< 0.2	36%	< 15	107%	< 2	79%
C-LR	UE	Blood	< 0.2	3%	< 15	92%	< 2	83%

[1] above the limit of detection, but below the limit of quantification. [2] false positive data (refer to text for explanation).

MC-LR was detected in all the tested specimens from exposed animals with exception of the hair (C-GR #1) and one blood sample (C-GR #2). The negative control specimens were all below detection for targeted MC-LR. The remaining MC-LR concentrations (metabolized specimens) only accounted for 0.5%–2.8% of ELISA data and 0.3–2.3% MMPB data.

The MMPB data was integral to results interpretation, as the data is representative of total Adda MCs (and nodularins, when present), regardless of form (free, bound, partially degraded). Since the approach to quantification was pre-oxidation spiking (standard addition), confidence in MMPB data was higher than that of ELISA. The spike returns in Table 3 are shown only for reference and are used to illustrate how the method is impacted by various complex matrices (as compared to MC-LR oxidized in water). A low level of detection (sub-ppb) was achieved for urine and blood samples. The strength of this approach was illustrated when 2.6 ng mL^{-1} total MCs was detected in the urine sample of C-Pom over 3 weeks after suspected exposure. While this level could not be confirmed using alternate techniques due to higher test MDLs, the MMPB urinalysis of another exposure case (C-GR#2) supported the continued excretion of Adda >60 days post exposure (Figure 4). In addition to renal elimination, it was determined that hair was a potential route of elimination. MMPB chromatograms of a dog hair sample collected 72 days post exposure with overlaid pre-oxidation spikes of MC-LR can be viewed in Figure 5, with a negative control sample. A total MCs of 180 ng g^{-1} (dry weight) was determined to be present in the C-GR#1 hair specimen.

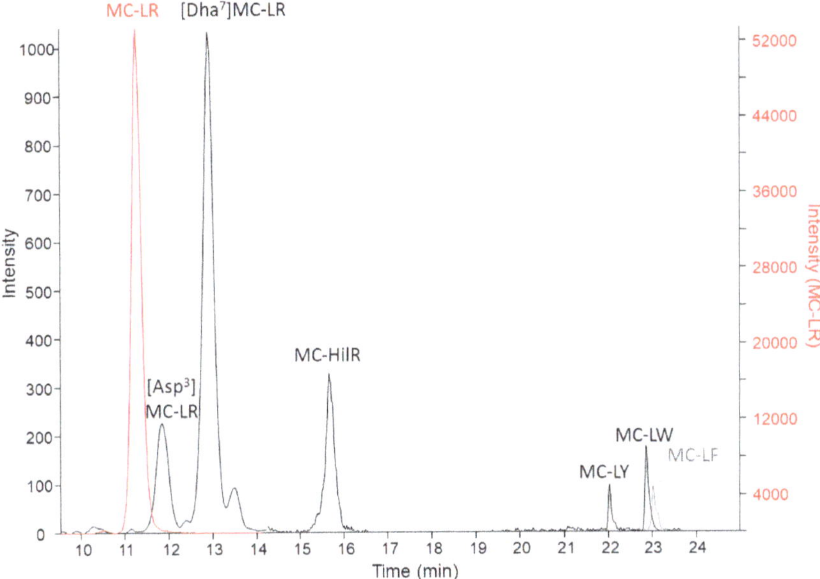

Figure 3. The LC-MS/MS chromatograms of MC variants confirmed present in the C-GR#2 vomit sample with a sum of 14,000 ng g^{-1} MCs. The MC-LR scale is on the right due to high levels detected in comparison to the other variants. MC-LR > [Dha^7]MC-LR > MC-HilR > [DAsp^3]MC-LR > MC-LY > MC-LW > MC-LF. Transitions monitored are reported in Table S1.

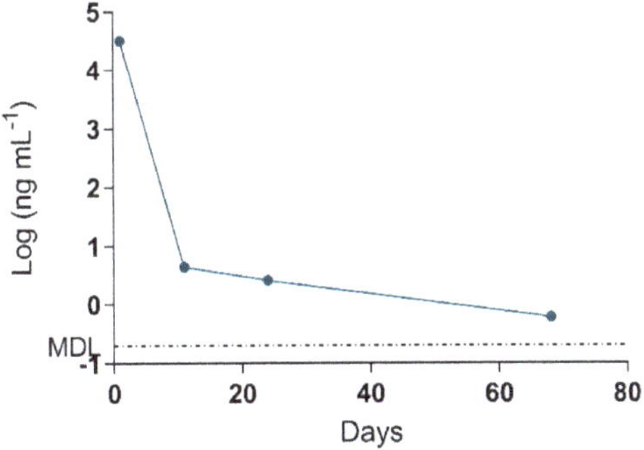

Figure 4. Data derived from the log of total Adda MCs (by MMPB) of the urine collected from one of the surviving dogs (C-GR#2) plotted against days post exposure. Urine was collected within 1 day from initial exposure event, and the animal continued to excrete MC metabolites >60 days post exposure. The MDL for total MCs in urine was determined to be 0.2 ng mL^{-1}.

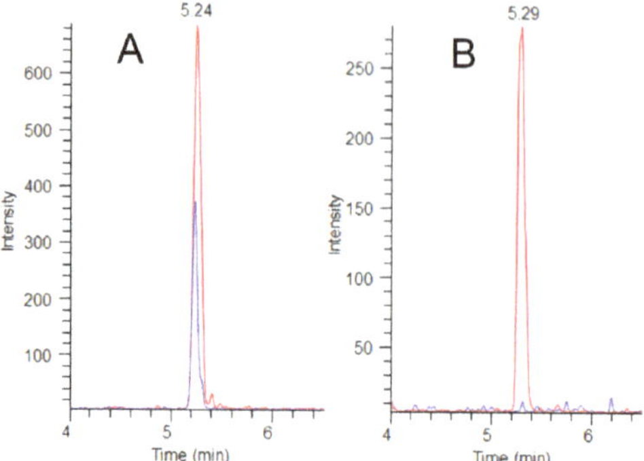

Figure 5. MMPB LC-MS/MS chromatograms (*m/z* 207→131) showing sample peaks (blue) overlaid with their paired pre-oxidation MC-LR spikes (red) at 200 ng g^{-1} for A (exposed dog C-GR#1) and 100 ng-g^{-1} for B (unexposed dog C-CBR). Total MCs detected in the C-GR#1 hair at 72 days post exposure was determined to be 180 ng g^{-1}, illustrating hair as a potential route of MC elimination.

The testing of the organs of the deceased canine (C-SP) revealed that the kidney had higher total Adda MCs (MMPB), free MCs (ELISA) and MC-LR when compared to the liver. In similar fashion, the urine was higher than the bile, with the least MCs measured in heart blood. Regardless of method used to test the specimens, the levels of MCs were as follows: urine > bile > kidney > liver > blood. Figure 6 illustrates the MCs detected using the 3-techniques in the specimens collected from the deceased dog (C-SP).

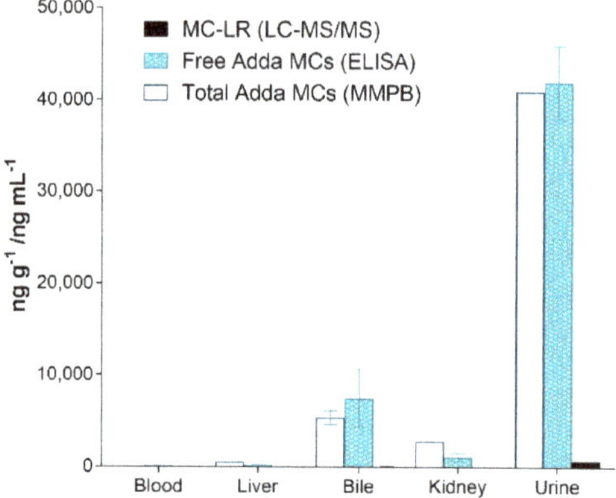

Figure 6. Results showing the different specimens collected from the dog (C-SP) that succumbed to intoxication 2 days post exposure and reported as ppb (ng mL^{-1} or ng g^{-1}). As illustrated, the urine contained the highest amounts of MCs (all methods), followed with the bile, kidney, liver and finally heart blood. All specimens were collected post-mortem.

It should be noted that the method of quantification for each technique is different, with interpretations of spike returns and final data essential to understanding results. The Adda ELISA method employs an external curve (certified reference material of MC-LR) for quantification, with the only assessable quality controls being replicate extraction data (reported as the average with standard deviations in Table 3) and spike returns. Both high (>130%) and low (<70%) spike returns were obtained from ELISA data. Exaggerated spike returns observed with ELISA may indicate overestimates of natively reported MCs, as supported by MMPB data in the bile sample. In contrast, low spike returns (e.g., hair and liver specimens), indicate matrix inhibition. The same extracts (and

MC-LR spikes) were analyzed by both Adda ELISA and LC-MS/MS and can be directly compared. The MC-LR spikes analyzed by LC-MS/MS returned 57% to 111%, while the same spikes were 0% to 211% when analyzed with the ELISA, further supporting that the ELISA resulted in both exaggerated and inhibited responses.

Further complicating ELISA data interpretation was the positive (≥15 ng g^{-1} MCs/NODs) response elicited from a negative control liver sample. The negative control specimen was collected from an animal without a recent exposure to waterways or cyanobacteria. Furthermore, MMPB analysis confirmed that Adda MCs/NODs were not present over 4 ng g^{-1}, which is more sensitive than the ELISA MDL of 15 ng g^{-1}. Sample extract clean-up (SPE), dilution (100-fold) and hexane washing were not sufficient in mitigating false positive data for the negative control liver sample. In comparison, prior to hexane washing, the positive liver and kidney samples were 260 and 1105 ng g^{-1}, respectively. Post hexane washing, the levels were similar at 258 and 1080 ng g^{-1}, with little change to spike returns (193→201%, 106→111%). Hexane washing did lower the negative control liver assay responses (analyzed with 100-fold dilution) from 1.78 to 0.52 ng mL^{-1} (correlating to 178 to 52 ng g^{-1}), and increased the spike return from 12% to 33%. However, this was not sufficient in preventing the false positive assay result.

3. Discussion

Harmful algal blooms (HABs) can be hazardous to humans, animals and the environment [43,44]. Reports of bloom formation with toxin production are becoming more frequent, likely due to anthropogenic influence leading to nutrient accumulation in waterways and changing global environmental conditions [45–47]. The 2018 *Microcystis* bloom in the St. Lucie River is a fitting example of the sublethal and lethal toxic effects that may be observed in dogs exposed to HABs.

The clinical signs and pathology of blue-green algal toxicity can mimic a handful of toxicoses, including, but not limited to, xylitol, sago palm, rodenticide, *Amanita*, and ricin. Rapid detection of the inciting toxin can be problematic in clinical settings. Samples of tissues may be difficult or dangerous to obtain and few laboratories provide cyanotoxin testing for animal tissues. In this work, the source water was not available for testing, but *Microcystis* and MCs were determined present in the source water during the event by the Florida Department of Environmental Protection [42]. The vomit from one of the six dogs did have measurable MCs, with MC-LR found to be 97% of the total targeted MCs. This observation is similar to other work analyzing MCs in the St. Lucie River [48], where MC-LR was 98% of the total targeted variants by LC-MS/MS. Specimens tested in past dog exposure events that succumbed to MC intoxication included the source water, feces [4], liver [5], and vomitus [5,6]. In this work, clinicopathological observations, post-mortem pathology, and MC testing were all used to confirm that the exposed dog morbidity was a result of MC intoxication and findings correlate with previous reports [4–6,49]. This work illustrates the most comprehensive multidisciplinary reporting on microcystin induced animal morbidity and mortality following a HAB exposure event. Confirmation of cyanobacteria intoxication is rare in a clinical setting due to a multitude of reasons. Improved antemortem methods for testing is essential for diagnosis and timely treatment recommendations, especially in the face of a multi-animal exposure event.

For the first time, urine was used to diagnose an ongoing MC intoxication event. The importance of this finding cannot be overstated, as urine is a specimen that can be easily, quickly and non-invasively collected by veterinarians faced with a suspected toxicosis, even if several days or weeks have passed. The depuration of MCs, even months after exposure, is likely due to the accumulation of MCs in tissues and bound to proteins. The MCs are then slowly released over time, with the Adda being conserved. The achieved sensitivity and broad specificity of the MMPB test, especially for urine, was integral in confirming exposure and metabolism. Figure 7 outlines the recommended oxidation and extraction approach for screening urine samples in the event of a suspected exposure case. This approach could be applied to other sentinel species, or even following human exposure events.

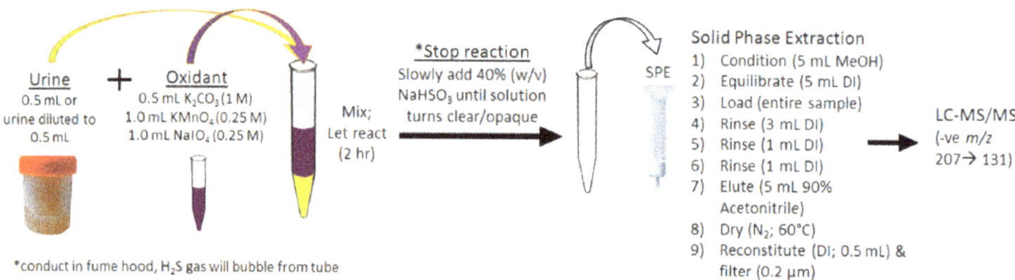

Figure 7. Schematic showing the recommended approach to the oxidation, extraction and clean-up of urine for the purposes of total Adda MCs by the MMPB method. In order to properly quantitate, it is essential that pre-oxidation spiking of MC-LR be conducted on duplicate sample.

With regard to more long term, chronic exposure, it has been demonstrated that hair might be a candidate for assessing exposure. This is the first use of MC testing on mammalian hair following an intoxication event and positive results were observed using the MMPB technique. This preliminary exercise shows promise with regard to potential avenues of testing in both acute and chronic mammalian exposure cases. However, confirmatory testing beyond the MMPB analysis did not detect free MC-LR in the hair above the established method detection limits. This area of research should be expanded in order to determine the applicability of this specimen for monitoring exposed populations of mammalian species.

Pharmacokinetic studies of orally ingested MCs in canines have not been conducted to date, making interpretation of measured toxin levels challenging. The amount of toxin needed to cause the observed gross and microscopic lesions has not been standardized in canine patients and extrapolation from lab animal models may not be representative. In experimental settings, the median lethal dose (LD_{50}) for MCs is typically derived from an intraperitoneal (i.p.) route, which has been reported as low as 36 μg per kg body weight for mice exposed to MC-LR [50]. The oral LD_{50} of MC-LR has been reported to be as high as 5,000 μg per kg body weight for mice, and even higher for rats, supporting interspecies variability [10]. The likely presence of other protein phosphate inhibitors concomitant with MC-LR, such as other MC variants and peptides (e.g., anabaenopeptins), can also complicate interpretations of toxin data in reference to toxicity [51,52]. Since other MC variants were detected in this event, and MMPB data indicated even more variants were present but not accounted for in targeted analyses, correlating MC-LR levels from previous toxicity studies would provide little benefit. The inter-/intra-species variability combined with complicated real-world exposure, results in a high level of uncertainty with regard to MC dose and observed toxicity.

In this study, the dose for each animal was unknown, but it can be inferred from MMPB analysis of the urine samples from a surviving individual (32,000 ng mL^{-1}) at 1-day post exposure and from the deceased animal (41,000 ng mL^{-1}) at 2-days post exposure, that doses were similarly high and caused the observed hepatoxicity. Additionally, the single collected vomit sample provided evidence that >46 μg kg^{-1} total Adda MCs was initially ingested, but it is impossible to determine what level entered metabolism. Post metabolism, several potential MC metabolites were observed in the urine employing an MS scan, but require more sophisticated analyses, such as high-resolution mass spectrometry and deconjugation experiments, to elucidate the profile. Work to identify these metabolites is ongoing.

Although the oral exposure route has been confirmed in at least one of these animals, the potential for compounded exposure through inhalation cannot be ruled out. Inhalation of MC-LR by mice has a reported LD_{50} of 43 μg kg^{-1} [53], similar to that reported for the i.p. route. The higher toxicity observed via inhalation coupled with evidence that MCs can become aerosolized during cyanobacteria blooms [54,55] may present additional health risks to those living in close proximity to an ongoing bloom. Research has shown that low dose, subchronic exposure can result in accumulation of MCs in the mammalian liver [56]. Since dogs are considered a proposed sentinel species and they intimately share the human environment, canine exposure events such as this support the importance of a pro-active approach with regard to HABs in relation to human and animal health.

Collection of appropriate specimen type and selection of optimal analytical test are of high importance for accurate diagnosis. It was illustrated that the Adda ELISA provided false positive data and exaggerated assay responses, likely due to non-specific binding to kit antibodies. This highlights the need to confirm any ELISA data, especially for the analysis of matrices more complicated than water, for which the assay was intended. Matrix effects and exaggerated ELISA responses have been observed in other work [36,40] and require, at minimum, a secondary method of confirmation when a positive assay response is observed. In this work, both LC-MS/MS of targeted MCs and total Adda MCs by MMPB were conducted in addition to the MC Adda ELISA. However, due to method specificity, targeting MC variants, such as the MC-LR targeted in this work, under-represents total MCs, both due to MC structural variability and metabolic transformations. Therefore, if only one method is to be used in initial screening of samples in a suspected exposure case, it is recommended that the MMPB method be used. The MMPB approach provided a value representative of total (e.g., bound, free, conjugated) Adda MCs for the evaluated samples with a low detection limit.

Although clinicopathological data and epidemiology supported involvement of six dogs in the described morbidity/mortality event, only four dogs were confirmed to have been exposed to MCs using analytical techniques. Due to monetary constraints and lack of published sampling/testing protocols for antemortem testing of cyanotoxin in canines, confirmation could not be made in the other three cases. In the present study, a viable test using free catch urine collection has been established. Furthermore, a set protocol for directed toxin testing is described and can hopefully be employed in similar events to quickly and accurately diagnose MC intoxication. The development of an antemortem assay using non-invasive collection techniques is sure to have a significant impact on the diagnosis, treatment, and exposure of animals to HABs.

4. Materials and Methods

4.1. Pathology and Specimen Collection

A thorough gross examination of the deceased patient was performed within 12 h of death. Throughout the procedure, aseptically collected sections of liver, kidney, lung, spleen, cerebral frontal lobe, small intestine, large intestine, gastric contents, feces, urine, bile, and, heart blood were collected for ancillary testing. Fresh tissues and fluids were individually placed into sterile bags or sterile syringes to be held in a −20 °C freezer for storage until testing could be performed.

Sections of all abdominal and thoracic organs along with brain, eyes, skin, skeletal muscle, sciatic nerve, and bone marrow were preserved in 10% neutral buffered formalin. Tissues were trimmed, placed into cassettes, and processed via routine paraffin embedding techniques. All tissues were made into 3–5 μm sections and stained with hematoxylin and eosin stain (H&E) for histological review. Microscopic review was performed by a board-certified veterinary pathologist using a Nikon eclipse 80i (Nikon, Minato, Tokyo, Japan) and photomicrographs were captured by an Accu-scope Excelis HD (Commack, NY, USA).

Antemortem sample collection from a subset of the surviving dogs (Table 1) including urine, blood, vomit, and hair was performed by owners or veterinary staff during and after the initial onset of clinical signs. Urine was collected via free catch method and stored in sterile jars or plastic collection containers without preservative. Liver, kidney, free catch urine, blood and hair samples from three control dogs not associated with the HAB event were additionally collected (one deceased and the other living). Tissue, hair, and fluid samples were chilled and/or frozen prior to overnight shipment or transfer. Sample collection dates in relation to exposure can be referenced in Table 1.

4.2. *Phycology of Vomitus Sample*

A ca 0.5 g subset of the C-GR#2 vomitus sample was suspended in 5 mL of deionized water (DI) and gently vortex mixed. Wet mounts of un-fixed sample were prepared (3x) and scanned at 100X for the presence of potentially toxigenic cyanobacteria using a Nikon TE200 inverted microscope equipped with phase contrast and epifluorescence (green light excitation, 510–560, FT580, LP590). A Nikon digital sight DS-Fi1 camera was used for micrographs. Higher magnification was used as necessary for identification and micrographs.

4.3. *Adda MC/NOD Analyses*

4.3.1. Specimen Homogenization

Subsets were taken from liver samples (composited 5 subsets; ca 1 g each) and the entire kidney (2.7 g) was cut into small pieces and placed into 30 mL homogenization vials with 10 mM phosphate buffer at pH 7 (1:3). The vomitus sample was emptied into a glass beaker, mixed with a spatula and subsampled (2 subsets; 1 g each) into individual 7 mL homogenization vials with extractant solution (75% acetonitrile in 100 mM acetic acid; 1:3). Acetonitrile (HPLC grade) and glacial acetic acid (>99%) were both from Thermo Fisher Scientific (Waltham, MA, USA). Ceramic beads (2.8 mm) were added and the samples were homogenized at 5 m/s for 30 seconds (1 cycle for the vomitus and 2 cycles for the tissues) using an Omni Bead Ruptor 24 (Omni International, Inc, Kennesaw, GA, USA). Hair samples were pulverized in 30 mL vials with stainless steel 2.8 mm beads dry at 6.8 m/s (30 seconds) and transferred to glass vials as 50 mg subsets. Aliquots of kidney and liver (400 μL) were transferred to glass vials for subsequent oxidations/extractions. The vomitus vials were centrifuged (1500 g; 10 min) and supernatants retained. The pellets were vortex mixed with additional extractant (1 mL), centrifuged as before and supernatants pooled and saved for extractions.

4.3.2. Total Adda MCs/NODs Oxidation, Extraction and Analysis (MMPB)

MMPB oxidations and extractions were conducted as previously described [40]. Briefly, samples were oxidized as 100 mg (liver & kidney; wet weight), 50 mg (hair; dry weight), 500–1500 μL (blood), 0.2 μL–500 μL (bile, urine) and 10 mg (vomit; wet weight) subsets. All samples were paired with at least one pre-oxidation matrix spike of certified reference material (CRM) MC-LR (National Research Council Canada; Halifax, NS, Canada) for quantification. Spikes ranged from 5 ng mL^{-1} (or ng g^{-1}) to 50,000 ng mL^{-1} and were added to approximately double native peak areas (positive samples) for quantification (Table S2). Subsets of liver, hair, kidney, vomit and blood were oxidized by adding 5 mL of oxidant (0.2 M K_2CO_3, 0.1 M $KMnO_4$ and 0.1 M $NaIO_4$), while urine and bile samples were oxidized with 2.5 mL oxidant. Reactions were stopped after 2 h with the addition of 40% sodium bisulfite (w/v) until solutions turned white/opaque. The liver, kidney, hair, and blood oxidized aliquots were cooled (10 min; −20 °C), centrifuged (1500 *g*; 10 min) and supernatants retained. The pellets were rinsed with DI (2 mL) by vortex mixing followed by centrifugation (1500× *g*; 10 min). Extracts were loaded onto preconditioned (MeOH→DI) 200 mg Strata X SPE (Phenomenex, Torrance, CA, USA), rinsed (DI; 3x) and eluted (5 mL; 90% acetonitrile). Extracts were blown to dryness (N_2, 40 °C) and samples (with exception to liver and blood) reconstituted in DI (0.5–1.0 mL). The liver and blood were reconstituted in 0.1 M HCl (1.5 mL) and loaded onto simplified liquid extraction (SLE) columns (12cc; Phenomenex, Torrance, CA, USA), allowed to sit (10 min) and eluted with 2x–5mL ethyl acetate. Elutions from SLE were blown to dryness (N_2, 40 °C) and reconstituted in DI (0.5–1.0 mL). All samples were filtered using 0.2 μm polyvinylidene difluoride (PVDF) syringe filters (Millipore Sigma, St. Louis MO, USA).

A Thermo Surveyor high performance liquid chromatography (HPLC) system coupled to a TSQ Quantum Access MAX Triple quadrupole mass spectrometer system coupled with a Kinetex C18 column (2.6 μm; 100 Å; 150 × 2.1 mm; Phenomenex; Torrance, CA, USA) were utilized as described previously [36]. The $[M-H]^-$ ion of MMPB (*m*/*z* 207) was fragmented and *m*/*z* 131 (CE = 12%) was monitored. A five-point standard curve (0.5–100 ng/mL of oxidized MC-LR) was used to interpolate spike returns. Quantification of samples was conducted using individually prepared pre-oxidation spikes of MC-LR (standard addition). The instrument detection limit coupled with spike responses were used to determine the method detection limits (MDLs; Table 4).

Table 4. Method detection limits (MDLs) achieved on each specimen type at lowest dilution analyzed. Reported as ng mL^{-1} unless otherwise specified.

Specimen	Total Adda MCs (MMPB)	Adda ELISA	MC-LR (LC-MS/MS)
Liver (ng g^{-1})	4.0	15	5.0
Kidney (ng g^{-1})	4.0	15	5.0
Bile	50	15	50
Blood	0.2	15	2.0
Urine	0.2	15	2.0
Hair (ng g^{-1})	20	30	30

4.3.3. Free Adda MCs/NODs Extraction and Analyses

Samples with total MCs >15 ppb (ng g^{-1}; ng mL^{-1}) via MMPB analysis and representative negative control samples (unexposed individuals) were extracted and analyzed for free MCs, unless exhausted. Subsets of liver (100 mg), kidney (100 mg), blood/serum (500 µL), and hair (50 mg) were suspended in 5 mL extractant (75% acidified acetonitrile in 100 mM acetic acid) and sonicated (bath, 25 min). The samples were cooled (10 min; −20 °C), centrifuged (1500× *g*; 10 min) and supernatants retained. The pellets were resuspended in extractant (1 mL), cooled (10 min; −20 °C) and re-centrifuged. The pooled supernatants (including previously homogenized vomitus) were diluted (70 mL DI). Samples of urine (ranging from 10 µL–500 µL) and bile (10 µL) were diluted (10 mL DI) prior to SPE. Preconditioned Oasis HLB (200 mg; Waters Corporation, Milford, MA) or Strata X were loaded with diluted sample, rinsed with DI (2×; 5 mL and 3 mL) and eluted (5 mL; 90% acetonitrile). Solutions were blown to dryness (N_2; 60 °C) and reconstituted in DI at sample concentrations within range of the calibration curves for each analysis used. All samples were filtered using 0.2 µm PVDF prior to analysis.

Final liver and kidney extracts were washed 1:1 (*v*/*v*) using hexane (ACS grade; Fisher Scientific, Waltham, MA, USA). Hexane was added to sample, vortex mixed, centrifuged (1500× *g*; 5 min) and the hexane layer discarded. Analysis on these extracts by ELISA was conducted both prior to and after hexane washing.

An MCs/NODs Adda ELISA (Abraxis; Warminster, PA) was used as previously described [40]. Dilutions were prepared using DI to achieve absorbance values within the range of the standard curve (0.15–4.0 ng mL^{-1}) and all solutions were analyzed in duplicate. The minimum method detection limit (MDL) was 15 ppb (ng g^{-1}/ng mL^{-1}) based on the 100-fold dilution factor (DF) used and assay sensitivity (0.15 ng mL^{-1}).

Extracts prepared for ELISA were post-spiked with the internal standard (IS) d_7-MC-LR. The vomitus sample extracts were spiked with a second IS (*d5*-MC-LF). Both IS were acquired from Abraxis Kits (Warminster, PA, USA). A targeted analysis for MC-LR was used on all extracts using a Thermo Surveyor HPLC system coupled with an LTQ XL Linear Ion Trap Mass Spectrometer. Separation was achieved using the same column used in MMPB analysis (Kinetex C18 column) with mobile phase C (2 mM formic acid and 3.6 mM ammonium formate in deionized water) and D (95% acetonitrile (*v*/*v*) in 2 mM formic acid and 3.6 mM ammonium formate). Acetonitrile (Optima LC/MS), water (HPLC), ammonium formate (ACS grade), and formic acid (98%) were from Thermo Fisher Scientific (Waltham, MA, USA). The gradient (0.2 mL min^{-1}) was as follows: solvent C 70–30% over 5 min, 30–70% C over 2 min, and held at 70% C for 3 min. Each chromatographic run was 10 min and 20 µL full loop injections were employed. The analysis of MC-LR (*m*/*z* 995.5→375.0, 553.4, 599.4, 866.6) was calibrated using a seven-point IS curve (0.5–100 ng mL^{-1}) with *d7*-MC-LR (*m*/*z* 1002.5→599.5).

The analysis of the vomit extracts was conducted using targeted LC-MS/MS (Table S1) and LC-MS scans (*m/z* 400–1800 (-ve and +ve)) with a PDA set to λ 238 nm. The gradient (0.2 mL min^{-1}) was as follows: solvent C held at 70% for 11 min, 70–30% C over 9 min, 30–70% C over 2 min, and held at 70% C for 5 min. XCalibur v 2.2 (Thermo Fisher Scientific, Waltham, MA, USA) was utilized for data processing. Standards used to calibrate the method included CRMs (NOD-R, MC-LR, MC-RR, [Dha^7]MC-LR) and the RM (MC-RY) from the National Research Council Canada (Halifax, NS, Canada), RMs (MC-WR, [Asp^3]MC-RR, [Asp^3]MC-LR, MC-HtyR, MC-LF, MC-LW, MC-HilR) from Enzo Biochem (Farmingdale, NY, USA) and RMs ([Leu^1]MC-LR, MC-YR, MC-LA, MC-LY) from GreenWater Laboratories (Palatka, FL, USA). Additional RMs ([$ADMAdda^5$]MC-LR, [$ADMAdda^5$]MC-LHar) were produced in-house through the extraction and purification of a strain of *Nostoc* sp. 152 generously supplied by Kaarina Sivonen (University of Helsinki) using methods previously described [38]. The desmethyl RMs [$DMAdda^5$]MC-LR and [$DMAdda^5$]MC-LHar were prepared by hydrolysis of the [$ADMAdda^5$]MCs [57]. Quantification of targeted MCs was conducted using the previously reported IS approach [36].

Supplementary Materials: The following are available online at http://www.mdpi.com/2072-6651/11/8/456/s1, Figure S1: A general representation of the heptapeptide microcystin and the formation of the MMPB molecule following oxidative cleavage of the Adda side chain; Figure S2: Significant gross lesions observed during autopsy of deceased dog (C-SP); Table S1: Multiple reaction monitoring (MRM) transitions for targeted analysis of MCs (19 variants), NOD-R; Table S2: Matrix spikes (of MC-LR) associated with returns reported in Table 3.

Author Contributions: Conceptualization, A.F. and S.B.; methodology, A.F. and S.B.; validation, A.F. and M.A.; formal analysis, A.F.; investigation, B.G., A.M.; resources, A.F., M.A., S.F., B.G., A.M.; writing—original draft preparation, A.F.; writing—review and editing, A.F., S.B., N.M.

Funding: The authors declare that GreenWater Laboratories received funding for MMPB analysis of one liver sample and ELISA analysis of the vomitus sample from Pet Emergency of Martin County (Stuart FL). All other research received no external funding.

Acknowledgments: The authors thank Kamil Cieslik for his assistance in laboratory (MC) testing and the pet owners for their assistance throughout this study.

Conflicts of Interest: The authors declare no conflict of interest

References

1. Backer, L.C.; Manassaram-Baptiste, D.; LePrell, R.; Bolton, B. Cyanobacteria and algae blooms: Review of health and environmental data from the harmful algal bloom-related illness surveillance system (HABISS) 2007–2011. *Toxins* **2015**, *7*, 1048–1064. [CrossRef] [PubMed]
2. Backer, L.C.; Landsberg, J.H.; Miller, M.; Keel, K.; Taylor, T.K. Canine cyanotoxin poisonings in the United States (1920s–2012): Review of suspected and confirmed cases from three data sources. *Toxins* **2013**, *5*, 1597–1628. [CrossRef] [PubMed]
3. Backer, L.C.; Grindem, C.B.; Corbett, W.T.; Cullins, L.; Hunter, J.L. Pet dogs as sentinels for environmental contamination. *Sci. Total Environ.* **2001**, *274*, 161–169. [CrossRef]
4. Rankin, K.A.; Alroy, K.A.; Kudela, R.M.; Oates, S.C.; Murray, M.J.; Miller, M.A. Treatment of cyanobacterial (microcystin) toxicosis using oral cholestyramine: Case report of a dog from Montana. *Toxins* **2013**, *5*, 1051–1063. [CrossRef] [PubMed]
5. Van der Merwe, D.; Sebbag, L.; Nietfeld, J.C.C.; Aubel, M.T.T.; Foss, A.; Carney, E. Investigation of a *Microcystis aeruginosa* cyanobacterial freshwater harmful algal bloom associated with acute microcystin toxicosis in a dog. *J. Vet. Diagnostic Investig.* **2012**, *24*, 679–687. [CrossRef] [PubMed]
6. Lürling, M.; Faassen, E.J. Dog poisonings associated with a *Microcystis aeruginosa* bloom in the Netherlands. *Toxins* **2013**, *5*, 556–567. [CrossRef] [PubMed]
7. Faassen, E.J.; Harkema, L.; Begeman, L.; Lürling, M. First report of (homo)anatoxin-a and dog neurotoxicosis after ingestion of benthic cyanobacteria in The Netherlands. *Toxicon* **2012**, *60*, 378–384. [CrossRef]
8. Stotts, R.R.; Twardock, A.R.; Koritz, G.D.; Haschek, W.M.; Manuel, R.K.; Hollis, W.B.; Beasley, V.R. Toxicokinetics of tritiated dihydromicrocystin-LR in swine. *Toxicon* **1997**, *35*, 455–465. [CrossRef]
9. Stotts, R.R.; Twardock, A.R.; Haschek, W.M.; Choi, B.W.; Rinehart, K.L.; Beasley, V.R. Distribution of tritiated dihydromicrocystin in swine. *Toxicon* **1997**, *35*, 937–953. [CrossRef]

10. Fawell, J.K.; Mitchell, R.E.; Everett, D.J.; Hill, R.E.; Everett, D.J. The toxicity of cyanobacterial toxins in the mouse: I microcystin-LR. *Hum. Exp. Toxicol.* **1999**, *18*, 168–173. [CrossRef]
11. Eriksson, J.E.; Grönberg, L.; Nygård, S.; Slotte, J.P.; Meriluoto, J. a Hepatocellular uptake of 3H-dihydromicrocystin-LR, a cyclic peptide toxin. *Biochim. Biophys. Acta* **1990**, *1025*, 60–66. [CrossRef]
12. Fischer, W.J.; Altheimer, S.; Cattori, V.; Meier, P.J.; Dietrich, D.R.; Hagenbuch, B. Organic anion transporting polypeptides expressed in liver and brain mediate uptake of microcystin. *Toxicol. Appl. Pharmacol.* **2005**, *203*, 257–263. [CrossRef]
13. Lee, W.; Glaeser, H.; Smith, L.H.; Roberts, R.L.; Moeckel, G.W.; Gervasini, G.; Leake, B.F.; Kim, R.B. Polymorphisms in human organic anion-transporting polypeptide 1A2 (OATP1A2): Implications for altered drug disposition and central nervous system drug entry. *J. Biol. Chem.* **2005**, *280*, 9610–9617. [CrossRef]
14. Li, X.-B.; Zhang, X.; Ju, J.; Li, Y.; Yin, L.; Pu, Y. Alterations in neurobehaviors and inflammation in hippocampus of rats induced by oral administration of microcystin-LR. *Environ. Sci. Pollut. Res.* **2014**, *21*, 12419–12425. [CrossRef]
15. Pouria, S.; De Andrade, A.; Barbosa, J.; Cavalcanti, R.L.; Barreto, V.T.S.; Ward, C.J.; Preiser, W.; Poon, G.K.; Neild, G.H.; Codd, G.A. Fatal microcystin intoxication in haemodialysis unit in Caruaru, Brazil. *Lancet* **1998**, *352*, 21–26. [CrossRef]
16. Wilby, A.J.; Maeda, K.; Courtney, P.F.; Debori, Y.; Webborn, P.J.H.; Kitamura, Y.; Kusuhara, H.; Riley, R.J.; Sugiyama, Y. Hepatic uptake in the dog: Comparison of uptake in hepatocytes and human embryonic kidney cells expressing dog organic anion-transporting polypeptide 1B4. *Drug Metab. Dispos.* **2011**, *39*, 2361–2369. [CrossRef]
17. König, J.; Cui, Y.; Nies, A.T.; Keppler, D. Localization and genomic organization of a new hepatocellular organic anion transporting polypeptide. *J. Biol. Chem.* **2000**, *275*, 23161–23168. [CrossRef]
18. Campos, A.; Vasconcelos, V. Molecular mechanisms of microcystin toxicity in animal cells. *Int. J. Mol. Sci.* **2010**, *11*, 268–287. [CrossRef]
19. MacKintosh, R.W.; Dalby, K.N.; Campbell, D.G.; Cohen, P.T.W.; Cohen, P.; MacKintosh, C. The cyanobacterial toxin microcystin binds covalently to cysteine-273 on protein phosphatase 1. *FEBS Lett.* **1995**, *371*, 236–240.
20. Honkanan, R.E.; Codispoti, B.A.; Tse, K.; Boynton, A.L. Characterization of natural toxins with inhibitory activity against serine/threonine protein phosphatases. *Toxicon* **1994**, *32*, 339–350. [CrossRef]
21. Mattos, L.J.; Valença, S.S.; Azevedo, S.M.F.O.; Soares, R.M. Dualistic evolution of liver damage in mice triggered by a single sublethal exposure to Microcystin-LR. *Toxicon* **2014**, *83*, 43–51. [CrossRef]
22. Ding, W.X.; Shen, H.M.; Ong, C.N. Critical role of reactive oxygen species formation in microcystin-induced cytoskeleton disruption in primary cultured hepatocytes. *J. Toxicol. Environ. Health. A* **2001**, *64*, 507–519. [CrossRef]
23. Falconer, I.R.; Yeung, D.S. Cytoskeletal changes in hepatocytes induced by *Microcystis* toxins and their relation to hyperphosphorylation of cell proteins. *Chem. Biol. Interact.* **1992**, *81*, 181–196. [CrossRef]
24. Kondo, F.; Ikai, Y.; Oka, H.; Okumura, M.; Ishikawa, N.; Harada, K.; Matsuura, K.; Murata, H.; Suzuki, M. Formation, characterization, and toxicity of the glutathione and cysteine conjugates of toxic heptapeptide microcystins. *Chem. Res. Toxicol.* **1992**, *5*, 591–596. [CrossRef]
25. Buratti, F.M.; Scardala, S.; Funari, E.; Testai, E. Human glutathione transferases catalyzing the conjugation of the hepatoxin microcystin-LR. *Chem. Res. Toxicol.* **2011**, *24*, 926–933. [CrossRef]
26. Kondo, F.; Matsumoto, H.; Yamada, S.; Ishikawa, N.; Ito, E.; Nagata, S.; Ueno, Y.; Suzuki, M.; Harada, K. Detection and identification of metabolites of microcystins formed in vivo in mouse and rat livers. *Chem. Res. Toxicol.* **1996**, *9*, 1355–1359. [CrossRef]
27. Ito, E.; Kondo, F.; Harada, K.I. First report on the distribution of orally administered microcystin-LR in mouse tissue using an immunostaining method. *Toxicon* **2000**, *38*, 37–48. [CrossRef]
28. Li, W.; He, J.; Chen, J.; Xie, P. Excretion pattern and dynamics of glutathione detoxification of microcystins in Sprague Dawley rat. *Chemosphere* **2018**, *191*, 357–364. [CrossRef]
29. Wang, Q.; Xie, P.; Chen, J.; Liang, G. Distribution of microcystins in various organs (heart, liver, intestine, gonad, brain, kidney and lung) of Wistar rat via intravenous injection. *Toxicon* **2008**, *52*, 721–727. [CrossRef]
30. Ito, E.; Takai, A.; Kondo, F.; Masui, H.; Imanishi, S.; Harada, K.I. Comparison of protein phosphatase inhibitory activity and apparent toxicity of microcystins and related compounds. *Toxicon* **2002**, *40*, 1017–1025. [CrossRef]

31. Guo, X.; Chen, L.; Chen, J.; Xie, P.; Li, S.; He, J.; Li, W.; Fan, H.; Yu, D.; Zeng, C. Quantitatively evaluating detoxification of the hepatotoxic microcystin-LR through the glutathione (GSH) pathway in SD rats. *Environ. Sci. Pollut. Res.* **2015**, *22*, 19273–19284. [CrossRef]
32. Milutinovic, A.; Sedmak, B.; Horvat-Znidarsic, I.; Suput, D. Renal injuries induced by chronic intoxication with microcystins. *Cell. Mol. Biol. Lett.* **2002**, *7*, 139–141.
33. Hooser, S.B.; Beasley, V.R.; Lovell, R.A.; Carmichael, W.W.; Haschek, W.M. Toxicity of microcystin LR, a cyclic heptapeptide hepatotoxin from *Microcystis aeruginosa*, to rats and mice. *Vet. Pathol.* **1989**, *26*, 246–252. [CrossRef]
34. Chorus, I.; Bartram, J.; World Health Organization. *Toxic Cyanobacteria in Water: A Guide to Their Public Health Consequences, Monitoring and Management*; CRC Press: London, UK, 1999; ISBN 0419239308.
35. Miles, C.O.; Stirling, D. Toxin Mass List, Version 15. Available online: https://www.researchgate.net/publication/316605326_Toxin_mass_list_version_15 (accessed on 1 May 2017). [CrossRef]
36. Brown, A.; Foss, A.; Miller, M.A.; Gibson, Q. Detection of cyanotoxins (microcystins/nodularins) in livers from estuarine and coastal bottlenose dolphins (*Tursiops truncatus*) from Northeast Florida. *Harmful Algae* **2018**, *76*, 22–34. [CrossRef]
37. Foss, A.J.; Aubel, M.T. Using the MMPB technique to confirm microcystin concentrations in water measured by ELISA and HPLC (UV, MS, MS/MS). *Toxicon* **2015**, *104*, 91–101. [CrossRef]
38. Foss, A.J.; Miles, C.O.; Samdal, I.A.; Løvberg, K.E.; Wilkins, A.L.; Rise, F.; Jaabæk, J.A.H.; McGowan, P.C.; Aubel, M.T. Analysis of free and metabolized microcystins in samples following a bird mortality event. *Harmful Algae* **2018**, *80*, 117–129. [CrossRef]
39. Sano, T.; Nohara, K.; Shiraishi, F.; Kaya, K. A method for micro-determination of total microcystin content in waterblooms of cyanobacteria (blue-green algae). *Int. J. Environ. Anal. Chem.* **1992**, *49*, 163–170. [CrossRef]
40. Foss, A.J.; Butt, J.; Fuller, S.; Cieslik, K.; Aubel, M.T.; Wertz, T. Nodularin from benthic freshwater periphyton and implications for trophic transfer. *Toxicon* **2017**, *140*, 45–59. [CrossRef]
41. German, A.J.; Holden, S.L.; Moxham, G.L.; Holmes, K.L.; Hackett, R.M.; Rawlings, J.M. A simple, reliable tool for owners to assess the body condition of their dog or cat. *J. Nutr.* **2006**, *136*, 2031S–2033S. [CrossRef]
42. Florida Department of Environmental Protection Algal Bloom Monitoring and Response. Available online: https://floridadep.gov/AlgalBloom (accessed on 1 June 2019).
43. Trevino-Garrison, I.; Dement, J.; Ahmed, F.S.; Haines-Lieber, P.; Langer, T.; Ménager, H.; Neff, J.; Van Der Merwe, D.; Carney, E. Human illnesses and animal deaths associated with freshwater harmful algal blooms—Kansas. *Toxins* **2015**, *7*, 353–366. [CrossRef]
44. Figgatt, M.; Hyde, J.; Dziewulski, D.; Wiegert, E.; Kishbaugh, S.; Zelin, G.; Wilson, L. Harmful Algal Bloom–Associated Illnesses in Humans and Dogs Identified Through a Pilot Surveillance System—New York, 2015. *MMWR. Morb. Mortal. Wkly. Rep.* **2017**, *66*, 1182–1184. [CrossRef]
45. Glibert, P.; Burford, M. Globally Changing Nutrient Loads and Harmful Algal Blooms: Recent Advances, New Paradigms, and Continuing Challenges. *Oceanography* **2017**, *30*, 58–69. [CrossRef]
46. Ndlela, L.L.; Oberholster, P.J.; Van Wyk, J.H.; Cheng, P.H. An overview of cyanobacterial bloom occurrences and research in Africa over the last decade. *Harmful Algae* **2016**, *60*, 11–26. [CrossRef]
47. Paerl, H.W.; Scott, J.T. Throwing fuel on the fire: Synergistic effects of excessive nitrogen inputs and global warming on harmful algal blooms. *Environ. Sci. Technol.* **2010**, 7756–7758. [CrossRef]
48. Oehrle, S.; Rodriguez-Matos, M.; Cartamil, M.; Zavala, C.; Rein, K.S. Toxin composition of the 2016 Microcystis aeruginosa bloom in the St. Lucie Estuary, Florida. *Toxicon* **2017**, *138*, 169–172. [CrossRef]
49. DeVries, S.E.; Galey, F.D.; Namikoshi, M.; Woo, J.C. Clinical and pathologic findings of blue-green algae (*Microcystis aeruginosa*) intoxication in a dog. *J. Vet. Diagn. Invest.* **1993**, *5*, 403–408. [CrossRef]
50. Stoner, R.D.; Adams, W.H.; Slatkin, D.N.; Siegelman, H.W. The effects of single L-amino acid substitutions on the lethal potencies of the microcystins. *Toxicon* **1989**, *27*, 825–828. [CrossRef]
51. Spoof, L.; Błaszczyk, A.; Meriluoto, J.; Cegłowska, M.; Mazur-Marzec, H. Structures and activity of new anabaenopeptins produced by Baltic Sea cyanobacteria. *Mar. Drugs* **2016**, *14*, 8. [CrossRef]
52. Elkobi-Peer, S.; Carmeli, S. New prenylated aeruginosin, microphycin, anabaenopeptin and micropeptin analogues from a Microcystis bloom material collected in Kibbutz Kfar Blum, Israel. *Mar. Drugs* **2015**, *13*, 2347–2375. [CrossRef]

53. Creasia, D.A. Acute inhalation toxicity of microcystin-LR with mice. In Proceedings of the Third Pan-American symposium on animal, plant and microbial toxins, Oaxtepec, Morelos State, Mexico -U.S. Army Medical Research Institute of Infectious Diseases, Frederick, MD, USA, 12 January 1990; p. 605.
54. Wood, S.A.; Dietrich, D.R. Quantitative assessment of aerosolized cyanobacterial toxins at two New Zealand lakes. *J. Environ. Monit.* **2011**, *13*, 1617–1624. [CrossRef]
55. Backer, L.C.; McNeel, S.V.; Barber, T.; Kirkpatrick, B.; Williams, C.; Irvin, M.; Zhou, Y.; Johnson, T.B.; Nierenberg, K.; Aubel, M.; et al. Recreational exposure to microcystins during algal blooms in two California lakes. *Toxicon* **2010**, *55*, 909–921. [CrossRef]
56. Greer, B.; Meneely, J.P.; Elliott, C.T. Uptake and accumulation of Microcystin-LR based on exposure through drinking water: An animal model assessing the human health risk. *Sci. Rep.* **2018**, *8*, 4913. [CrossRef]
57. Ballot, A.; Sandvik, M.; Rundberget, T.; Botha, C.J.; Miles, C.O. Diversity of cyanobacteria and cyanotoxins in Hartbeespoort Dam, South Africa. *Mar. Freshw. Res.* **2014**, *65*, 175–189. [CrossRef]

toxins

Article

Cylindrospermopsin-Microcystin-LR Combinations May Induce Genotoxic and Histopathological Damage in Rats

Leticia Díez-Quijada [1], Concepción Medrano-Padial [1], María Llana-Ruiz-Cabello [1], Giorgiana M. Cătunescu [2], Rosario Moyano [3], Maria A. Risalde [4,5], Ana M. Cameán [1,*] and Ángeles Jos [1]

1 Area of Toxicology, Faculty of Pharmacy, University of Sevilla, Profesor García González n2, 41012 Sevilla, Spain; ldiezquijada@us.es (L.D.-Q.); cmpadial@us.es (C.M.-P.); mllana@us.es (M.L.-R.-C.); angelesjos@us.es (Á.J.)
2 University of Agricultural Sciences and Veterinary Medicine Cluj-Napoca, Calea Mănăștur 3-5, 400372 Cluj-Napoca, Romania; giorgiana.catunescu@usamvcluj.ro
3 Department of Pharmacology, Toxicology and Legal and Forensic Medicine, Faculty of Veterinary Medicine, University of Córdoba, Campus de Rabanales, 14014 Córdoba, Spain; ft1mosam@uco.es
4 Animal Pathology Department. Faculty of Veterinary Medicine, University of Córdoba, Campus Universitario de Rabanales s/n, 14014 Cordoba, Spain; maria.risalde@uco.es
5 Instituto Maimonides de Investigación Biomédica de Córdoba (IMIBIC)-Hospital Universitario Reina Sofía de Córdoba-Universidad de Córdoba, Avenida Menendez Pidal s/n, 14006 Cordoba, Spain
* Correspondence: camean@us.es; Tel.: +34-954-556762

Received: 8 April 2020; Accepted: 23 May 2020; Published: 26 May 2020

Abstract: Cylindrospermopsin (CYN) and microcystins (MC) are cyanotoxins that can occur simultaneously in contaminated water and food. CYN/MC-LR mixtures previously investigated in vitro showed an induction of micronucleus (MN) formation only in the presence of the metabolic fraction S9. When this is the case, the European Food Safety Authority recommends a follow up to in vivo testing. Thus, rats were orally exposed to 7.5 + 75, 23.7 + 237, and 75 + 750 µg CYN/MC-LR/kg body weight (b.w.). The MN test in bone marrow was performed, and the standard and modified comet assays were carried out to measure DNA strand breaks or oxidative DNA damage in stomach, liver, and blood cells. The results revealed an increase in MN formation in bone marrow, at all the assayed doses. However, no DNA strand breaks nor oxidative DNA damage were induced, as shown in the comet assays. The histopathological study indicated alterations only in the highest dose group. Liver was the target organ showing fatty degeneration and necrotic hepatocytes in centrilobular areas, as well as a light mononuclear inflammatory periportal infiltrate. Additionally, the stomach had flaking epithelium and mild necrosis of epithelial cells. Therefore, the combined exposure to cyanotoxins may induce genotoxic and histopathological damage in vivo.

Keywords: in vivo; genotoxicity; cylindrospermopsin; microcystin-LR; micronucleus; comet assay; enzyme-modified comet assay; rats

Key Contribution: CYN/MC-LR combinations induced in rats increases MN formation in bone marrow. No DNA strand breaks were induced and the DNA was not oxidatively damaged as detected by the comet assays.

---

1. Introduction

Climate and nutrient changes are contributing to global eutrophication and global expansion of harmful algal blooms [1], including the proliferation of cyanobacterial blooms [2]. Cyanobacteria are producers of a broad group of secondary metabolites called cyanotoxins [3]. The main human exposure route to cyanotoxins is oral intake, primarily from drinking water. However, the consumption of contaminated food and dietary supplements with cyanotoxins cannot be disregarded, although some of them (such as cylindrospermopsin) have not been found in commercially-available blue-green algal supplements until recently [2,4,5]. The most studied cyanotoxins are microcystins (MCs) and cylindrospermopsin (CYN) as a consequence of their toxicity and wide distribution. The severity of these intoxications depends on several factors, such as their chemical structure, mechanisms of action, and concentrations of cyanotoxins involved [2].

MCs are cyclic heptapeptides that are mainly hepatotoxic [6], including as primary target organs the liver and kidney, as well as several secondary targets such as testes, ovaries, heart, lung, and the CNS; to date 246 MC congeners have been identified [7]. Microcystin-LR (MC-LR) is the most studied MC variant, because of its higher toxicity and wider distribution than other MC congeners [8,9]. Hepatotoxicity of MC-LR is mediated by the presence of organic anion transport polypeptides (OATPs) which are expressed mostly in the liver and are responsible for its uptake into the hepatocytes [10]. MC-LR is a specific inhibitor of protein serine/threonine phosphatases 1 and 2A (PP1, PP2A) [11]. It induces a hyperphosphorylation of cytoskeletal proteins affecting cell morphology and cellular adhesion, leading to necrosis [12,13]. Oxidative stress is another mechanism involved in MC toxicity [14–17]. Therefore, MC-LR could have carcinogenic or genotoxic properties because of its inhibitory action on PP1 and PP2A, acting as a liver tumor-promoter [18–20]. In fact, MC-LR was classified only as a possible human carcinogen (Group 2B) by the International Agency of Research on Cancer (IARC) [21], mainly because of the lack of sufficient evidence for its direct carcinogenicity in both humans and experimental animals [2].

The genotoxicity of pure MC-LR cyanotoxin has been widely studied and reviewed in the scientific literature [20,22]. Globally, contradictory results were reported following in vitro and in vivo experimental systems, and the mechanisms involved are not yet fully understood. Pioneering studies indicated that MC-LR damaged the DNA in the liver of male and female Swiss albino mice dependent upon the assayed dose and on the duration of exposure after intraperitoneal injection (i.p.) [23]. This effect was observed after just a single administration of a LD_{50} dose of MC-LR [6,24]. Likewise, DNA damage was observed in the blood cells of Swiss albino mice after oral exposure to a single dose of MC-LR. Higher damage was found in different organs after i.p. injection [25]. Moreover, MC-LR induced a rapid increase of the amount of DNA in the tail of the comets and increased micronucleus (MN) frequencies in male mice injected i.p. [26] and in Balb/c mice exposed repeatedly to MC-LR for 30 d [27]. By contrast, other studies reported that MC-LR did not induce DNA damage in vivo. Thus, no DNA damage was observed in rat liver after a single intravenous administration (i.v.) of MC-LR [28], and negative results were observed in the MN assay in male transgenic mice after intragastric administration [29], and in male CBA mice after i.p. injection of MC-LR [19].

CYN is a stable tricyclic alkaloid consisting of a guanidine moiety combined with a hydroxymethyluracil [30], whose occurrence in aquatic systems is increasing. This cyanotoxin can be produced by different cyanobacterial genera such as *Aphanizomenon, Cylindrospermopsis, Lyngbya, Oscilliatoria, Raphidiopsis and Umezaki*, but *Cylindrospermopsis raciborskii* is its main producer [31,32]. CYN has cytotoxic activity and its effects on the liver and other organs have been proven [2,33]. The main mechanism of CYN toxicity is the inhibition of both protein and glutathione synthesis [34–36], although oxidative stress is also involved [37]. Several studies have shown that cytochrome P450 enzymes are necessary in the metabolic activation of CYN, and consequently in its toxicity [38,39].

At present, the mechanisms of pro-genotoxicity and potential carcinogenic activity of CYN (not yet classified by the IARC) are still not completely described and, thus, further investigations are needed. To this end, several in vivo studies in rodents were performed to clarify the genotoxicity of CYN. Thus, CYN induced DNA strand breaks (sb) in the liver of mice after i.p. administration [40]. Similarly, DNA damage was observed in the colon of mice after i.p. injection of the toxin, and both in the colon and bone marrow after oral (gavage) administration [41]. Dordevic et al. [42] demonstrated DNA damage (comet assay) in the liver of rats exposed i.p. to CYN and to an extract of *Cylindrospermopsis raciborskii*. Recently, Diez-Quijada et al. [43] exposed rats to pure CYN (7.5–75.0 µg/kg b.w) and performed a battery of assays consisting of MN in bone marrow, as well as the standard and modified comet assays in stomach, liver and blood. The DNA seemed to be damaged only in the bone marrow of rats regardless of concentration. By contrast, neither DNA strand breaks nor oxidative DNA damage was observed in the comet assays in any of the investigated tissues.

Both MC-LR and CYN can be found at the same time in the environment and their simultaneous presence was previously described [44–46]. Thus, the European Food Safety Authority (EFSA) documented and stated the importance of studying the effects of their combined exposure [47]. A few studies dealt with the toxicological profile of CYN and MC-LR combinations. Thus, the in vitro assessments of the potential interactions of CYN and MC-LR are very scarce [48–51], and only two focus on their genotoxicity. Hercog et al. [48] described an induction of DNA sb by MN and comet assays at 24 h after the treatment of HepG2 cells with CYN/MC-LR combinations, but to a lesser extent for only CYN. The genomic instability detected by the cytokinesis block micronucleus assay was, however, comparable to the individual CYN. Recently, Diez-Quijada et al. [49] applied a battery of in vitro tests in several cell lines, including bacterial systems, and they described genotoxic effects only in the MN test, when the metabolic fraction S9 was used. To the best of our knowledge, no in vivo studies have been yet performed to assess the genotoxicity of CYN/MC-LR mixtures and they are necessary to elucidate the contradictory results obtained in vitro.

Thus, according to the recommendations of the EFSA [52], the purpose of this research was to investigate, for the first time, the potential in vivo genotoxicity of CYN/MC-LR combinations in rats, as an experimental model, after oral administration (gavage) of relevant environmental concentrations. A combined MN—standard and modified comet assay was applied. The enzyme-modified comet assay was performed with Endonuclease-III (Endo-III) and Formamidopyrimidine glycosilase (Fpg) enzymes. Bone marrow was the selected tissue for the MN test, Organisation for Economic Co-operation and Development (OECD)474 [53], and stomach, liver OECD 489 [54] and blood cells for the standard and enzyme-modified comet assays. Additionally, the potential histopathological alterations were assessed in stomach and liver.

2. Results

2.1. Micronucleus Assay

This assay was conducted following the recommendations of OECD guideline 474 [53], and it is especially relevant for assessing genotoxicity because, although they may vary among species, factors of in vivo metabolism, pharmacokinetics, and DNA repair processes are active and contribute to the responses. Its purpose is to identify cytogenetic damage which results in the formation of micronuclei (MN) containing lagging chromosome fragments or whole chromosomes. As positive control, ethylmethanesulfonate (EMS) was chosen according to this guideline. Results are measured as the polychromatic erythrocytes (PCE) out of total erythrocytes (normochromatic erythrocytes (NCE) + (PCE)), and the PCE/NCE ratios, which were calculated by counting 500 erythrocytes per animal. An increase in the frequency (%) of micro-nucleated polychromatic erythrocytes (%MN-PCEs) in treated animals is an indication of induced chromosome damage.

The results obtained for the MN test in rats exposed to CYN/MC-LR mixtures are shown in Table 1, with individual data shown in the Supplementary Materials (Table S1). Significant differences versus the negative and solvent control groups were found in the PCE/total erythrocytes and PCE/NCE ratios in male and female rats treated with the highest assessed dose (75 + 750 µg/kg b.w. CYN/MC-LR; ** $p < 0.01$). Treatment with the positive control, ethylmethanesulfonate (EMS), produced similar significant decreases in the PCE/total erythrocytes and PCE/NCE ratios. Furthermore, significant increases in the percentage of MN in immature erythrocytes were observed in all treated groups of both sexes, when compared with the negative and solvent control groups.

Table 1. Micronucleus assays results. Bone marrow cytotoxicity expressed as polychromatic erythrocytes (PCE) out of total erythrocytes (normochromatic erythrocytes (NCE) + PCE), ratio of PCE out of NCE, and the micronuclei induction expressed as % MN-PCE´s. The values are expressed as mean ± SD. Significantly different from the negative and solvent control (** $p < 0.01$).

Groups	Sex	n	Doses	PCE/Total	% MN-PCE´s	PCE/NCE
Negative Control (water)	♂	5		0.49 ± 0.02	0.45 ± 0.64	0.95 ± 0.08
	♀	5		0.49 ± 0.03	0.66 ± 0.32	0.96 ± 0.14
Solvent Control (0.5% MeOH)	♂	5		0.50 ± 0.02	0.83 ± 0.51	0.98 ± 0.08
	♀	5		0.50 ± 0.03	0.58 ± 0.24	1.00 ± 0.11
Positive Control (EMS *)	♂	3	200 mg/kg b.w.	0.34 ± 0.02 **	1.93 ± 0.15 **	0.51 ± 0.05 **
	♀	3		0.36 ± 0.03 **	2.47 ± 0.9 **	0.57 ± 0.08 **
CYN/MC-LR	♂	5	7.5 + 75 µg/kg b.w.	0.51 ± 0.02	1.75 ± 0.42 **	1.03 ± 0.07
	♀	5		0.50 ± 0.03	1.98 ± 0.21 **	1.05 ± 0.13
	♂	5	23.7 + 237 µg/kg b.w.	0.46 ± 0.04	1.83 ± 0.41 **	0.86 ± 0.14
	♀	5		0.47 ± 0.03	2.09 ± 0.15 **	0.88 ± 0.09
	♂	5	75 + 750 µg/kg b.w.	0.37 ± 0.03 **	1.88 ± 0.51 **	0.60 ± 0.08 **
	♀	5		0.34 ± 0.06 **	2.22 ± 0.27 **	0.51 ± 0.12 **

* EMS: ethylmethanesulfonate.

2.2. Standard and Enzyme-Modified Comet Assay

The standard comet assay can detect single and double stranded breaks (SBs), resulting, for example, from direct interactions with DNA, alkali labile sites or as a consequence of transient DNA strand breaks resulting from DNA excision repair [54]. Moreover, the enzyme-modified comet assay allowed for detection of oxidative DNA damage using the enzymes Endonuclease III (EndoIII) and Formamidopyrimidine DNA glycosylase (Fpg), which detect oxidized pyrimidines and purines, respectively. The results are expressed as the DNA content in the tail (% of DNA in the tail), which is the intensity of the comet tail relative to the total intensity. As positive control, ethylmethanesulfonate (EMS) was chosen.

No DNA strand breaks were induced in the standard comet assay at any assessed dose in liver, stomach, and blood cells, with the exception of rats treated with the positive control (EMS) (Figure 1A,B). Furthermore, no increases were observed in the % of DNA in the tail of the comets in liver, stomach and blood cells of both sexes at any exposure assayed after post-treatment with Endo-III (Figure 2). Similarly, no differences were found in the tissues from the treated and control groups after Fpg post-exposure (Figure 3). Significant DNA damage was observed in the studied tissues obtained in all the experiments in the positive control group treated with 200 mg/kg b.w. of EMS, (Figures 1–3). Individual data of alkaline and enzyme-modified comet assays are shown in the Supplementary Materials (Table S1). Summary statistics for the treatment groups (male + female) of both assays are also included in the Supplementary Materials (Table S2).

2.3. Clinical and Histopathological Analysis

No clinical signs of toxicity were detected during the experiment at any assayed dose of CYN/MC-LR combinations. No macroscopic changes were observed after necropsy in the gastric mucosa or liver samples of treated groups of both sexes, whereas the stomach had an increased size in the positive control groups. The relative weight (RW) of liver (excised wet liver weight/animal weight) and stomach collected from the treated animals with cyanotoxins was similar to that of the control groups (see Tables S3 and S4 in Supplementary Materials).

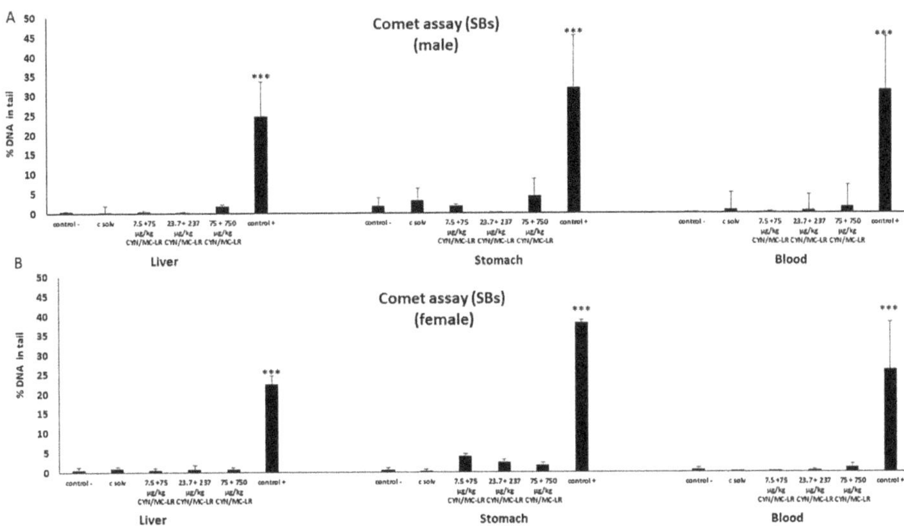

Figure 1. The level of DNA damage in cells isolated from liver, stomach, and blood of male (**A**) and female (**B**) rats exposed to cylindrospermopsin/microcystins (CYN/MC-LR) mixtures as the formation of strand breaks (SBs) detected by the standard comet assay. The levels of DNA strand breaks are expressed as % DNA in the tail of the comets. All values are represented as mean ± SD. Significantly different from control (*** $p < 0.001$).

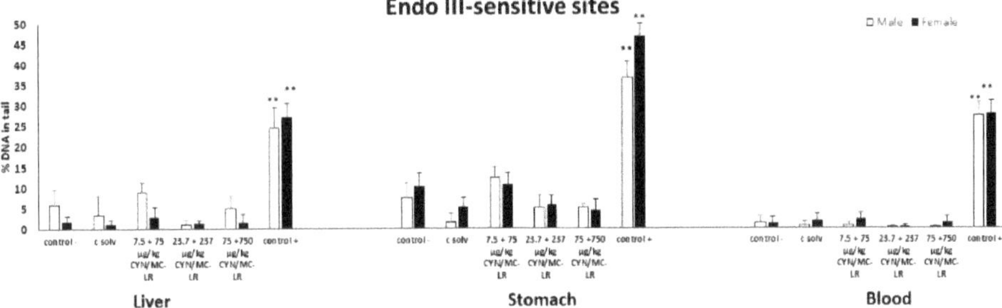

Figure 2. The level of DNA damage in cells isolated from liver, stomach and blood of male and female rats exposed to CYN/MC-LR combinations as the formation of oxidative DNA damage in the form of Endo-III sensitive sites. The levels of oxidized pyrimidines are expressed as % DNA in the tail of the comets. All values are represented as mean ± SD. Significantly different from control (** $p < 0.01$).

The liver and stomach from the negative and solvent control groups showed normal histology without pathological lesions (Figure 4A,B). Degenerate hepatocytes in centrilobular areas together with small, round, and clear vacuoles (microvesicular vacuolation) inside hepatocytes with a diffuse distribution were observed in the liver from rats exposed to medium-lower doses (23.7 + 237 and 7.5 + 75 µg/kg b.w. CYN/MC-LR), (Figure 4C,D); however, they did not reach statistical significance with regard to the negative and solvent controls when quantified (Figure 5).

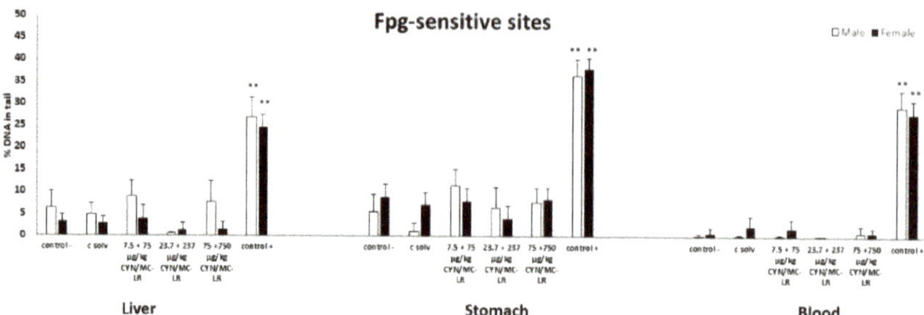

Figure 3. The level of DNA damage in cells isolated from liver, stomach, and blood of male and female rats exposed to CYN/MC-LR mixtures as the formation of oxidative DNA damage in the form of Formamidopyrimidine glycosilase (Fpg)-sensitives sites. The levels of oxidized purines are expressed as % DNA in the tail of the comets. All values are represented as mean ± SD. Significantly different from control (** $p < 0.01$).

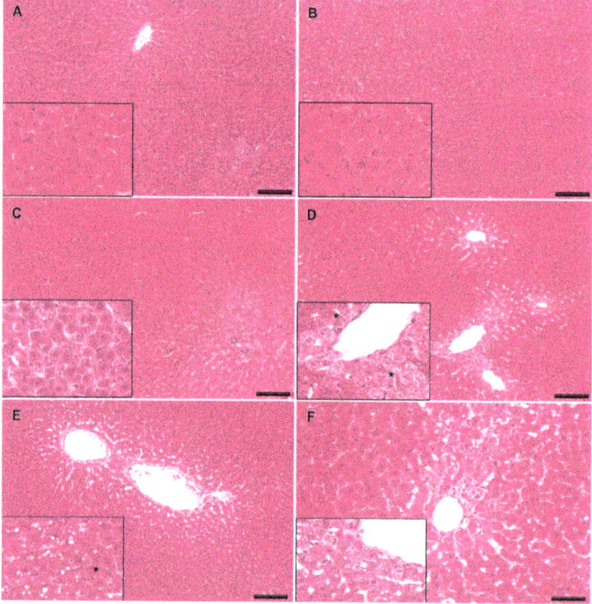

Figure 4. Representative histopathological changes in the liver of rats exposed to CYN/MC-LR. Normal hepatic parenchyma is observed in negative (**A**) and solvent (**B**) control groups. Details of normal hepatocytes are observed in the insets (**A**,**B**). Rats exposed to 7.5 + 75 µg/kg b.w. CYN/MC-LR showed a diffuse distribution of mild degenerate hepatocytes with micro-vesicular lipid vacuolation (**C**). There are details of intracellular accumulation of small, round and clear vacuoles in hepatocytes (**C**, inset). Rats exposed to 23.7 + 237 µg/kg b.w. CYN/MC-LR presented mild degenerate and necrotic hepatocytes in centrilobular areas (**D**). There are details of degenerate hepatocytes, some of them multinucleated (*) (**D**, inset). Rats exposed to 75 + 750 µg/kg b.w. CYN/MC-LR showed moderate degenerate and necrotic hepatocytes in centrilobular areas with macro-vesicular lipid vacuolation (**E**). There are details of intracellular accumulation of large, round, and clear vacuoles inside hepatocytes, some of them multinucleated (*) (**E**, inset). The positive control group showed hepatocyte degeneration with lipid accumulation and centrilobular necrosis (**F**). There are details of the degeneration of hepatocytes with macro-vesicular lipid vacuolation (**F**, inset). Hematoxylin and eosin staining; bars = 100 µm (**A**–**E**) and 50 µm (**F**).

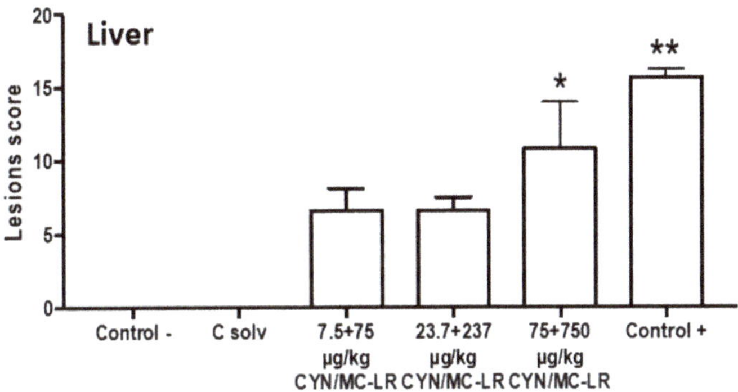

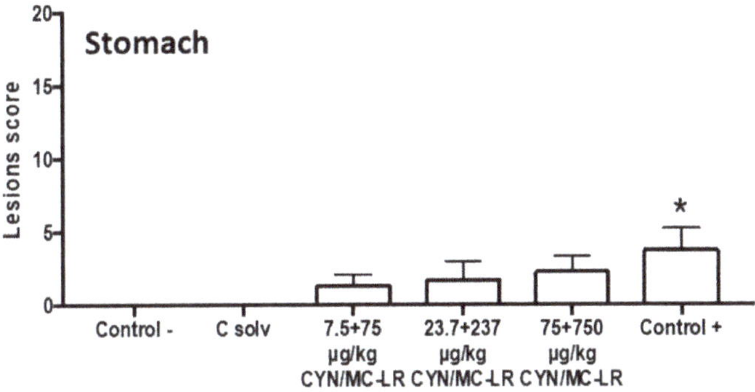

Figure 5. Mean ± standard error (SE) of the histopathological lesion score from liver and stomach of rats exposed to CYN/MC-LR. Significantly different from control (* $p < 0.05$; ** $p < 0.01$).

Similar, but more severe liver lesions than in the lower doses were observed in the highest exposed group (75 + 750 µg/kg b.w. CYN/MC-LR). In this case, degenerate and necrotic hepatocytes in centrilobular areas were observed, as well as ample intracellular lipid accumulation in hepatocytes in the form of round and clear vacuoles (macrovesicular vacuolation). These had a multifocal distribution together with mild presence of multinuclear hepatocytes (Figure 4E). This group showed significant liver damage compared to negative and solvent control groups (Figure 5).

The liver pathology of the positive controls was characterized by severe centrilobular necrosis, hepatocyte degeneration with lipid accumulation, and mild mononuclear inflammatory periportal infiltrates (Figure 4F). This group showed significant differences in the lesion scores compared to negative and solvent control groups (Figure 5).

On the other hand, the main damage in the stomach included flaking epithelium and minimal-mild necrosis of epithelial cells with different severity, depending on the groups. These lesions were moderate in the positive controls accompanied by hyperplasia of the gastric glands, while being mild in the highest exposed group, minimal in the medium-lower doses, and totally absent in negative and solvent controls (Figure 6A–F). Statistical differences were observed only between positive control and the rest of the studied groups (Figure 5).

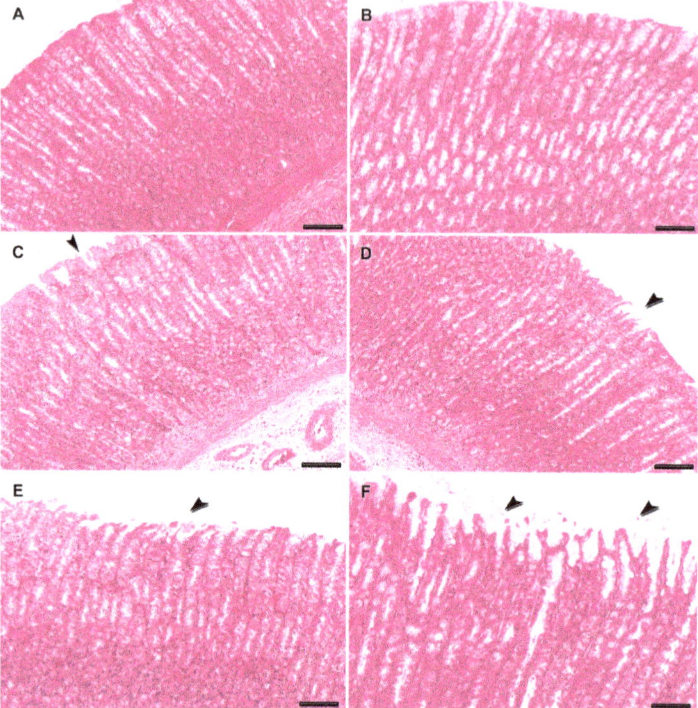

Figure 6. Representative histopathological changes in the stomach of rats exposed to CYN/MC-LR. Normal gastric mucosa is observed in negative (**A**) and solvent (**B**) control groups. Rats exposed to 7.5 + 75 and 23.7 + 237 μg/kg b.w. CYN/MC-LR showed minimal flaking gastric epithelium (arrows) (**C**,**D**, respectively). Rats exposed to 75 + 750 μg/kg b.w. CYN/MC-LR showed a mild flaking gastric epithelium with occasional necrosis of epithelial cells and a mild increase of the secretory epithelium (arrow) (**E**). Positive control group presented a moderate flaking gastric epithelium with mild necrosis of epithelial cells and hyperplasia of the gastric glands (arrows) (**F**). Hematoxylin and eosin staining; bars = 100 μm (**A**,**C**,**D**), 75 μm (**B**,**E**) and 50 μm (**F**). The lesions were independently examined by two experienced assessors: a veterinary pathologist and an investigator (M.A.R. and R.M.) in a single-blinded approach. These findings were evaluated in all the fields of one slide with a section of 1 × 1 cm from each organ studied in all the animals.

3. Discussion

Although the simultaneous occurrence of cyanotoxins such as MCs and CYN is being reported more and more frequently [44,55], toxicological studies focusing on their potential interaction are very scarce [50,56]. Particularly, the genotoxicity of mixtures is of great interest for humans exposed to contaminated waters and food with cyanotoxins, due to the possible carcinogenic effects of MC-LR and pro-genotoxic activity of CYN. Moreover, as EFSA recommends [47], further studies are needed to characterize the hazard for a more realistic and reliable risk assessment.

Only two in vitro studies evaluated the genotoxic effects of CYN/MC-LR mixtures. The first was carried out on the HepG2 cell line exposed to MC-LR (1 μg/mL), CYN (0.01, 0.05, 0.1, and 0.5 μg/mL) and their mixtures by comet and MN assay [48]. The authors indicated that CYN might have a higher genotoxic potential than MC-LR and the genotoxic potential of CYN/MC-LR combination was similar to CYN alone. More recently, the mutagenicity and genotoxicity of 1:10 CYN/MC-LR mixtures were assayed through a complete battery of in vitro tests including the MN test on L5178Y Tk$^{\pm}$ cells and the standard and enzyme-modified comet assays in Caco-2 cells [49]. Genotoxicity was observed only for the combination in the MN test with S9 metabolic fraction, in agreement with the previous reports for only CYN [57]. Together, both in vitro studies suggested the predominance of the pro-genotoxic activity of CYN in the combinations, hence the necessity to evaluate in vivo the genotoxicity of CYN/MC-LR mixtures by application of international guidelines such as the OECD guidelines.

This is the first in vivo study which reported the genotoxicity of CYN/MC-LR in rats orally exposed (by gavage). It combines and relates the results of the two types of comet assays (standard and enzyme-modified) on cells isolated from stomach, liver, and blood, with the MN test in bone marrow cells. The assays were performed according to the OECD 489 and 474 guidelines, respectively [53,54] with some alterations, as described by Diez-Quijada et al. [43]. The combined comet-MN assay [58] reduces the use of animals according to the 3Rs principles (Replace, Reduce, and Refine), and it increases the sensitivity and specificity of the assays, decreasing the number of false negative results [59]. Moreover, in this case, the use of DNA repair enzymes (Endo III and Fpg) was also included, increasing the sensitivity of the in vivo comet assay [60,61].

In this study, the combined MN-comet assay was performed in both sexes, because differences in toxicity linked to gender have previously been described. This is in agreement with the OECD 489 [54] that encourages the use of both sexes when relevant differences between males and females are reported (e.g. differences in systemic toxicity, metabolism, bioavailability, etc. including those in a range-finding study). In this sense, although no sex-related differences were observed in rats after MC-LR i.p. exposure [62], male mice were more sensitive than females to CYN administered i.p. [63]. CYN exposure doses (75, 23.7, and 7.5 μg/kg b.w.) were selected according to a previous in vivo study performed with the individual toxin [43]. The doses of MC-LR were 10 times higher than CYN because of its proportionally higher abundance in the environment [64]. Overall in this study, using rats as experimental model, the CYN/MC-LR combinations increased the % MN-PCE in bone marrow cells, although it induced no DNA strand breaks nor any oxidative DNA damage in any of the investigated tissues. These results are in accordance with the only two previous genotoxicity in vivo assay carried out in rats with pure individual toxins: CYN alone under the same combined assays [43] and MC-LR administered i.v. through the comet assay [28]. Generally, the results of this study show no synergism or antagonism between CYN and MC-LR in the assayed combinations and concentrations in rats, because the effects are similar to the in vivo exposure to CYN alone [43]. Moreover, these findings confirm the in vitro experiments performed with the CYN/MC-LR mixture through a battery of tests, in which MN and comet assays were included [49].

In the present study, the significant decreases in the PCE/total erythrocytes and PCE/NCE ratios in the highest dose groups (75 + 750 μg/kg b.w. CYN/MC-LR) in comparison to the controls were consistent with results observed for CYN tested individually at the same doses (7.5–75.0 μg/kg b.w.) [43]. Previously, in the case of MC-LR no in vivo effects were reported for polychromatic erythrocytes (PCE) from peripheral blood of male mice after i.p. administration of the toxin (0–55 μg/kg b.w.) nor induction of MN [19].

Moreover, the induction of MN by the CYN/MC-LR mixture in rats contrasts with the contradictory results obtained in previous assays performed in mice as experimental model. Thus, negative results were reported in the colonic cells of mice orally exposed (1–4 mg/kg b.w.) and in bone marrow after i.v. administration [41]. In contrast, positive results were found in the blood cells of mice exposed i.p. to 37.5 μg/kg b.w. MC-LR, with significant increases in the frequency of MN at 48–72 h after treatment [26]. Additionally, only a rapid and temporary two-fold increase had been detected in the amounts of DNA in the tail of the comets after 30 min of MC-LR exposure was detected in the comet assay. The authors suggested that although MC-LR induced DNA damage, the leukocytes might repair the lesions, prior to the genotoxicity assessment by the comet assay. They concluded that the MN assay could be more sensitive than the comet assay to evaluate the genotoxicity of MC-LR. This hypothesis could also explain the results obtained in the present work. Other studies showed a significant dose-dependent increase in MN frequency of bone marrow cells of Balb/C mice i.p. injected with MC-LR (0.5–8 μg/kg b.w.) every 48 h for 30 d [27]. In contrast, an in vivo negative MN induction was found in CBA mice exposed i.p. to pure MC-LR [19]. Moreover, MC-LR did not show mutagenicity (in lung and liver), and it did not induce gene mutation nor MN in peripheral blood cells of transgenic λ/lacZ mice at 48 h after exposure to 1 mg/kg b.w. MC-LR [29]. The authors suggested that the tumor-promoting effect of MC-LR is independent of its interactions with DNA.

Overall, the in vivo induction of MN by the CYN/MC-LR combinations observed in this study confirms the genotoxicity demonstrated previously in vitro, and corroborates the reported in vitro pro-genotoxic activity of CYN [39,57,65–67], and to a lesser degree of MC-LR [26,68]. Moreover, previous in vivo studies resulted in similar effects for CYN whereas the diversity of results obtained for MC-LR could be attributed to different animal species, exposure routes, and doses used. Our results were obtained on rats, the experimental model recommended by the OECD guidelines on genotoxicity [53,54] and toxicity studies; and they clarify previous contradictory findings reported in mice.

The application of the alkaline comet assay in stomach, liver, and blood cells is a good complement to the bone-marrow and peripheral blood MN test [58] because this combination allows assessment of the DNA damage in various potential target tissues (site of contact, metabolism and peripheral distribution) and it can detect multi-endpoint genotoxic effects [59]. In the present study, the absence of effects is in consensus with a previous in vivo combined study carried out with CYN alone in male rats under similar experimental conditions [43].

In contrast, to the best of our knowledge, the few in vivo comet assays available revealed positive results [40–42]. However, it must be emphasized that CYN was administered i.p., and this could influence the higher intensity of the results obtained: induction of DNA sb in the liver of mice [40], in the colon of mice [41], and in the liver of rats [42]. Nonetheless, the oral route is the most representative for human exposure to cyanotoxins [2,5,33]. In this sense, only Bazin et al. [41] found DNA damage in colon and bone marrow samples from mice exposed by gavage to CYN in a range of 1–4 mg/kg b.w., doses much higher than in this study.

Previous in vivo studies reported contradictory results for individual and pure MC-LR from the alkaline comet assay. Thus, there were DNA lesions in the blood cells (strand breaks, labile sites, etc.) of mice at 3 h after oral exposure to MC-LR (4 mg/kg b.w.). However, DNA lesions were observed mostly in the liver after i.p. administration, although they were also induced in the kidney, intestines, and colon [25]. The authors concluded that the DNA damage induced in the organs was probably due to oxidation, and it could be attributed to the more sensitive i.p. route of administration. Similarly, the DNA breaks were reported in blood cells of mice exposed i.p. to MC-LR (37.5 μg /kg b.w. MC-LR) [26]. In contrast, MC-LR did not induce DNA damage in rat hepatocytes at 2–4 h or 12–16 h after exposure to a single sublethal doses (ranging between 12.5–50 μg/kg b.w.) administered i.v. [28]. The authors suggested that the administration route influences the results from the in vivo comet assay and changes the MC-LR kinetic parameters. Further in vivo combined MN-comet assay for MC-LR should be carried out in rats, in order to know the genotoxic profile of this toxin.

Moreover, the oxidative stress could be one of the genotoxic mechanisms of CYN and MC cyanotoxins [20]. Thus, the modified-comet assay was also performed in this study. No oxidative damage was observed in the pyrimidine and purine bases of DNA, in agreement with the negative results found in the in vivo experiment performed at the same doses with CYN alone in male rats [43]. Moreover, it corroborates the results reported in vitro on Caco2 cells exposed to CYN/MC-LR mixtures [49], and the scarce results obtained from in vitro enzyme-modified comet assay for CYN in HepG2 cells [69] and in Caco-2 cells [57]. This suggests that oxidative stress could have a minor role in the CYN mediated genotoxicity [22,69].

In contrast to CYN, MC-LR tested in vitro showed increased DNA strand breaks by oxidation of pyrimidine and purine bases in HepG2 cells [16,70], providing evidence that the toxin induced strand breaks from excision of oxidative DNA adducts. Thus, reactive oxygen species (ROS) are involved in this type of DNA damage. The same authors confirmed that MC-LR displayed oxidation of purine bases in human peripheral blood lymphocytes when DNA was digested with purified Fpg. Moreover, oxidative stress-responsive genes were up-regulated at the molecular level after 24 h, supporting the hypothesis that MC-LR is an indirect genotoxic agent, acting via induction of oxidative stress [71].

The discrepancy between the negative results of the present study on the DNA oxidative damage induced by the mixtures and previous reports on individual MC-LR suggests that the genotoxic activity of CYN rather than MC-LR is responsible of the effects of CYN/MC-LR mixtures in rats. In fact, Hercog et al. [48] who studied CYN/MC-LR combinations in HepG2 cells concluded that MC-LR did not seem to deregulate the investigated genes, and they suggested the higher pro-genotoxic potential of CYN. However, MC-LR was classified in the group 2B by the IARC [21] due to its tumor promotion mechanism. Thus, caution is required when assessing its toxicity when in the mixture.

The liver was the main target organ of the CYN/MC-LR mixtures as showed by the histopathologic evaluation. It was also the most severely affected organ in MC-exposed fish [72], rats [73,74], and mice [75,76]. The major findings of oral toxicity in the present experiment were fatty generation and necrotic hepatocytes in centrilobular areas, as well as a light mononuclear inflammatory periportal infiltrate, specially noted in the highest doses of toxins (75 + 750 µg/kg b.w. CYN/MC-LR). This degeneration and the necrotic processes of hepatocytes were similar to those reported in experimental administrations of MC-LR in mice [75,77,78] and rats [74]. Thus, our results indicate that MC-LR is undeniably incorporated into the liver, resulting in the characteristic hepatotoxicity and confirming the sensitivity of this organ to the oral administration of MC-LR. Further studies with additional inflammatory and hepatotoxic markers (gene and/or protein expression) should be performed to support the histopathological analysis. These in vivo results are in accord with other in vivo and in vitro studies where MC-LR induced liver injury through the production of ROS among other mechanisms [74,79,80]. Hepatocyte injury was also observed after CYN exposure in rats [42,43,81] and mice [41,63,82,83], with its severity increasing with the dose [37,63,82–84]. On the other hand, a variable number of multinucleated hepatocytes was observed in the high and medium-dosed groups (23.7 + 237 or 75 + 750 µg/kg b.w. CYN/MC-LR) implying the presence of liver injuries. However, this histological change was not observed in the low-dosed group. An increase in the mitotic frequency was also described in rats exposed only to the highest CYN dose (75 µg/kg b.w.) [43]. Nevertheless, other liver injuries associated with MC-LR, such as fibrosis, were not observed upon histological examination. MC-LR is known to induce liver fibrosis in pre-clinical models and in people after acute exposure to MC [14,85,86]. However, these findings are normally associated with a more severe liver damage and advanced disease [74].

In the present study, mild histopathological findings were recorded in the stomach of all exposed groups. In contrast, significantly altered gastric mucus secretions were found previously after oral administration of MC-LR in mice [87] or pure CYN in rats [43]. These lesions were associated with an irritant effect of the toxins since the gastric mucus is the first line of defense against luminal irritants [43,88]. However, the discrepancies could be associated with the age of the animals [87] or the time of exposure, which was longer in the study of Diez-Quijada et al. [43].

4. Conclusions

The study performed provided evidence of an enhancement of MN formation in the bone marrow of rats subsequent to an oral exposure to CYN/MC-LR combinations in the range 7.5 + 75 to 75 + 750 µg/kg b.w. However, no genotoxic damage was observed in other organs such as liver, stomach, and blood as evaluated by the standard and enzyme-modified comet assay. Therefore, the combined exposure to cyanotoxins may induce genotoxic damage in vivo, although there is no evidence of synergistic or additive effects due to their combination. Histopathological lesions were observed mainly in the liver, in agreement with the well-known hepatotoxicity of these cyanotoxins. These results support the demand for further confirmatory studies needed for a thorough risk assessment of cyanotoxins and their mixtures, and consequently for possible risk management measures limiting human exposure if required.

5. Materials and Methods

5.1. Chemicals and Reagents

Microcystin-LR standard (99% purity) and Cylindrospermopsin standard (95% purity) were acquired from Alexis Corporation (Lausen, Switzerland). All assay chemicals were obtained from C-Viral S.L. (Seville, Spain), Gibco (Biomol, Seville, Spain), Moltox (Trinova, Biochem, Germany) and Sigma-Aldrich (Madrid, Spain).

5.2. Animal Housing and Feeding Conditions

The Ethics Committee on Animal Experimentation of the University of Sevilla approved this in vivo experiment (09/03/2016/027). In addition, all animals received care following the Directive 2010/63/UE for the protection of animals used for scientific purposes.

Male and female Wistar rats, strain RjHan:WI (type outbred rats), between seven and eight weeks old were purchased from Animal Production and Experimentation Service (SEPA, University of Cádiz). Animals were weighed on the arrival day (weight variation did not exceed ± 20%) and housed into polycarbonate cages with stainless steel covers. Afterwards, the animals were fed during one week before the experiments with standard laboratory diet (Harlan, 2014; Harlan Laboratories, Barcelona, Spain) and water *ad libitum*. During this time, animals were acclimatized to the environmental conditions with a 12 h dark/light cycle, temperature 23 ± 1 °C, relative humidity (55 ± 10)%, and free from any type of chemical contamination.

5.3. Experimental Design and Treatment

The treatment doses of CYN were selected based on our previous study of genotoxicity following oral administration of 75 µg CYN /kg b.w. for three days [43]. The chosen concentrations of MC-LR were 10 times higher than CYN because MC are proportionally more abundant in nature than CYN [2,89]. Even though occurrence data show that MC-RR and other minority MCs are distributed worldwide becoming sometimes predominant, the congener MC-LR is widely distributed and the main focus of toxicological studies [8,64]. The ratio CYN/MC concentrations could oscillate between countries and continents, climatic conditions, or composition of the cyanobacteria communities [44,45], and in this study the ratio 1:10 CYN/MC-LR was chosen to compare or confirm previous toxicity studies in which combinations of CYN/MC-LR were assayed [49–51]. Furthermore, the International Conference on Harmonisation (ICH) S2 guidelines [90] and the OECD 474 [53] and OECD 489 [54] guidelines for the MN and Comet assay, respectively, recommend for the combined MN-comet assay the use of the highest dose and two additional lower doses [58] appropriately separated by less than $\sqrt{10}$ to prove dose-related responses [54]. Thus, increasing concentrations of CYN/MC-LR mixtures were selected: 7.5 + 75, 23.7 + 237, and 75 + 750 µg/kg b.w. CYN/MC-LR, respectively, according to Diez-Quijada et al. [43]. These doses are not only experimentally relevant (according to the OECD guidelines), but also environmentally significant, especially the lower ones. Thus, lower doses of 23.7 or 7.5 µg CYN/kg, b.w. are very relevant, being the equivalent to an exposure of 3.4 µg or 1.0 µg CYN/rat/day, respectively.

In this study, 28 male and 28 female rats were randomly divided into six groups, three controls and three treatment groups: the negative control group (C-) (five male and five female rats) administered with water by gavage; the solvent control group (C solv) (five male and five female rats) treated with 0.5% Methanol (MeOH); the positive control group (C+) (three males and three females rats) exposed to 200 mg/kg b.w. ethylmethanesulfonate (EMS). The three exposed groups (five males and five female rats per group) were treated with 7.5 + 75, 23.7 + 237, or 75 + 750 µg/kg b.w. CYN/MC-LR. All doses were prepared from a concentrate stock solution to a final volume of 1 mL with 0.5% MeOH. Although some organic solvents may affect the activity of the cytochrome P450, a 0.5% concentration of MeOH is lower than the 2% indicated to impact the activity of CYP450s [91] The number of animals included in each group was based on the OECD 474 [53] and OECD 489 [54] guidelines. Both indicate five animals of one sex per group, or five of each if both sexes are used, as an adequate number of rats. The OECD 489 [54] permits the use of a minimum of three animals of one sex, or three of each if both sexes are used, treated with a positive control. The animals for combined MN and comet assay need to be dosed at 0, 24, and 45 h, and sacrificed at 3 h after the last administration [58]. The rats were treated by gavage using an enteral feeding tube (Vygon, Ecouen, France). Clinical signs (e.g. mobility, activity, posture, blood around nose and eyes, dyspnea and piloerection) and body weight were recorded during the exposure period.

5.4. Sample Collection

The liver and stomach were extracted, dissected, washed with cold saline solution, and weighed. Sections of each were rapidly processed for the standard and enzyme modified comet assay as explained in Section 5.6. Furthermore, blood samples were collected and conserved in Vacutainer®sodium Heparin Tubes (Becton Dickinson, Rutherford, NJ, USA). Samples were collected from the bone marrow of both femurs of each animal for the MN assay and immediately spread on slides. The smear was then allowed to air dry, fixed with absolute methanol and stained with 10% Giemsa. Sections of liver and stomach were processed according to Diez-Quijada et al. [43] to investigate potential histopathological changes.

5.5. Micronucleus Assay

For this assay, two smeared glass slides (one per femur of each animal) were prepared with the bone marrow cells re-suspended in a drop of fetal bovine serum. After allowing the smear to air-dry, it was fixed in absolute methanol for five minutes and then air-dried and stained with 10% Giemsa for 10 min. The polychromatic erythrocytes (PCE)/total erythrocytes (normochromatic erythrocytes (NCE) + (PCE)) ratio and the PCE/NCE ratio were calculated by counting 500 erythrocytes per animal. The frequency of micro-nucleated immature erythrocytes (MNPCE) was expressed as % MN-PCE´s and it was determined by counting a total of 5000 PCE per animal, following Diez-Quijada et al. [43].

5.6. Standard and Enzyme-Modified Comet Assay

Single cell suspensions from liver and stomach were isolated according to Corcuera et al. [92] and Diez-Quijada et al. [43] for the standard and enzyme-modified comet assay. Liver and stomach were quickly rinsed with Merchant´s buffer (MB) (0.14 M NaCl, 1.47 nM KH_2PO_4, 2.7 mM KCl, 8.1 mM Na_2HPO_4, 10 mM Na_2EDTA, with pH 7.4). Then, a section of each tissue was homogenized in the cold by immersing it in an ice-filled beaker, and the homogenates were centrifuged, filtered, and mixed with 5 mL MB buffer before slide preparation. Heparinized blood samples were mixed with phosphate buffered saline solution (PBS) v/v (1/1), and afterwards, the lymphocytes were isolated with Histopaque® (Sigma-Aldrich, Madrid, Spain) and centrifuged (30 min, 400 G). The cells were washed with PBS twice and re-suspended at a concentration of 2×10^5 cells/mL in PBS.

Thirty μL of blood cell suspensions were mixed with 140 μL pre-warmed 0.5% low-melting point agarose, and 12 drops of 5 μL of each cell suspension were placed on microscope slides pre-coated with agarose. Cell suspensions of liver and stomach were mixed with 1% low-melting point agarose, and the mixtures were placed on microscope slides precoated with agarose, similar to blood samples. The standard and enzyme-modified comet assays were carried out as previously described by Diez-Quijada et al. [43]. The slides were cleaned up three times for 5 min with enzyme buffer (40 mM HEPES; 0.1 M KCl; 0.5 mM EDTA; 0.2 mg/mL bovine serum albumin; pH 8) subsequent to lysis at 4 °C. Later, two gels in each slide were exposed successively to 30 μL of each of the following: lysis solution; enzyme buffer alone (buffer F); buffer F containing Endo III; and buffer F containing Fpg in a metal box at 37 °C for 30 min. Then, the nuclei were denatured by electrophoresis carried out for 20 min, 0.81 V/cm up to 400 mA. The DNA was neutralized in PBS, washed with water and fixed with 70% and absolute ethanol. Finally, once the slides were dried, nuclei were stained with SYBR Gold and visualized with an Olympus BX61 fluorescence microscope coupled via a CCD camera to an image-analysis system (DP controller-DP manager). Images of at least 150 randomly selected nuclei per animal were analyzed with the image analysis software Comet Assay IV (Perceptive Instruments, UK). Percentages of tail DNA (% DNA in tail), automatically obtained by the software, were used to describe each of the nuclei/comets analyzed and the medians of the scored comets were obtained to describe each animal.

Endo III and Fpg sensitive sites were determined by subtracting the % of DNA in tail after repair enzymes incubation.

5.7. Histopathological Analysis

Tissue samples from liver and stomach were fixed in 10% phosphate-buffered formalin for 24 h, and then immediately dehydrated in ethanol, immersed in xylol, and embedded in paraffin wax employing an automatic processor. Sections of 4 μm were stained with hematoxylin and eosin and examined microscopically, with a Modular Microscopy BX43 (Olympus, Shinjuku, Tokyo, Japan). For the histopathological study, males and females were evaluated and both presented similar lesions, but the lesion score was performed only in males as they have been reported to be more sensitive to cylindrospermopsin effects [43,63]. A semiquantitative evaluation of the severity of lesions was scored. These were independently examined by two blinded and experienced observers, a veterinary pathologist and an investigator (M.A.R. and R.M.), in all the fields of one slide with a section of 1 × 1 cm from each organ studied in all the animals. The lesions scored in the liver were multinucleated hepatocytes, hepatocyte degeneration with lipid accumulation, hepatocellular necrosis, and mononuclear inflammatory periportal infiltrate, while those scored in stomach were flaking epithelium, necrosis of epithelial cells and hyperplasia of the gastric glands. Pathology scores were as follows: 0, no significant lesions (0%); 1, minimal (<10%); 2, mild (11–25%); 3, moderate (26–50%); 4, marked (51–75%); 5, severe (>75%).

5.8. Statistical Analysis

The results of the MN test are expressed as mean ± standard deviation (SD) for each group of animals, and a statistical analysis was carried out using the analysis of variance (ANOVA) followed by Dunnett's multiple comparison test. The results of the standard and enzyme-modified comet assays were calculated for each group as mean ± SD of the medians. The distribution of the results was verified for normality utilizing the Kolmogorov- Smirnov test and total scores of the different groups were compared using the non-parametric Kruskal-Wallis test followed by Dunn's multiple comparison test. Analyses were conducted using Graph-Pad InStat software (Graph-Pad Software Inc., La Jolla, San Diego, CA, USA).

Supplementary Materials: The following are available online at http://www.mdpi.com/2072-6651/12/6/348/s1, Table S1: Measurements (MN, alkaline comet and enzyme-modified comet assays) obtained in the target tissues (liver, stomach and blood) by dose and treatment assayed in male and female rats, Table S2: Summary statistic for treatment groups including both sexes (male and female) for the standard and enzyme-modified comet assay in liver, stomach and blood cells, Table S3: Relative weight (RW) (excised wet organ weight /animal weight) of liver and stomach from exposed male rats, Table S4: Relative weight (RW) (excised wet organ weight /animal weight) of liver and stomach from exposed female rats.

Author Contributions: Conceptualization, design of the study, A.M.C. and Á.J.; methodology, L.D.-Q., C.M.-P., M.L.-R.-C. and G.M.C.; software L.D.-Q., C.M.-P., M.L.-R.-C. and G.M.C.; formal analysis, L.D.-Q., C.M.-P., M.L-R.-C. and G.M.C.; investigation, L.D.-Q., C.M.-P., M.L.-R.-C., G.M.C., M.A.R., and R.M.; writing—original draft preparation, L-D.-Q., A.M.C., Á.J. and M.A.R.; writing—review and editing, A.M.C., Á.J, M.A.R. and G.M.C.; visualization, L.D.-Q., C.M.-P., M.L-R.-C., G.M.C., R.M. and M.A.R.; supervision, A.M.C. and Á.J.; project administration, A.M.C. and Á.J.; funding acquisition, A.M.C. and Á.J. All authors have read and agreed to the published version of the manuscript.

Funding: This research was funded by the SPANISH MINISTERIO DE ECONOMÍA Y COMPETITIVIDAD (AGL2015-64558-R, MINECO/FEDER, EU), and the grant FPI (BES-2016-078773) awarded to Leticia Díez-Quijada Jiménez. Also, María Llana-Ruiz-Cabello is grateful to the Junta de Andalucía for her postdoctoral grant associated to AGR7252 project and Giorgiana M. Cătunescu gratefully acknowledges the EFSA funding under the EU-FORA Programme for her contribution in the present work.

Acknowledgments: Spanish Ministerio de Economía y Competitividad for the project AGL2015-64558-R, (MINECO/FEDER, EU); and for the FPI grant number BES-2016-078773 awarded to Leticia Díez-Quijada Jiménez. Junta de Andalucía for the postdoctoral grant awarded to María Llana-Ruiz-Cabello associated to AGR7252 project, and the EFSA funding under the EU-FORA Programme of Giorgiana M. Cătunescu.

Conflicts of Interest: The authors declare no conflict of interest.

References

1. Glibert, P.M. Harmful algae at the complex nexus of eutrophication and climate change. *Harmful Algae* **2020**, *91*, 101583. [CrossRef] [PubMed]
2. Buratti, F.M.; Manganelli, M.; Vichi, S.; Stefanelli, M.; Scardala, S.; Testai, E.; Funari, E. Cyanotoxins: Producing organisms, occurrence, toxicity, mechanism of action and human health toxicological risk evaluation. *Arch. Toxicol.* **2017**, *91*, 1049–1130. [CrossRef] [PubMed]
3. Carmichael, W.W. Cyanobacteria secondary metabolites – the cyanotoxins. *J. Appl. Bacteriol.* **1992**, *72*, 445–459. [CrossRef]
4. Environmental Protection Agency. Health Effects Support Document for the Cyanobacterial Toxin Cylindrospermopsin. 2015. Available online: https://www.epa.gov/sites/production/files/2017-06/documents/cylindrospermopsin-support-report-2015.pdf (accessed on 4 May 2020).
5. Gutiérrez-Praena, D.; Jos, Á.; Pichardo, S.; Moreno, I.M.; Cameán, A.M. Presence and bioaccumulation of microcystins and cylindrospermopsin in food and the effectiveness of some cooking techniques at decreasing their concentrations: A review. *Food Chem. Toxicol.* **2013**, *53*, 139–152. [CrossRef]
6. Gupta, N.; Pant, S.C.; Vijayaraghavan, R.; Lakshmana Rao, P.V. Comparative toxicity evaluation of cyanobacterial cyclic peptide toxin microcystin variants (LR, RR, YR) in mice. *Toxicology* **2003**, *188*, 285–296. [CrossRef]
7. Spoof, L.; Catherine, A. Appendix 3: Tables of microcystins and nodularins. *Handb. Cyanobact. Monit. Cyanotoxin Anal.* **2017**, 526–537. [CrossRef]
8. Diez-Quijada, L.; Prieto, A.I.; Guzmán-Guillen, R.; Jos, Á.; Cameán, A.M. Occurrence and toxicity of microcystin congeners other than MC-LR and MC-RR: A review. *Food Chem. Toxicol.* **2019**, *125*, 106–132. [CrossRef]
9. Puerto, M.; Pichardo, S.; Jos, Á.; Cameán, A.M. Comparison of the toxicity induced by microcystin-RR and microcystin-YR in differentiated and undifferentiated Caco-2 cells. *Toxicon* **2009**, *54*, 161–169. [CrossRef] [PubMed]
10. Fischer, W.J.; Altheimer, S.; Cattori, V.; Meier, P.J.; Dietrich, D.R.; Hagenbuch, B. Organic anion transporting polypeptides expressed in liver and brain mediate uptake of microcystin. *Toxicol. Appl. Pharmacol.* **2005**, *203*, 257–263. [CrossRef]
11. MacKintosh, C.; Beattie, K.A.; Klumpp, S.; Cohen, P.; Codd, G.A. Cyanobacterial microcystin-LR is a potent and specific inhibitor of protein phosphatases 1 and 2A from both mammals and higher plants. *FEBS Lett.* **1990**, *264*, 187–192. [CrossRef]
12. Falconer, I.R.; Yeung, D.S.K. Cytoskeletal changes in hepatocytes induced by Microcystis toxins and their relation to hyperphosphorylation of cell proteins. *Chem. Biol. Interact.* **1992**, *81*, 181–196. [CrossRef]
13. Chorus, I.; Bartram, J. *Toxic Cyanobacteria in Water: A Guide to Their Public Health Consequences, Monitoring and Management*; Chorus, I., Bertram, J., Eds.; E & FN Spon: London, UK, 1999; pp. 1–400. ISBN 0-419-23930-8.
14. Guzman, R.E.; Solter, P.F. Hepatic oxidative stress following prolonged subletal Microcystin-LR exposure. *Toxicol. Pathol.* **1999**, *27*, 582–588. [CrossRef] [PubMed]
15. Ding, W.-X.; Shen, H.-M.; Ong, C.-N. Critical role of reactive oxygen species formation in microcystin-induced cytoskeleton disruption in primary cultured hepatocytes. *J. Toxicol. Environ. Health A* **2001**, *64*, 507–519. [CrossRef] [PubMed]
16. Žegura, B.; Lah, T.T.; Filipič, M. The role of reactive oxygen species in microcystin-LR induced DNA damage. *Toxicology* **2004**, *200*, 59–68. [CrossRef] [PubMed]

17. Puerto, M.; Pichardo, S.; Jos, Á.; Prieto, A.I.; Sevilla, E.; Frías, J.E.; Cameán, A.M. Differential oxidative stress response to pure Microcystin-LR and Microcystin-containing and non-containing cyanobacterial crude extracts on Caco-2 cells. *Toxicon* **2010**, *55*, 514–522. [CrossRef]
18. Nishiwaki-Matsushima, R.; Ohta, T.; Nishiwaki, S.; Suganuma, M.; Kohyama, K.; Ishikawa, T.; Carmichael, W.W.; Fujiki, H. Liver tumor promotion by the cyanobacterial cyclic peptide toxin microcystin-LR. *J. Cancer Res. Clin. Oncol.* **1992**, *118*, 420–424. [CrossRef]
19. Abramsson-Zetterberg, L.; Sundh, U.B.; Mattsson, R. Cyanobacterial extracts and microcystin-LR are inactive in the micronucleus assay in vivo and in vitro. *Mutat. Res.* **2010**, *699*, 5–10. [CrossRef]
20. Žegura, B. An overview of the mechanisms of microcystin-LR genotoxicity and potential carcinogenicity. *Mini Rev. Med. Chem.* **2016**, *16*, 1042–1062. [CrossRef]
21. IARC Working Group on the Evaluation of Carcinogenic Risks to Humans. Ingested Nitrate and Nitrite, and Cyanobacterial Peptide Toxins. 2010; World Health Organization, International Agency for Research on Cancer. Available online: https://www.ncbi.nlm.nih.gov/books/NBK326544/pdf/Bookshelf_NBK326544.pdf (accessed on 6 April 2020).
22. Žegura, B.; Štraser, A.; Filipič, M. Genotoxicity and potential carcinogenicity of cyanobacterial toxins – a review. *Mutat. Res.-Rev. Mutat. Res.* **2011**, *727*, 16–41. [CrossRef]
23. Rao, P.V.L.; Bhattacharya, R. The cyanobacterial toxin Microcystin-LR induced DNA damage in mouse liver in vivo. *Toxicology* **1996**, *114*, 29–36. [CrossRef]
24. Rao, P.V.L.; Gupta, N.; Jayaraj, R.; Bhaskar, A.S.B.; Jatav, P.C. Age-dependent effects on biochemical variables and toxicity induced by cyclic peptide toxin microcystin-LR in mice. *Comp. Biochem. Physiol. C Toxicol. Pharmacol.* **2005**, *140*, 11–19. [CrossRef] [PubMed]
25. Gaudin, J.; Huet, S.; Jarry, G.; Fessard, V. in vivo DNA damage induced by the cyanotoxin microcystin-LR: Comparison of intra-peritoneal and oral administrations by use of the comet assay. *Mutat. Res.* **2008**, *652*, 65–71. [CrossRef] [PubMed]
26. Dias, E.; Louro, H.; Pinto, M.; Santos, T.; Antunes, S.; Pereira, P.; Silva, M.J. Genotoxicity of microcystin-LR in in vitro and in vivo experimental models. *BioMed. Res. Int.* **2014**, *2014*, 1–9. [CrossRef]
27. Zhou, W.; Zhang, X.; Xie, P.; Liang, H.; Zhang, X. The suppression of hematopoiesis function in Balb/c mice induced by prolonged exposure of microcystin-LR. *Toxicol. Lett.* **2013**, *219*, 194–201. [CrossRef]
28. Gaudin, J.; Le Hegarat, L.; Nesslany, F.; Marzin, D.; Fessard, V. In vivo genotoxic potential of microcystin-LR: A cyanobacterial toxin, investigated both by the unscheduled DNA synthesis (UDS) and the comet assays after intravenous administration. *Environ. Toxicol.* **2009**, *24*, 200–209. [CrossRef]
29. Zhan, L.; Honma, M.; Wang, L.; Hayashi, M.; Wu, D.-S.; Zhang, L.-S.; Rajaguru, P.; Suzuki, T. Microcystin-LR is not Mutagenic in vivo in the *λ/lacZ* Transgenic Mouse (Muta^{TM}Mouse). *Gene. Environ.* **2006**, *28*, 68–73. [CrossRef]
30. Ohtani, I.; Moore, R.E.; Runnegar, M.T.C. Cylindrospermopsin: A potent hepatotoxin from the blue-green alga *Cylindrospermopsis raciborskii*. *J. Am. Chem. Soc.* **1992**, *114*, 7941–7942. [CrossRef]
31. Kinnear, S. Cylindrospermopsin: A decade of progress on bioaccumulation research. *Mar. Drugs* **2010**, *8*, 542–564. [CrossRef]
32. Manning, S.R.; Nobles, D.R. Impact of global warming on water toxicity: Cyanotoxins. *Curr. Opin. Food Sci.* **2017**, *18*, 14–20. [CrossRef]
33. Pichardo, S.; Cameán, A.M.; Jos, Á. In vitro toxicological assessment of Cylindrospermopsin: A review. *Toxins* **2017**, *9*, 402. [CrossRef]
34. Terao, K.; Ohmori, S.; Igarashi, K.; Ohtani, I.; Watanabe, M.F.; Harada, K.I.; Ito, E.; Watanabe, M. Electron microscopic studies on experimental poisoning in mice induced by cylindrospermopsin isolated from blue-green alga *Umezakia natans*. *Toxicon* **1994**, *32*, 833–843. [CrossRef]
35. Runnegar, M.T.; Kong, S.-M.; Zhong, Y.-Z.; Lu, S.C. Inhibition of reduced glutathione synthesis by cyanobacterial alkaloid cylindrospermopsin in cultured rat hepatocytes. *Biochem. Pharmacol.* **1995**, *49*, 219–225. [CrossRef]
36. Froscio, S.M.; Humpage, A.R.; Burcham, P.C.; Falconer, I.R. Cylindrospermopsin-induced protein synthesis inhibition and its dissociation from acute toxicity in mouse hepatocytes. *Environ. Toxicol. Int. J.* **2003**, *18*, 243–251. [CrossRef] [PubMed]

37. Puerto, M.; Jos, Á.; Pichardo, S.; Moyano, R.; Blanco, A.; Cameán, A.M. Acute exposure to pure Cylindrospermopsin results in oxidative stress and pathological alterations in Tilapia (*Oreochromis niloticus*). *Environ. Toxicol.* **2014**, *29*, 371–385. [CrossRef] [PubMed]
38. Humpage, A.R.; Fontaine, F.; Froscio, S.; Burcham, P.; Falconer, I.R. Cylindrospermopsin genotoxicity and cytotoxicity: Role of cytochrome P-450 and oxidative stress. *J. Toxicol. Environ. Health A* **2005**, *68*, 739–753. [CrossRef]
39. Bazin, E.; Mourot, A.; Humpage, A.R.; Fessard, V. Genotoxicity of a freshwater cyanotoxin, cylindrospermopsin, in two human cell lines: Caco-2 and HepaRG. *Environ. Mol. Mutat.* **2010**, *51*, 251–259. [CrossRef]
40. Shen, X.; Lam, P.K.S.; Shaw, G.R.; Wickramasinghe, W. Genotoxity investigation of a cyanobacterial toxin, cylindrospermopsin. *Toxicon* **2002**, *40*, 1499–1501. [CrossRef]
41. Bazin, E.; Huet, S.; Jarry, G.; Le Hégarat, L.; Munday, J.S.; Humpage, A.R.; Fessard, V. Cytotoxic and genotoxic effects of cylindrospermopsin in mice treated by gavage or intraperitoneal injection. *Environ. Toxicol.* **2012**, *27*, 277–284. [CrossRef]
42. Dordevic, N.B.; Matíc, S.L.J.; Simić, S.B.; Stanić, S.M.; Mihailivić, V.B.; Stancović, N.M.; Stancović, V.D.; Cirić, A.R. Impact of the toxicity of *Cylindrospermopsis raciboskii* (Woloszynska) Seenayya & Subba Raju on laboratory rats in vivo. *Environ. Sci. Pollut. Res.* **2017**, *24*, 14259–14272. [CrossRef]
43. Diez-Quijada, L.; Llana-Ruiz-Cabello, M.; Cătunescu, M.G.; Puerto, M.; Moyano, R.; Jos, Á.; Cameán, A.M. In vivo genotoxicity evaluation of cylindrospermopsin in rats using a combined micronucleus and comet assay. *Food. Chem. Toxicol.* **2019**, *132*, 1–9. [CrossRef]
44. Bittencourt-Oliveira, M.; Carmo, D.; Piccin-Santos, V.; Moura, A.N.; Aragão-Tavares, N.K.; Cordeiro-Araújo, M.K. Cyanobacteria, microcystins and cylindrospermopsin in public drinking supply reservoirs of Brazil. *An. Acad. Bras. Cienc.* **2014**, *86*, 297–310. [CrossRef] [PubMed]
45. Jančula, D.; Straková, L.; Sadílek, J.; Maršálek, B.; Babica, P. Survey of cyanobacterial toxins in Czech water reservoirs—The first observation of neurotoxic saxitoxins. *Environ. Sci. Pollut. Res. Int.* **2014**, *21*, 8006–8015. [CrossRef] [PubMed]
46. León, C.; Peñuela, G.A. Detected cyanotoxins by UHPLC MS/MS technique in tropical reservoirs of northeastern Colombia. *Toxicon* **2019**, *167*, 38–48. [CrossRef]
47. Testai, E.; Buratti, F.M.; Funari, E.; Manganelli, M.; Vichi, S.; Arnich, N.; Biré, R.; Fessard, V.; Sialehaamoa, A. Review and analysis of occurrence, exposure and toxicity of cyanobacteria toxins in food. *EFSA Support. Publ.* **2016**, *13*, 1–309. [CrossRef]
48. Hercog, K.; Maisanaba, S.; Filipič, M.; Jos, Á.; Cameán, A.M.; Žegura, B. Genotoxic potential of the binary mixture of cyanotoxins microcystin-LR and cylindrospermopsin. *Chemosphere* **2017**, *189*, 319–329. [CrossRef]
49. Diez-Quijada, L.; Prieto, A.I.; Jos, Á.; Cameán, A.M. in vitro mutagenic and genotoxic assessment of a mixture of the cyanotoxins Microcystin-LR and Cylindrospermopsin. *Toxins* **2019**, *11*, 318. [CrossRef]
50. Gutiérrez-Praena, D.; Guzmán-Guillén, R.; Pichardo, S.; Moreno, F.J.; Vasconcelos, V.; Jos, Á.; Cameán, A.M. Cytotoxic and morphological effects of microcystin-LR, cylindrospermopsin, and their combinations on the human hepatic cell line HepG2. *Environ. Toxicol.* **2018**, *34*, 240–251. [CrossRef] [PubMed]
51. Hinojosa, M.G.; Prieto, A.I.; Gutiérrez-Praena, D.; Moreno, F.J.; Cameán, A.M.; Jos, Á. Neurotoxic assessment of Microcystin-LR, cylindrospermopsin and their combination on the human neuroblastoma SH-SY5Y cell line. *Chemosphere* **2019**, *224*, 751–764. [CrossRef] [PubMed]
52. EFSA Scientific Committee. Scientific opinion on genotoxicity testing strategies applicable to food and feed safety assessment. *EFSA J.* **2011**, *9*, 2379. [CrossRef]
53. OECD Guidelines for the Testing of Chemicals: Mammalian Erythrocyte Micronucleus Test. 2016. Guideline 474. pp. 1–21. Available online: https://www.oecd.org/env/test-no-474-mammalian-erythrocyte-micronucleus-test-9789264264762-en.htm (accessed on 6 April 2020).
54. OECD Guideline for the Testing of Chemicals: In vivo Mammalian Alkaline Comet Assay. 2016. Guideline 489. pp. 1–27. Available online: https://www.oecd.org/env/test-no-489-in-vivo-mammalian-alkaline-comet-assay-9789264264885-en.htm (accessed on 6 April 2020).
55. Zervou, S.K.; Christophoridis, C.; Kaloudis, T.; Triantis, T.M.; Hiskia, A. New SPE-LC-MS/MS method for simultaneous determination of multi-class cyanobacterial and algal toxins. *J. Hazard. Mater.* **2017**, *323*, 56–66. [CrossRef]

56. Pinheiro, C.; Azevedo, J.; Campos, A.; Vasconcelos, V.; Loureiro, S. The interactive effects of microcystin-LR and cylindrospermopsin on the growth rate of the freshwater algae *Chlorella vulgaris*. *Ecotoxicology* **2016**, *25*, 745–758. [CrossRef] [PubMed]
57. Puerto, M.; Prieto, A.I.; Maisanaba, S.; Gutiérrez-Praena, D.; Mellado-García, P.; Jos, Á.; Cameán, A.M. Mutagenic and genotoxic potential of pure Cylindrospermopsin by a battery of in vitro tests. *Food Chem. Toxicol.* **2018**, *121*, 413–422. [CrossRef] [PubMed]
58. Bowen, D.E.; Whitwell, J.H.; Lillford, L.; Henderson, D.; Kidd, D.; McGarry, S.; Pearce, G.; Beevers, C.; Kirkland, D.J. Evaluation of a multi-endpoint assay in rats, combining the bone-marrow micronucleus test, the comet assay and the flow-cytometric peripheral blood micronucleus test. *Mutat. Res. Genet. Toxicol. Environ. Mutagen* **2011**, *722*, 7–19. [CrossRef] [PubMed]
59. Kirkland, D.; Levy, D.D.; LeBaron, M.J.; Aardema, M.J.; Beevers, C.; Bhalli, J.; Douglas, G.R.; Escobar, P.A.; Farabaugh, C.S.; Guerard, M.; et al. A comparison of transgenic rodent mutation and in vivo comet assay responses for 91 chemicals. *Mutat. Res. Genet.Toxicol. Environ. Mutagen.* **2019**, *839*, 21–35. [CrossRef]
60. Azqueta, A.; Shaposhnikov, S.; Collins, A.R. DNA oxidation: Investigating its key role in environmental mutagenesis with the comet assay. *Mutat. Res. Genet. Toxicol. Environ. Mutagen.* **2009**, *674*, 101–108. [CrossRef]
61. Llana-Ruiz-Cabello, M.; Maisanaba, S.; Puerto, M.; Prieto, A.I.; Pichardo, S.; Moyano, R.; González-Pérez, J.A.; Cameán, A.M. Genotoxicity evaluation of carvacrol in rats using a combined micronucleus and comet assay. *Food Chem. Toxicol.* **2016**, *98*, 240–250. [CrossRef]
62. Hooser, S.B.; Beasley, V.R.; Lovell, R.A.; Carmichael, W.W.; Haschek, W.M. Toxicity of Microcystin LR, a Cyclic Heptapeptide Hepatotoxin from *Microcystis aeruginosa*, to Rats and Mice. *Vet. Pathol.* **1989**, *26*, 246–252. [CrossRef]
63. Chernoff, N.; Hill, D.J.; Chorus, I.; Diggs, D.L.; Huang, H.; King, D.; Lang, J.R.; Le, T.T.; Schmid, J.E.; Travlos, G.S.; et al. Cylindrospermopsin toxicity in mice following a 90-d oral exposure. *J. Toxicol. Environ. Health A* **2018**, *81*, 549–566. [CrossRef]
64. Diez-Quijada, L.; Puerto, M.; Gutiérrez-Praena, D.; Llana-Ruiz-Cabello, M.; Jos, Á.; Cameán, A.M. Microcystin-RR: Occurrence, content in water and food and toxicological studies. A review. *Environ. Res.* **2019**, *168*, 467–489. [CrossRef]
65. Humpage, A.R.; Fenech, M.; Thomas, P.; Falconer, I.R. Micronucleus induction and chromosome loss in transformed human white cells indicate clastogenic and aneugenic action of the cyanobacterial toxin, cylindrospermopsin. *Mutat. Res. Genet. Toxicol. Environ. Mutagen.* **2000**, *472*, 155–161. [CrossRef]
66. Sieroslawska, A.; Rymuszka, A. Cylindrospermopsin induces oxidative stress and genotoxic effects in the fish CLC cell line. *J. Appl. Toxicol.* **2015**, *35*, 426–433. [CrossRef]
67. Štraser, A.; Žegura, B.; Filipič, M. Genotoxic effects of the cyanobacterial hepatotoxin cylindrospermopsin in the HepG2 cell line. *Arch. Toxicol.* **2011**, *85*, 1617–1626. [CrossRef]
68. Zhan, L.; Sakamoto, H.; Sakuraba, M.; Wu, D.-S.; Zhang, L.-S.; Suzuki, T.; Hayashi, M.; Honma, M. Genotoxicity of microcystin-LR in human lymphoblastoid TK6 cells. *Mutat. Res. Genet. Toxicol. Environ. Mutagen.* **2004**, *557*, 1–6. [CrossRef]
69. Štraser, A.; Filipič, M.; Gorenc, I.; Žegura, B. The influence of cylindrospermopsin on oxidative DNA damage and apoptosis induction in HepG2 cells. *Chemosphere* **2013**, *92*, 24–30. [CrossRef]
70. Žegura, B.; Sedmak, B.; Filipič, M. Microcystin-LR induces oxidative DNA damage in human hepatoma cell line HepG2. *Toxicon* **2003**, *41*, 41–48. [CrossRef]
71. Žegura, B.; Gajski, G.; Štraser, A.; Garaj-Vrhovac, V.; Filipič, M. Microcystin-LR induced DNA damage in human peripheral blood lymphocytes. *Mutat. Res. Genet. Toxicol. Environ. Mutagen.* **2011**, *726*, 116–122. [CrossRef]
72. Amé, M.V.; Baroni, M.V.; Galanti, L.N.; Bocco, J.L.; Wunderlin, D.A. Effects of microcystin-LR on the expression of P-glycoprotein in *Jenynsia multidentata*. *Chemosphere* **2009**, *74*, 1179–1186. [CrossRef]
73. Qiu, T.; Xie, P.; Liu, Y.; Li, G.; Xiong, Q.; Hao, L.; Li, H. The profound effects of microcystin on cardiac antioxidant enzymes, mitochondrial function and cardiac toxicity in rat. *Toxicology* **2009**, *257*, 86–94. [CrossRef]
74. Arman, T.; Lynch, K.D.; Montonye, M.L.; Goedken, M.; Clarke, J.D. Sub-Chronic Microcystin-LR Liver Toxicity in Preexisting Diet-Induced Nonalcoholic Steatohepatitis in Rats. *Toxins* **2019**, *11*, 398. [CrossRef] [PubMed]

75. Yoshida, T.; Makita, Y.; Nagata, S.; Tsutsumi, T.; Yoshida, F.; Sekijima, M.; Tamura, S.I.; Ueno, Y. Acute oral toxicity of microcystin-LR, a cyanobacterial hepatotoxin, in mice. *Nat Toxins* **1997**, *5*, 91–95. [CrossRef] [PubMed]
76. Sedan, D.; Andrinolo, D.; Telese, L.; Giannuzzi, L.; de Alaniz, M.J.; Marra, C.A. Alteration and recovery of the antioxidant system induced by sub-chronic exposure to microcystin-LR in mice: Its relation to liver lipid composition. *Toxicon* **2010**, *55*, 333–342. [CrossRef]
77. Zhang, X.X.; Zhang, Z.; Fu, Z.; Wang, T.; Qin, W.; Xu, L.; Cheng, S.; Yang, L. Stimulation effect of microcystin-LR on matrix metalloproteinase-2/-9 expression in mouse liver. *Toxicol. Lett.* **2010**, *199*, 377–382. [CrossRef]
78. Sun, X.; Mi, L.; Liu, J.; Song, L.; Chung, F.L.; Gan, N. Sulforaphane prevents microcystin-LR-induced oxidative damage and apoptosis in BALB/c mice. *Toxicol. Appl. Pharmacol.* **2011**, *255*, 9–17. [CrossRef]
79. Weng, D.; Lu, Y.; Wei, Y.; Liu, Y.; Shen, P. The role of ROS in microcystin-LR-induced hepatocyte apoptosis and liver injury in mice. *Toxicology* **2007**, *232*, 15–23. [CrossRef]
80. Han, Z.X.; Yang, L.; Zhang, L.; Xu, C.; Shu, W.Q. The antagonistic action of epigallocatechin-3-gallate on microcystin LR-induced oxidative damage on hepatocytes of mice and the expression of cytochrome P450 2E1. *Zhonghua Yu Fang Yi Xue Za Zhi* **2010**, *44*, 24–29. [CrossRef]
81. Sibaldo de Almeida, C.; Costa de Arruda, A.C.; Caldas de Queiroz, E.; Matias de Lima Costa, H.T.; Fernandes Barbosa, P.; Araújo Moura Lemos, T.M.; Nunes Oliveira, C.; Pinto, E.; Schwarz, A.; Kujbida, P. Oral exposure to cylindrospermopsin in pregnant rats: Reproduction and foetal toxicity studies. *Toxicon* **2013**, *74*, 127–129. [CrossRef]
82. Humpage, A.R.; Falconer, I.R. Oral toxicity of the cyanobacterial toxin cylindrospermopsin in male Swiss albino mice: Determination of no observed adverse effect level for deriving a drinking water guideline value. *Environ. Toxicol.* **2003**, *18*, 94–103. [CrossRef]
83. Chernoff, N.; Rogers, E.H.; Zehr, R.D.; Gage, M.I.; Malarkey, D.E.; Bradfield, C.A.; Liu, Y.; Schmid, J.E.; Jaskot, R.H.; Richards, J.H.; et al. Toxicity and recovery in the pregnant mouse after gestational exposure to the cyanobacterial toxin, cylindrospermopsin. *J. Appl. Toxicol.* **2010**, *31*, 242–254. [CrossRef]
84. Seawright, A.A.; Nolan, C.C.; Shaw, G.R.; Chiswell, R.K.; Norris, R.L.; Moore, M.R.; Smith, M.J. The oral toxicity for mice of the tropical cyanobacterium *Cylindrospermopsis raciborskii* (Wolonszynska). *Environ. Toxicol.* **1999**, *14*, 135–142. [CrossRef]
85. Pouria, S.; de Andrade, A.; Barbosa, J.; Cavalcanti, R.L.; Barreto, V.T.; Ward, C.J.; Preiser, W.; Poon, G.K.; Neild, G.H.; Codd, G.A. Fatal microcystin intoxication in haemodialysis unit in Caruaru, Brazil. *Lancet* **1998**, *352*, 21–26. [CrossRef]
86. Frangež, R.; Kosec, M.; Sedmak, B.; Beravs, K.; Demsar, F.; Juntes, P.; Pogačnik, M.; Šuput, D. Subchronic liver injuries caused by microcystins. *Pflügers Arch.* **2000**, *440*, 103–104. [CrossRef]
87. Ito, E.; Kondo, F.; Harada, K.I. Hepatic necrosis in aged mice by oral administration of microcystin-LR. *Toxicon* **1997**, *35*, 231–239. [CrossRef]
88. Iwabuchi, T.; Iijima, K.; Ara, N.; Koike, T.; Shinkai, H.; Ichikawa, T.; Kamata, Y.; Ishihara, K.; Shimosegawa, T. Increased gastric mucus secretion alleviates non-steroidal anti-inflammatory drug-induced abdominal pain. *Tohoku J. Exp. Med.* **2013**, *231*, 29–36. [CrossRef]
89. De La Cruz, A.A.; Hiskia, A.; Kaloudis, T.; Chernoff, N.; Hill, D.; Antoniou, M.G.; He, X.; Loftin, K.; O'Shea, K.; Zhao, C.; et al. A review on cylindrospermopsin: The global occurrence, detection, toxicity and degradation of a potent cyanotoxin. *Environ. Sci. Process. Impacts* **2013**, *15*, 1979–2003. [CrossRef]
90. International Conference of Harmonisation (ICH). Guidance on Genotoxicity Testing and Data Interpretation for Pharmaceuticals Intended for Human Use S2 (R1). 2012. Available online: https://www.fda.gov/regulatory-information/search-fda-guidance-documents/s2r1-genotoxicity-testing-and-data-interpretation-pharmaceuticals-intended-human-use (accessed on 6 April 2020).

91. Easterbrook, J.; Lu, C.; Sakai, Y.; Li, A.P. Effects of organic solvents on the activities of cytochrome P450 isoforms, UDP-dependent glucuronyl transferase, and phenol sulfotransferase in human hepatocytes. *Drug Metab. Dispos.* **2001**, *29*, 141–144.
92. Corcuera, L.A.; Vettorazzi, A.; Arbillaga, L.; Pérez, N.; Gil, A.G.; Azqueta, A.; González-Peñas, E.; García-Jalón, J.A.; López de Cerain, A. Genotoxicity of Aflatoxin B1 and Ochratoxin A after simultaneous application of the in vivo micronucleus and comet assay. *Food Chem. Toxicol.* **2015**, *76*, 116–124. [CrossRef]

Article

Chronic Low Dose Oral Exposure to Microcystin-LR Exacerbates Hepatic Injury in a Murine Model of Non-Alcoholic Fatty Liver Disease

Apurva Lad [1], Robin C. Su [1], Joshua D. Breidenbach [1], Paul M. Stemmer [2], Nicholas J. Carruthers [2], Nayeli K. Sanchez [1], Fatimah K. Khalaf [1], Shungang Zhang [1], Andrew L. Kleinhenz [1], Prabhatchandra Dube [1], Chrysan J. Mohammed [1], Judy A. Westrick [3], Erin L. Crawford [1], Dilrukshika Palagama [4], David Baliu-Rodriguez [4], Dragan Isailovic [4], Bruce Levison [5], Nikolai Modyanov [6], Amira F. Gohara [7], Deepak Malhotra [1], Steven T. Haller [1,*] and David J. Kennedy [1,*]

1 Department of Medicine, University of Toledo, Toledo, OH 43614, USA
2 Institute of Environmental Health Sciences, Eugene Applebaum College of Pharmacy and Health Sciences, Wayne State University, Detroit, MI 48201, USA
3 Department of Chemistry, Wayne State University, Detroit, MI 48202, USA
4 Department of Chemistry and Biochemistry, University of Toledo, Toledo, OH 43606, USA
5 Department of Pediatrics, University of Toledo, Toledo, OH 43614, USA
6 Department of Physiology and Pharmacology, University of Toledo, Toledo, OH 43614, USA
7 Department of Pathology, University of Toledo, Toledo, OH 43614, USA
* Correspondence: Steven.Haller@UToledo.edu (S.T.H.); David.Kennedy@UToledo.edu (D.J.K.); Tel.: +1-(419)-383-6859 (S.T.H.); +1-(419)-383-6822 (D.J.K.)

Received: 2 July 2019; Accepted: 19 August 2019; Published: 23 August 2019

Abstract: Microcystins are potent hepatotoxins that have become a global health concern in recent years. Their actions in at-risk populations with pre-existing liver disease is unknown. We tested the hypothesis that the No Observed Adverse Effect Level (NOAEL) of Microcystin-LR (MC-LR) established in healthy mice would cause exacerbation of hepatic injury in a murine model ($Lepr^{db}$/J) of Non-alcoholic Fatty Liver Disease (NAFLD). Ten-week-old male $Lepr^{db}$/J mice were gavaged with 50 μg/kg, 100 μg/kg MC-LR or vehicle every 48 h for 4 weeks (n = 15–17 mice/group). Early mortality was observed in both the 50 μg/kg (1/17, 6%), and 100 μg/kg (3/17, 18%) MC-LR exposed mice. MC-LR exposure resulted in significant increases in circulating alkaline phosphatase levels, and histopathological markers of hepatic injury as well as significant upregulation of genes associated with hepatotoxicity, necrosis, nongenotoxic hepatocarcinogenicity and oxidative stress response. In addition, we observed exposure dependent changes in protein phosphorylation sites in pathways involved in inflammation, immune function, and response to oxidative stress. These results demonstrate that exposure to MC-LR at levels that are below the NOAEL established in healthy animals results in significant exacerbation of hepatic injury that is accompanied by genetic and phosphoproteomic dysregulation in key signaling pathways in the livers of NAFLD mice.

Keywords: Microcystin-LR; Non-alcoholic Fatty Liver Disease; No Observed Adverse Effect Level; $Lepr^{db}$/J mice; hepatotoxicity; oxidative stress; TiO_2 enriched phosphopeptides

Key Contribution: This is the first study to investigate the effect of the permissible levels of the cyanotoxin on the susceptibility to MC-LR toxicity in a model of a pre-existing liver condition.

1. Introduction

Cyanobacteria are a natural component of the freshwater phytoplankton, but their overgrowth has increasingly been recognized as a potential health hazard due to the release of toxins by selected species of algae, such as *Microcystis aeruginosa, Raphidiopsis mediterranea, Cylindrospermopsin raciborskii, Anabaena circinalis, Planktothrix rubescens and Planktothrix agardhii* [1–5]. Their ability to adapt and flourish in a wide variety of climate conditions, including extremes, contributes to their successful occurrence in a variety of ecosystems. An extensive study done by Mantzouki et al. in 2018, showed that global warming has direct and indirect effects on the temperature and nutrient gradient in the freshwater systems, thus, affecting the distribution of the cyanotoxins [5]. Therefore, changing climatic conditions and increased eutrophication of freshwater ecosystems are two main factors that favor the growth of cyanobacterial blooms and allow them to flourish, increasing the duration and intensity on a global scale [2,6].

Toxin production is a common characteristic of many of the cyanobacterial species. Of the many toxins produced by cyanobacteria, microcystins (MCs) are among the most widely distributed, and microcystin-LR (MC-LR) is one of the most commonly produced [7]. MC-LR is a cyclic heptapeptide characterized structurally by leucine and arginine amino acids at positions 2 and 4 within a cyclo-(D-alanine-1-X2-D-MeAsp3-Y4-Adda 5-d-glutamate 6-Mdha7) structure [8]. It is usually taken up into the cells through organic anion transporting polypeptides (OATPs) [9]. MC-LR then exhibits its deleterious effects by inhibiting the activity of protein phosphatases 1 (PP1) and 2A (PP2A) and by increasing the production of reactive oxygen species (ROS) [10,11]. As a result, there is a hyperphosphorylation and pro-oxidative state induced by MC-LR, which leads to alterations in the cytoskeleton, OATP expression, mitogen-activated protein kinase (MAPK) activity, glycogen storage, and mitochondrial structure and function in addition to DNA damage, cellular disruption, endoplasmic reticulum dysfunction, inflammation and tumor growth [12–20].

MC-LR has become a global health concern due to its potent toxicity in humans and animals alike. In addition to its well-known hepatotoxic effects, MCs also cause reproductive, developmental and, immune toxicities [9,21–28]. Humans can be exposed to MC-LR most commonly through oral ingestion of contaminated water, or through recreational activities like swimming or boating in contaminated waters, as well as intake of contaminated fish or algal supplements [29,30]. Once ingested, the toxin is absorbed into the tissues, especially into hepatic tissue via the bile acid carrier system [1].

The current guidelines for safe exposure to the toxin in humans have been extrapolated from experiments performed in healthy animal models. In a 13-week study performed by Fawell et al., the No Observed Adverse Effect Level (NOAEL) was 40 μg/kg and the Low Observed Adverse Effect Level (LOAEL) was 200 μg/kg of body weight [1]. In accordance with these findings, the World Health Organization (WHO) established the provisional guideline value of 1 μg/L of MC-LR in drinking water. In 2015, as part of the Safe Drinking Water Act, the United States Environmental Protection Agency developed a Health Advisory for microcystins of 0.3 μg/L for bottle-fed infants and pre-school children and 1.6 μg/L for School-age children and adults [31]. Nevertheless, the effects of MC-LR in the setting of pre-existing liver disease remain unknown.

Non-alcoholic Fatty Liver Disease (NAFLD) is one of the most common liver conditions in the United States. The incidence of NAFLD is increasing concurrent with the rise in obesity and diabetes [32–34]. NAFLD is characterized by the presence of steatosis/ inflammation with ≥ 5% fat infiltration into the liver and potentially elevated liver enzymes [35]. In the past two decades, the prevalence of this condition has nearly doubled, affecting about 20–30% of the population in the Western countries [36]. Given the increase in the occurrence of both NAFLD and cyanobacteria blooms globally, we evaluated the effects of MC-LR in the setting of pre-existing liver disease using the well-established Lepr^{db}/J murine model of NAFLD [37]. Given NAFLD's growing prevalence worldwide, several studies have recently investigated MC-LR's effects in pre-existing NAFLD [18,28,38]. These studies have provided valuable insight, alerting us to the fact that individuals with pre-existing NAFLD may be more susceptible to MC-LR's toxic effects. However, these studies utilize intraperitoneal (IP) and intravenous (IV) methods of MC-LR exposure. In our current study, we provide additional insight by using a robust genetic model of NAFLD and expose mice to MC-LR by gavage, an exposure route more relevant to environmental methods of exposure. We tested the hypothesis that chronic low dose oral exposure to MC-LR at levels below the NOAEL established in healthy animal models would exacerbate the hepatotoxicity seen in a NAFLD murine model.

2. Results

2.1. *Survival, Appearance and Weight*

No visible symptoms of sickness or discomfort were observed in $Lepr^{db}$/J mice from either the 50 μg/kg or 100 μg/kg MC-LR exposed groups. However, early mortality was observed in 1/17 mice (95% survival) from the 50 μg/kg MC-LR exposed group and 3/17 mice (83% survival) from the 100 μg/kg MC-LR exposed group (Kaplan-Meier log-rank $p = 0.07$, Figure S1A). In each case the animals died overnight and there was no observed acute trauma (e.g., tracheal rupture resulting in immediate death) or other signs of improper gavage technique such as visible signs of discomfort or bloating in the time preceding death although we were not able to fully exclude mortality attributable to the gavage procedure in these animals. It should be noted that the $Lepr^{db}$/J mice are more susceptible to comorbidities including compromised immune system [39]. No deaths were observed in either the vehicle exposed $Lepr^{db}$/J group ($n = 13$) or in either of the vehicle or 100 μg/kg MC-LR exposed healthy background strain control C57Bl/6J (Wild Type - WT) mice study groups ($n = 5$ mice/group each). The body weight gain throughout the exposure in $Lepr^{db}$/J MC-LR exposed groups was not significantly different from the respective vehicle exposed group (Figure 1A,B). The physical appearance (Figure S1B–D) as well as the weight of the livers (Figure 1C) were not significantly different from the MC-LR-exposed groups in either the $Lepr^{db}$/J or WT mice (data not shown). Other organs including heart, lung, and kidney were collected from both the $Lepr^{db}$/J and C57Bl/6J control and MC-LR-exposed groups. There was no significant difference in organ weight between WT C57Bl/6J control and C57Bl/6J MC-LR-exposed mice as well as no significant differences among $Lepr^{db}$/J vehicle control and $Lepr^{db}$/J MC-LR exposed mice. (Tables S1 and S2).

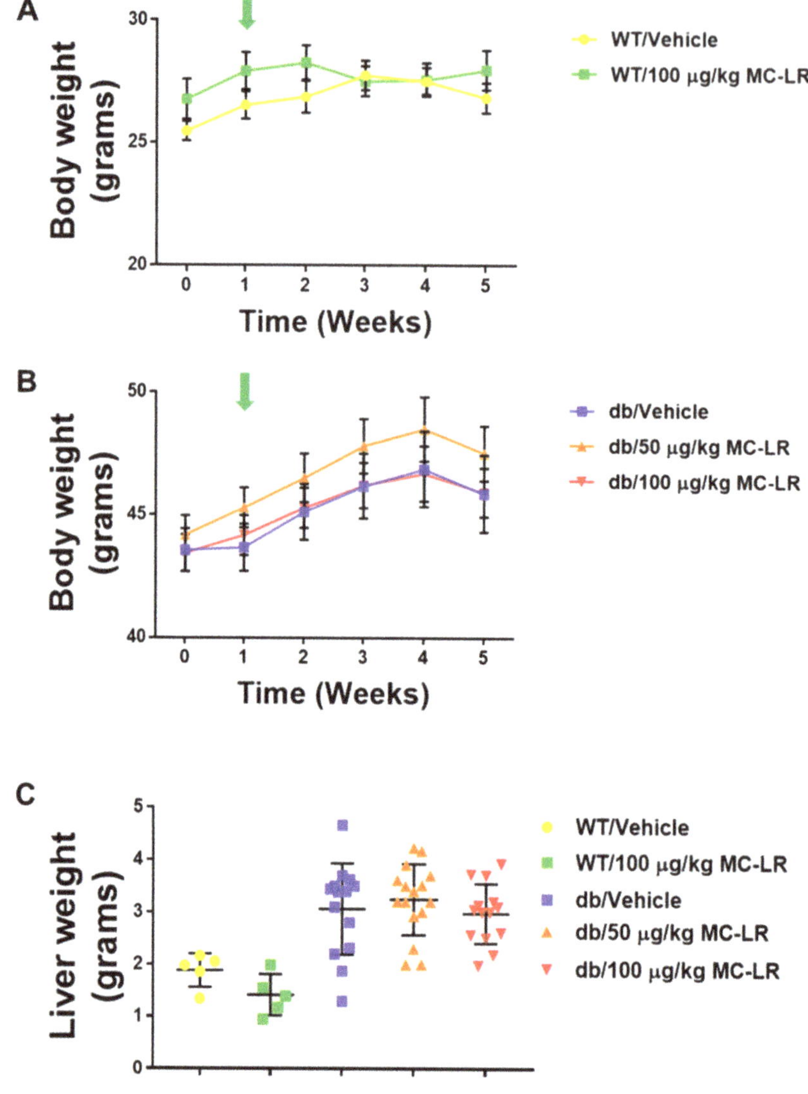

Figure 1. Effect of Microcystin-LR (MC-LR) on body and liver weights in both healthy (C57Bl/6J) and NAFLD ($Lepr^{db}$/J) mice. (**A** and **B**) Total body weights. (A) C57Bl6/J (WT) mice exposed to vehicle (n = 5) or 100 µg/kg MC-LR (n = 5) and (B) $Lepr^{db}$/J (db) mice exposed to vehicle (n = 13), 50 µg/kg MC-LR (n = 16), or 100 µg/kg MC-LR (n = 14). Mean and S.E.M. are indicated. (**C**) liver weights of animals from all groups. Individual, mean, and S.E.M. values indicated. NOTE: The green arrow over Week 1 represents the initiation of exposure to MC-LR or vehicle.

2.2. Detection of MC-LR in Plasma and Urine

To determine circulating levels and urinary excretion of MC-LR resulting from this exposure regimen, we collected plasma and 24-h urine samples at the end of the study prior to euthanasia. We then used Ultra-high-performance liquid chromatography coupled to triple quadrupole mass spectrometry (UHPLC-QqQ-MS/MS) and High-Performance Liquid chromatography–orbitrap mass spectrometry (HPLC-orbitrap-MS) to detect and quantify the MC-LR levels in animals which met the sample volume requirements for the analysis. MC-LR was detected in the plasma (Figure 2A) and urine (Figure 2B) of exposed Lepr^{db}/J mice. Significantly more MC-LR was detected in the 100 μg/kg MC-LR exposed Lepr^{db}/J mice ($p < 0.001$ for plasma and $p < 0.01$ for urine, respectively) as compared to those exposed to 50 μg/kg MC-LR. Interestingly, the toxin was undetectable in the plasma of C57Bl6/J mice exposed to the toxin (Figure 2A), whereas the amounts detected in the 24-h urine excretion samples were significantly elevated (nearly 100-fold) as compared to the Lepr^{db}/J mice exposed to the same amount of the toxin (Figure 2B).

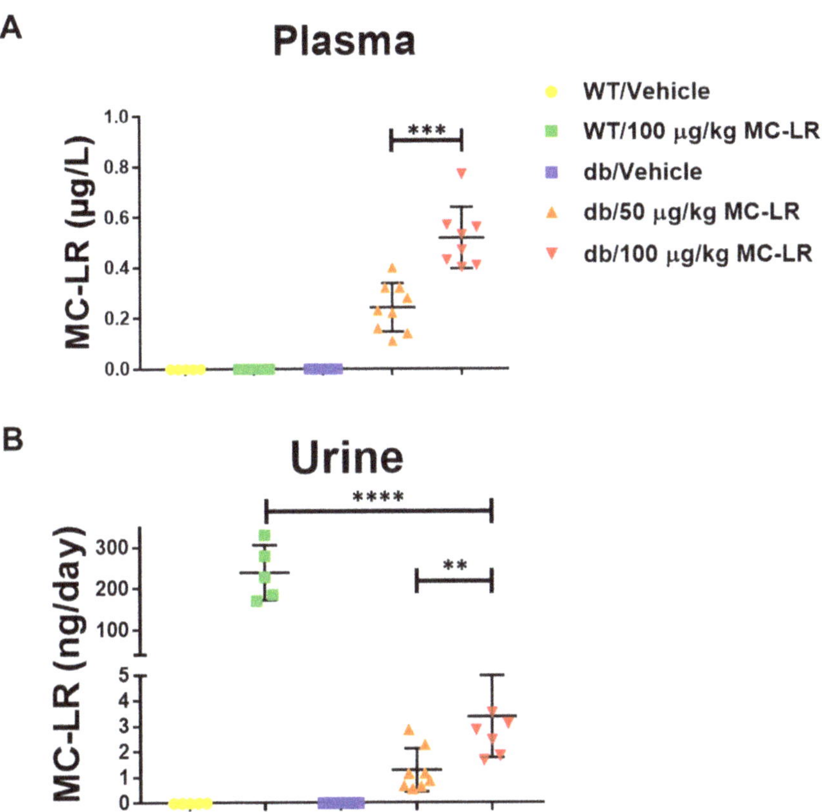

Figure 2. MC-LR determination in plasma and urine of both healthy (C57Bl/6J) and NAFLD (Lepr^{db}/J) mice. (**A**) Plasma and (**B**) 24-h urine excretion levels of MC-LR in C57Bl/6J (WT) Leprdb/J (db) mice as assessed by Ultra-high-performance liquid chromatography coupled to triple quadrupole mass spectrometry (UHPLC-QqQ-MS/MS) and High-Performance Liquid chromatography–orbitrap mass spectrometry (HPLC-orbitrap-MS) at the end of the 4–week exposure protocol. For plasma analysis, WT/vehicle $n = 5$, WT/100 μg/kg MC-LR $n = 5$, db/vehicle $n = 8$, db/50 μg/kg MC-LR $n = 9$, db/100 μg/kg MC-LR $n = 8$. For 24-h urine excretion analysis, WT/vehicle $n = 5$, WT/100 μg/kg MC-LR $n = 5$, db/vehicle $n = 7$, db/50 μg/kg MC-LR $n = 8$, db/100 μg/kg MC-LR $n = 8$. The number of samples shown in the figure for the db mice vary due to insufficient amount of plasma or urine needed for the analysis. Significance calculated by unpaired Student's t-test for the indicated comparisons. **, $p < 0.01$; ***, $p < 0.001$; ****, $p < 0.0001$.

2.3. Blood Biochemistry

To determine the levels of liver injury enzymes, namely alanine aminotransferase (ALT) and alkaline phosphatase (ALP), as well as to study the factors of blood biochemistry, we conducted a comprehensive diagnostic analysis using whole blood collected via retro-orbital bleed at days 15 and 30 during the exposure. Results from the comprehensive diagnostic analysis are shown in Tables S3 and S4. In paired analysis of the Lepr^{db}/J mice, the levels of ALT were non-specifically elevated (15 vs. 30-day time points) in the vehicle exposed mice ($p < 0.001$), 50 μg/kg MC-LR exposed mice ($p < 0.001$), as well as the 100 μg/kg MC-LR exposed mice ($p < 0.05$, Figure S2A). On the other hand, the levels of ALP were only significantly elevated (15 vs. 30-day time points) in the 50 μg/kg MC-LR exposed mice ($p < 0.05$) as well as in the 100 μg/kg MC-LR exposed mice ($p < 0.01$), but not in the vehicle exposed mice (Figure S2B). In the unpaired, between group analysis (Figure 3), the only significant change in blood chemistry between vehicle and MC-LR exposed mice was that ALP levels were elevated in the 100 μg/kg MC-LR exposed mice vs. vehicle ($p < 0.01$) at the 30-day time point (Figure 3B). All other blood biochemistry markers did not show any significant change between vehicle or MC-LR exposed groups for either the Lepr^{db}/J mice (Figure 3, Table S3) or the C57Bl/6J mice (Figure 3, Table S4).

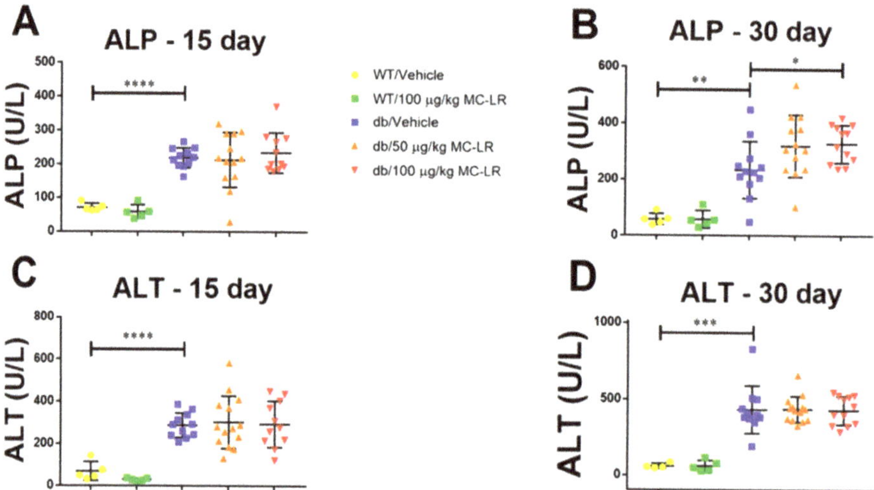

Figure 3. Effect of MC-LR on levels of liver injury enzymes in both healthy (C57Bl/6J) and NAFLD (Lepr^{db}/J) mice. Circulating levels of Alkaline Phosphatase (ALP) (**A** and **B**) and Alanine aminotransferase (ALT) (**C** and **D**) were measured across all exposure groups at 15 days and 30 days after initial exposure. Significance between groups tested by unpaired Student's t-test for indicated comparisons. For 15-day analysis, WT/vehicle $n = 5$, WT/100 μg/kg MC-LR $n = 5$, db/vehicle $n = 11$, db/50 μg/kg MC-LR $n = 13$, db/100 μg/kg MC-LR $n = 11$. For 30-day analysis, WT/vehicle $n = 5$, WT/100 μg/kg MC-LR $n = 5$, db/vehicle $n = 12$, db/50 μg/kg MC-LR $n = 13$, db/100 μg/kg MC-LR $n = 12$, *, $p < 0.05$, **, $p < 0.01$; ***, $p < 0.001$; ****, $p < 0.0001$.

2.4. Liver Histology

Histopathological analysis was performed using Hematoxylin & Eosin (H&E) and Periodic Acid-Schiff (PAS) staining of the liver sections. The slides were assessed by a pathologist, who was blinded to the group assignments, and histopathologic scoring was assessed based on the severity of fat infiltration, hepatic damage, glycogen content, and micro- and macro-vesicular lipid accumulation within the hepatocytes. As shown in Figure 4, significant hepatic injury was noted in both 50 μg/kg and 100 μg/kg MC-LR exposed Lepr^{db}/J mice as compared to the vehicle control Lepr^{db}/J mice. This was noted by the presence of fat vacuoles (Figure 4A, top panel, yellow arrows) as well as macro- and micro-vesicular fat infiltration and accumulation inside the hepatocytes (Figure 4A, top panels, red arrows). Furthermore, PAS staining demonstrated reduction in the glycogen content in Lepr^{db}/J MC-LR exposed mice as compared to vehicle control, although this appeared to be more pronounced in the 50 μg/kg group (Figure 4A, bottom panels). Based on histopathologic scoring, liver injury was significantly increased in MC-LR exposed Lepr^{db}/J mice as compared to the vehicle control Lepr^{db}/J mice ($p < 0.001$ in 50 μg/kg and $p < 0.05$ in 100 μg/kg, Figure 4C). However, no significant changes were observed in the liver sections of MC-LR exposed C57Bl/6J mice compared to vehicle control C57Bl/6J mice (Figure 4B,C). The H&E and PAS derived injury scores for each individual animal are presented in Table S5.

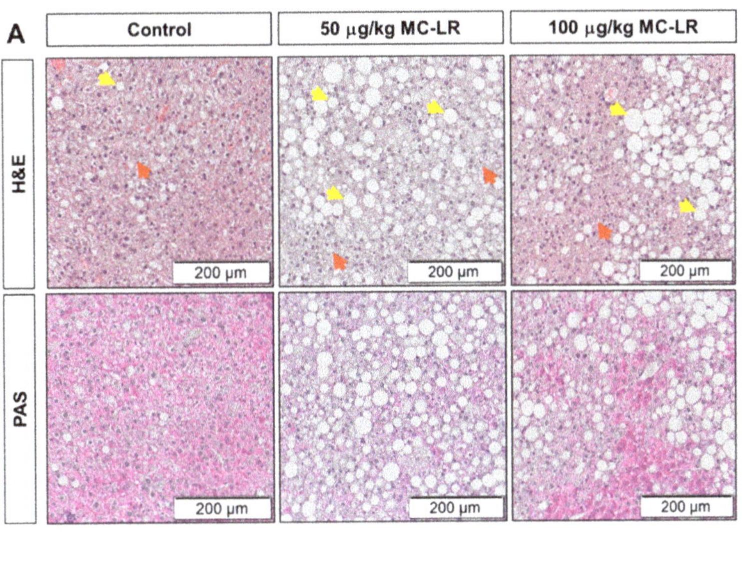

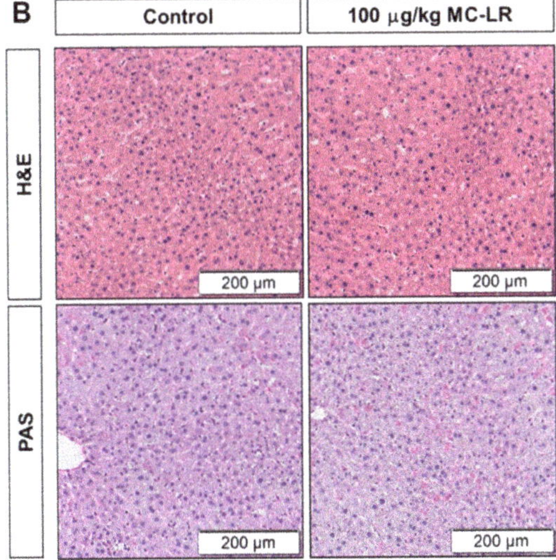

Figure 4. *Cont.*

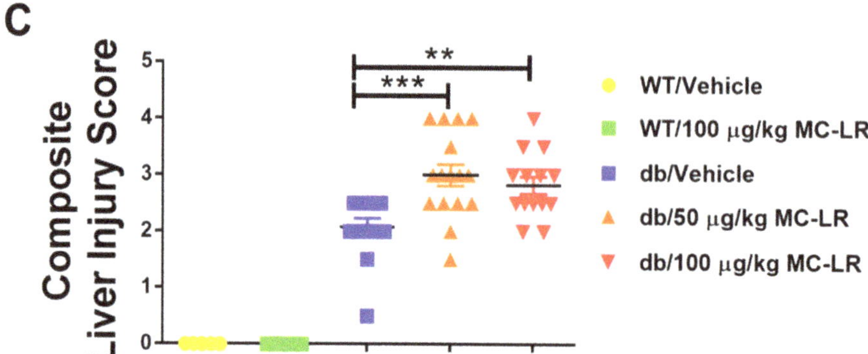

Figure 4. Histological analysis of hepatic injury in both healthy (C57Bl/6J) and NAFLD ($Lepr^{db}$/J) mice. (A and B) Representative images of liver tissue sections from different exposure groups of $Lepr^{db}$/J (db) mice (**A**) and C57Bl6/J (WT) mice (**B**) stained with Hematoxylin & Eosin (H&E) (top panels) and Periodic Acid-Schiff (PAS) (bottom panels) staining; yellow arrows indicate fat vacuoles, red arrows indicate fat infiltration inside hepatocytes; scale bar = 200 μm. (**C**) Composite liver injury score for C57Bl6/J mice (n = 5/group) and $Lepr^{db}$/J mice receiving vehicle (n = 13), 50 μg/kg MC-LR (n = 16; ***, $p < 0.0006$ vs. vehicle) and 100 μg/kg MC-LR (n = 14; *, $p = 0.0023$ vs. vehicle) as determined by Kruskal–Wallis ANOVA and Mann–Whitney test. No significant injury was determined in either the vehicle or 100 μg/kg MC-LR exposed C57Bl6/J mice.

2.5. Genetic Analysis Hepatotoxicity and Oxidative Stress

Since micro-vesicle lipid accumulation is associated with hepatic inflammation and oxidative stress, we assessed markers of inflammation and oxidative stress in the livers of $Lepr^{db}$/J mice after microcystin exposure. Interestingly, we observed that several genetic markers associated with hepatotoxicity and oxidative stress response were significantly upregulated as compared to the respective db/vehicle control (Tables S6 and S7). Significantly ($p < 0.05$) upregulated markers of hepatotoxicity (*Cxcl12*, 170.3 fold upregulated; *Casp3*, 15.6 fold upregulated; *Cyp1a2*, 187.2 fold upregulated; *Rb1*, 2.5 fold upregulated), necrosis (*Fam214a*, 15.6 fold upregulated; *Mlxip1*, 97.1 fold upregulated; *Col4a1*, 33.2 fold upregulated), nongenotoxic hepatocarcinogenicity (*Ccng1*, 19.3 fold upregulated) as well as oxidative stress response (*Prdx2*, 2876.3 fold upregulated; *Gpx5*, 4.7 fold upregulated) were noted in the 50 μg/kg MC-LR exposed db mice vs. db/vehicle controls. Limited analysis of targeted genes associated with hepatotoxicity (*Cyp1a2*, $p = 0.7$; *Slc17a3*, $p = 0.9$), necrosis (*OSMR*, $p = 0.9$), steatosis (*CD36*, $p = 0.5$), phospholipidosis (*Serpina3n*, $p = 0.07$) and cholestasis (*Abcb4*, $p = 0.6$) were tested by RT-PCR in the C57Bl/6J groups and were not found to have any change in the gene expression between vehicle and MC-LR exposed groups.

2.6. Phosphoproteomic Analysis

Based on the above results and the fact that MC-LR is a known protein phosphatase inhibitor, we next examined phosphoproteomic pathways, which were affected in the livers of MC-LR exposed $Lepr^{db}$/J mice, to better understand the possible disruptions in signaling pathways associated with the hepatotoxicity in this model. Phosphoproteomic analysis with mass spectrometry-based label-free quantification of TiO_2 enriched samples was used to determine the phosphorylation sites and signaling pathways affected by MC-LR exposure.

Data analysis using Maxquant identified 9616 phosphorylation sites in total. A moderated t-test indicated that 10 phosphorylation sites were affected by 50 μg/kg microcystin exposure compared to vehicle control samples ($q < 0.1$, n = 8, Table 1, Figure S3A,B). Similar comparison was done, to indicate changes in phosphorylation sites, between 100 μg/kg microcystin exposure and Vehicle control samples

(Table 1, Figure S3C,D). An alternative analysis using linear regression of site intensity vs microcystin dose identified 25 sites with a linear relationship between dose and intensity ($q < 0.1$, $n = 7$, Table 1).

Table 1. Effect of microcystin on phosphorylation sites in Lepr^{db}/J liver protein. Number of sites affected by microcystin exposure in Lepr^{db}/J livers using moderated t-tests or linear regression. Results were corrected for multiple testing, $n = 8$ for 50 μg/kg and $n = 7$ for 100 μg/kg microcystin.

Test	$q < 0.10$	$q < 0.30$
50 μg/kg vs. Ctrl	10	66
100 μg/kg vs. Ctrl	93	459
Linear regression	25	368

Enrichment analysis of Gene Ontology (GO) biological processes, using Platform for Integrative Analysis of Omics (PIANO, v2.0.2., Bioconductor, Package for R, Sweden) software [40], was used to identify molecular pathways that were affected by microcystin exposure. Here we tested whether the mean t-statistic for each pathway was different from 0 indicating that the process components taken together were affected by MC-LR exposure. Exposure to 50 μg/kg MC-LR resulted in decreased phosphosite abundance of 91 pathways which represented 16 biological processes including urogenital system development, regulation of T cell proliferation and response to oxidative stress as well as positive regulation of cell cycle proliferation. The 16 processes selected to reduce redundancy are shown in Table 2. No GO processes were hyperphosphorylated in response to 50 μg/kg MC-LR (PIANO, FDR < 0.1). An analysis to identify specific classes of kinases that were affected by microcystin exposure failed to identify any significant kinase enrichment (PIANO, FDR < 0.1). Pathways in the Reactome pathway database [41] were also tested for changes in response to microcystin exposure. All GO and Reactome results are shown in Tables S8–S10.

In order to identify which microcystin-affected systems were the most sensitive to MC-LR exposure, fuzzy c-means clustering was used to identify common phosphorylation site abundance profiles across the MC-LR exposure groups. Among them, 4259 sites were assigned to one of 6 clusters (Figure S4). Biological process categories that were identified as affected by 50 μg/kg microcystin at an FDR < 0.03 were tested for enrichment in the c-means clusters using a Fisher's exact test. The results of this enrichment analysis demonstrated significant enrichment of 11 biological processes (Fishers exact test, $p < 0.02$) including several processes related to immune cell function as well as kidney/urogenital system development (Table 3). Cluster 3 contains sites that were dephosphorylated in both the 50 and 100 μg/kg samples compared to controls. It had the greatest number of enriched categories including several processes related to immune cell function as well as kidney/urogenital system development, suggesting that proteins in those categories are dephosphorylated by MC-LR exposure. Equivalent analysis using pathways that were affected by 100 μg/kg microcystin (FDR < 0.03) did not identify any cluster-enriched pathways. Cluster enrichment of Reactome pathways were also conducted as previously described [41] and are shown in Tables S8–S10. Analysis using the UniProt database was performed to summarize all the proteins with their phosphorylation sites affected by either 50 or 100 μg/kg MC-LR exposure (Table S11).

Table 2. Gene Ontology Biological Processes showing decreased phosphorylation on exposure with 50 μg/kg and 100 μg/kg MC-LR in Lepr^{db}/J mouse liver samples relative to vehicle (PIANO—Platform for Integrative Analysis of Omics data, FDR < 0.1). Processes with decreased phosphorylation site abundance in microcystin exposed Lepr^{db}/J liver samples relative to control (PIANO—Platform for Integrative Analysis of Omics data, FDR, False Discovery Rate < 0.1). A subset of the affected processes is shown here selected to reduce redundancy Results were summarized from n = 7–8 mice/group; 50 μg/kg MC-LR group (n = 8), 100 μg/kg MC-LR group (n = 7), vehicle (n = 8), for all proteomics analyses.

Pathway	Similar Pathways [a]	Mean *t*-Statistic	Sites	FDR
Vehicle vs. 50 μg/kg MC-LR				
Urogenital system development	18	−0.493	71	0.029
Regulation of T cell proliferation	9	−0.730	35	0.029
Appendage morphogenesis	9	−0.813	30	0.029
Regulation of DNA-binding transcription factor activity	7	−0.368	102	0.036
Regulation of striated muscle tissue development	7	−0.590	36	0.058
Regulation of cellular response to oxidative stress	6	−0.677	30	0.046
Energy derivation by oxidation of organic compounds	6	−0.460	92	0.029
Positive regulation of cell cycle process	6	−0.464	86	0.029
Defense response to other organism	5	−0.431	85	0.029
Response to wounding	4	−0.367	95	0.056
Cognition	4	−0.414	76	0.056
Regulation of microtubule cytoskeleton organization	4	−0.345	102	0.046
Monosaccharide metabolic process	3	−0.479	66	0.045
Protein autophosphorylation	3	−0.346	95	0.068
Lung alveolus development	3	−0.736	24	0.070
Coronary vasculature development	4	−1.032	17	0.029
Vehicle vs. 100 μg/kg MC-LR				
Toll-like receptor TLR6:TLR2 cascade	17	−1.073	32	0.0292
Transport of mature mRNA derived from an intron-less transcript	6	−1.257	21	0.0292
Cell cycle, mitotic	5	−0.555	132	0.0292
Resolution of sister chromatid cohesion	5	−1.111	30	0.0292
Mitotic prometaphase	4	−0.797	49	0.0426
Cytokine signaling in immune system	3	−0.580	144	0.0200
Carbohydrate metabolism	3	−0.717	105	0.0200
L1CAM interactions	2	−0.846	55	0.0292

[a] To reduce redundancy categories were clustered and one pathway per cluster listed. This indicates the size of the cluster.

Table 3. Identification of the clusters of pathways affected in 50 and 100 µg/kg MC-LR versus vehicle exposed Lepr^{db}/J mice using Reactome database: Reactome pathways in liver were identified as affected by 50 as well as 100 µg/kg MC-LR versus db/vehicle (False Discovery Rate, FDR<0.2) and enriched in a c-means cluster (Fisher's exact test $p < 0.02$). Results were summarized from $n = 7$–8 mice/group; 50 µg/kg MC-LR group ($n = 8$), 100 µg/kg MC-LR group ($n = 7$), Vehicle ($n = 8$), for all proteomics analyses.

Process	Mean *t*-Statistic	Sites	FDR
Vehicle vs. 50 µg/kg MC-LR			
Cluster 3			
Renal system development	−0.506	68	0.029
Regulation of stem cell proliferation	−0.838	31	0.029
Regulation of epithelial cell proliferation	−0.414	111	0.029
Regulation of mononuclear cell proliferation	−0.689	37	0.029
Regulation of leukocyte proliferation	−0.689	37	0.029
Regulation of T cell proliferation	−0.730	35	0.029
Positive regulation of cell cycle process	−0.464	86	0.029
Regulation of lymphocyte proliferation	−0.689	37	0.029
Urogenital system development	−0.493	71	0.029
Kidney development	−0.566	62	0.029
Coronary vasculature development	−1.032	17	0.029
Cluster 4			
Defense response to another organism	−0.431	85	0.029
Cluster 6			
Generation of precursor metabolites and energy	−0.375	133	0.029
Vehicle vs. 100 µg/kg MC-LR			
Cluster 1			
Pre-mRNA splicing	0.246	223	0.0601
mRNA splicing	0.246	223	0.0601
Cluster 2			
Post-translational protein modification	−0.31193	326	0.080192
Axon guidance	−0.30646	203	0.18749
Cluster 3			
Recruitment of mitotic centrosome proteins and complexes	−0.78878	24	0.14638
Cell-cell junction organization	−0.56824	45	0.15862
Centrosome maturation	−0.78878	24	0.14638
Toll-like receptor 4 (TLR4) cascade	−0.86804	41	0.042602
Signaling by interleukins	−0.57231	103	0.033413
Innate immune system	−0.37959	287	0.044106
Cytokine signaling in immune system	−0.58043	144	0.020048
Cluster 4			
Cellular senescence	−0.65136	49	0.082148

3. Discussion

In the current study, we used Lepr^{db}/J mice, a well-established genetic model for NAFLD [42], to examine the effect of prolonged low dose oral exposure to MC-LR in mice with pre-existing liver disease. The dosages used in the current study approximated the 40 μg/kg NOAEL [1] albeit with a reduced dosing regimen (every other day vs. daily) and total study duration (4 weeks vs. 13 weeks) [1,2]. This chronic low dose regimen resulted in significant increases in both circulating plasma levels of MC-LR as well as increased 24 h urinary excretion of MC-LR in the Lepr^{db}/J mice, as measured by UHPLC-QqQ-MS/MS and HPLC-orbitrap MS as we have described [43]. Interestingly, while there was no mortality or clinically observable MC-LR induced liver injury in the non-NALFD wild type C57Bl/6J group in our study, we did note a decreasing, albeit non-significant trend in survival in the MC-LR exposed Lepr^{db}/J mice, despite similar body and liver weights, as well as the gross morphology of the livers measured in the surviving mice. Additionally, non-NALFD wild type C57Bl/6J had undetectable circulating plasma levels of MC-LR and a highly significant increase in the 24-h urine excretion of the toxin as compared to the Lepr^{db}/J mice. This finding may be an indication of differing metabolic and excretory mechanisms between the healthy vs. NALFD diseased models and is likely relevant to the increased pathology and gene expression associated with hepatotoxicity and oxidative stress in the Lepr^{db}/J mice.

Similarly, we noted in the Lepr^{db}/J NAFLD model that while ALT levels were elevated non-specifically, ALP levels had a modest but significant increase specifically in the MC-LR exposed Lepr^{db}/J mice. Neither ALT or ALP levels were significantly elevated in the wild type C57Bl/6J mice exposed to MC-LR at the highest dose studied, 100 μg/kg. While ALT is a marker of both acute or chronic liver injury, ALP is a marker of cholestatic predominance [44]. This indicates that ALT may not be a clinically useful marker of MC-LR induced liver injury in the setting of NAFLD. On the other hand, the fact that ALP levels were significantly elevated in MC-LR exposed Lepr^{db}/J mice suggests that MC-LR may exacerbate liver damage in a cholestatic and obstructive pattern.

There are different patterns of hepatic-injury including hepatocyte-degeneration, intracellular fat accumulation, apoptosis, inflammation, regeneration and fibrosis [45]. Indeed, our histological analysis confirmed that hepatic injury in the form of hepatic micro-vesicular lipid accumulation, ballooned hepatocytes and reduced glycogen content was significantly higher in the MC-LR exposed Lepr^{db}/J mice. This finding indicates that exposure to MC-LR affects lipid metabolism and leads to increased fat accumulation in the livers of NAFLD mice and supports the observation that MC-LR exposure may promote cholestasis in this setting. Recent studies have investigated MC-LR toxicity in the setting of pre-existing NAFLD [18,28,38]. Clarke et al. studied the acute effects of MC-LR by a single-dose intravenous (IV) injection (20 μg/kg) and a single-dose of intraperitoneal (IP) injection (60 μg/kg) in two separate rat models of NAFLD: a methionine and choline deficient (MCD) diet and a high fat/high cholesterol (HFHC) diet [18]. Observed findings were informative but different between the two NAFLD models. While plasma MC-LR area under the concentration-time curve (AUC) was doubled and biliary clearance (Cl_{bil}) was unchanged in the MCD rats, AUC was unchanged and CL_{bil} was doubled in the HFHC rats. In addition, while liver pathology decreased in the MCD rats due to decreased binding to PP2A, hepatic inflammation, plasma cholesterol, proteinuria, and urinary KIM1 were increased in HFHC rats. These results reveal that pre-existing NAFLD potentially affects subject susceptibility to MC-LR toxicity, but there is variation to exact manifestations due to differences in NAFLD induction. Our current study provides a robust genetic model of NAFLD in mice that helps further the investigation of MC-LR toxicity in pre-existing NAFLD. In addition, our study provides a chronic exposure model by gavage, a more physiologically relevant method of exposure, mimicking a more typical exposure route in animals and humans.

Other studies have investigated the potential of NOX2 as a mediator of MC-LR toxicity in the setting of pre-existing NAFLD [28,38]. Albadrani et al. have shown that in mice with pre-existing NAFLD, intraperitonial exposure of MC-LR potentially activates hepatic Kupffer cells and stellate cells through the NOX2 pathway, causing an exacerbation of pre-existing liver pathology through increased

CD68 and increased pro-inflammatory cytokines. Sarkar et al. has shown that intraperitonial exposure of MC-LR in NAFLD mice alters the gut microbiome, leading to inflammatory processes within the intestines. Again, while these studies have provided invaluable insight using intraperitoneal-delivered MC-LR in diet induced NAFLD mice, our study builds upon these results by utilizing a genetic model of NAFLD with a physiologically relevant oral exposure of MC-LR.

MC-LR is a potent inhibitor of protein phosphatases 1 (PP1) and 2A (PP2A), which are responsible for dephosphorylating serine and threonine residues on regulatory and structural proteins. The inhibition of PP1 and PP2A leads to increased protein phosphorylation, which may alter the structure of the cell cytoskeleton [46]. It has been demonstrated that oxidative stress may contribute to the toxicity of MC-LR and that the formation of excessive free radical species from oxidative lipid alterations causes an increase in lipid peroxidation in mice serum [11,46]. Since PP2A regulates several mitogen-activated protein kinases (MAPKs), its inhibition by MC-LR may have secondary effects on downstream MAPK signaling pathways involved in essential cellular processes [47]. Additionally, many inflammatory and immune responses are regulated by several MAPK signaling pathways, making the release of pro-inflammatory cytokines a consequence of MAPK disruptions caused by MC-LR [48].

Our genetic and phosphoproteomic analysis of livers from MC-LR exposed NAFLD mice is in agreement with the aforementioned studies. In the present study, we observed that genes associated with hepatotoxicity (*Cxcl12, Casp3, Cyp1a2, Rb1*), necrosis (*Fam214a, Mlxipl, Col4a1*) and nongenotoxic hepatocarcinogenicity (*Ccng1*) were significantly upregulated in the mice that were exposed to MC-LR. Oxidative stress is another mechanism associated with MC-LR toxicity [49]. Many in vitro studies have shown that exposure to MC-LR can cause oxidative stress in cells, further leading to apoptosis [50–53]. Our study shows that exposure to MC-LR significantly upregulated the expression of oxidative stress response genes such as *Prdx2 and Gpx5*. As mentioned earlier, limited analysis of targeted genes associated with hepatotoxicity (*Cyp1a2, Slc17a3*), necrosis (*OSMR*), steatosis (*CD36*), phospholipidosis (*Serpina3n*) and cholestasis (*Abcb4*) were tested for the C57Bl/6J groups and were not found to show any change in gene expression between vehicle and MC-LR exposed groups.

Our phosphoproteomic analysis revealed that MC-LR exposure significantly affected Reactome pathways related to immune function, urogenital system development, T cell proliferation, cellular response to oxidative stress, and cell cycle regulation pathways. Studies have shown that MC-LR exposure in mice increases the levels of superoxide dismutase and catalase as well as increases the expression of *TNF-α, NFκB, IL-6* and *IL-1β* in HepG2 cells exposed with MC-LR [53,54]. Another study showed an increase in transcription of IL-8, a chemokine involved in the recruitment of monocytes and neutrophils during inflammation, in fish that were exposed to MC-LR [55]. In the present study MC-LR exposure regulated pathways related to Toll-like receptor (TLR) 2 and 6 signaling as well as those related to cell cycle and cell division regulatory functions. TLR2 activation has been shown to activate the PI3K/AKT/NF-kB pathway in immunological response to MC-LR in testicular cells [56]. These studies performed in various in vivo and in vitro models, complement our findings about the various pathways that are affected by exposure with MC-LR.

The fact that we did not note significant liver injury in the C57Bl/6J mice in the current study is in contrast with other studies where damage to mice was observed after oral exposure of MC-LR at similar doses of MC-LR by Zhang et al. and He et al. [6,57]. However, it should be noted that Zhang et al. gavaged the mice every 24 h for 28 days while in our study the mice were only gavaged every 48 h. Therefore, the total dosing in these mice was doubled compared to our study and this may account for the differences observed. In the He et al. study, the researchers used the Balb/c strain, whereas we used the C57Bl/6J because this strain is a better background control for the Leprdb/J mice. Importantly strain differences exist between Balb/c and C57Bl/6J strains where Balb/c mice have been shown to be more susceptible and C57Bl/6J mice less susceptible to hepatic injury [58,59].

MC-LR has been observed in freshwater environments around the world at varying levels. In the USA, 1000 μg/L of MC-LR was detected in the Lake Erie region [60]. In August 2014, elevated microcystin levels in the US city of Toledo, Ohio, led to a drinking water crisis when the city was left without drinking water for >2 days [61]. MC-LR not only accumulates within the water, but also accumulates within the aquatic ecosystem as extensively reviewed by Pham et al. [62]. MC-LR within the aquatic food chain, as well as within aquatic environments, pose potential risk of exposure in humans. In fact, human cases of exposure have been extensively reviewed by Svircev et al. [63]. A study in Australia reported MC-LR levels measured at 1 and 12 μg/L on two separate occasions within recreational locations (lakes and rivers in southern Queensland and in the Myall Lakes area of New South Wales), with surveyed individuals reporting gastrointestinal and respiratory symptoms [64]. In China, levels of MC-LR in Lake Chaohu have been reported to range from 2.2–3.9 μg/L [29], with measurable liver damage found in fishermen that work on the lake. In Sweden, 121 individuals from three villages reported abdominal pain, nausea, vomiting, diarrhea, fever, headaches, and muscle pain after their drinking water became discolored [63]. It was later found that the raw lake water contained 1 μg/L of MC-LR.

Taken together, our study provides insight into the potential increase in susceptibility of individuals with pre-existing liver conditions such as NAFLD, to the harmful effects of MC-LR exposure. By using exposure levels below the established NOAEL value of 40 μg/kg, our study reveals that MC-LR can exacerbate the liver damage in the setting of pre-existing liver disease. Such liver damage can be attributed to excess micro-vesicular lipid accumulation in the liver, upregulation of genes associated with hepatotoxicity and oxidative stress as well as changes in phosphorylation of the biological pathways related to cell cycle and immune response regulation. The results of this study suggest a need to review the preventative guidelines for safe exposure to MC-LR in at-risk settings, such as those with a pre-existing NAFLD. Indeed, we have recently demonstrated that MC-LR not only prolongs, but also worsens the severity of pre-existing colitis, supporting the notion that the effects of cyanotoxins such as MC-LR may be magnified in certain at-risk populations [65]. We have also demonstrated that the levels of the cyanotoxin can be detected in both plasma and urine samples of mice that were gavaged with low doses of the toxin, using UHPLC-QqQ-MS/MS and HPLC-orbitrap-MS methods [43]. This suggests that mass spectrometric analysis can be used in a clinical setting to assess the exposure to the cyanotoxins in humans. Further studies are warranted to examine the mechanistic basis of MC-LR hepatotoxicity in conditions such as NAFLD in order to assess diagnostic and therapeutic strategies for these potentially vulnerable populations. Additionally, epidemiological observations are needed to provide clinical context for the potential effects of cyanotoxins such as MC-LR in vulnerable patient populations.

Limitations

An important limitation of the current study was the fact that, because of the premature deaths in the MC-LR exposed NAFLD groups, not all animals were available for histopathological, biochemical, and phosphoproteomic examination. This may have affected the significance and dose dependence of the histopathological, biochemical, and phosphoproteomic changes we reported. It should also be noted that the doses used in this study are likely above doses to which humans, including those with NALFD, are exposed. Another limitation of the current study is that we were not able to measure MC-LR levels in the livers of the exposed mice nor did we perform analysis of activities or expression of PP1 and PP2A. However, despite these limitations, our study is the first to report the effects of chronic low dose oral exposure to MC-LR in a model of NAFLD. Our study suggests that in settings of pre-existing liver disease such as NAFLD, there may be an increase in the susceptibility to cyanotoxin toxicity.

4. Materials and Methods

4.1. Mice

Eight-week-old male B6.BKS(D)-Lepr^{db}/J (JAX Stock No. 000697, B6 db) mice (hereafter referred to as Lepr^{db}/J mice) on the C57Bl/6J background and wild type C57Bl/6J (JAX Stock No. 000664, Black 6) healthy background strain control mice were obtained from The Jackson Laboratory (Bar Harbor, Maine, USA) and maintained in the Department of Laboratory Animal Research at University of Toledo. All the mice were specific-pathogen free status and were housed in plastic cages (five mice per cage) and fed ad libitum on balanced rodent diet (Teklad global 16% protein diet, Envigo, Indianapolis, IN, USA) and water. The mice were kept in a well-ventilated room maintained at 23 ± 1 °C on a 12-h light and dark cycle. The mice were allowed to acclimatize for a week, before the beginning of the study. All the protocols were approved by the University of Toledo Institutional Animal Care and Use Committee (IACUC protocol number 108663, approval date February 9, 2016) and conducted in accordance with the National Institutes of Health (NIH) Guide for the Care and Use of Laboratory Animals.

4.2. Exposure and Experimental Design

At 10-weeks of age, Lepr^{db}/J mice were randomly divided into three groups with 13–17 mice per group (Figure 5). Group 1 was the Lepr^{db}/J vehicle control group which was exposed to water with volume equivalent to that of the MC-LR exposed mice (300 µL). Group 2 and 3 consisted of Lepr^{db}/J mice that were exposed to 50 µg/kg and 100 µg/kg of body weight of MC-LR (Item No. 10007188, Cayman Chemicals, Ann Arbor, MI, USA) respectively. MC-LR or vehicle control was administered to the mice via gavage every 48 h for a period of 4 weeks (15 total administrations over 4 weeks). These doses approximate the currently accepted NOAEL (40 µg/kg) established over a 13 week study [1]. The toxin was freshly prepared by dissolving it in purified water at a working concentration of 0.5 mg/mL. The mice were weighed once every week. Blood samples were collected at day 15 (midpoint) and 30 (endpoint) via retro-orbital bleeding (after isoflurane anesthesia) for analyzing blood chemistry, and at study termination, via intracardiac puncture after euthanasia. At the end of the study, mice were placed in metabolic cages for a 24-h urine collection immediately following the final dose of MC-LR or vehicle. 24-h urine samples were collected and stored at −80 °C. Mice were then removed from metabolic cages and euthanized for final blood and organ collection after another 24 h (i.e., a total of 48 h after last MC-LR/vehicle gavage). Blood samples were collected at day 15 (midpoint) and 30 (endpoint) via retro-orbital bleeding (after isoflurane anesthesia) for analyzing blood chemistry and at study termination via intracardiac puncture after euthanasia.

The animals were humanely euthanized and flushed with 10 ml of 1X Phosphate Buffered Saline (PBS) (Fisher Scientific, Pittsburgh, PA, USA). Liver tissues from individual animals were collected and weighed. Part of the liver tissue was frozen in OCT embedding media (Fisher Scientific, Pittsburgh, PA, USA) and another section was fixed with 4% *w/v* Formalin (Fisher Scientific, Pittsburgh, PA, USA) for histological purposes. The remainder of the tissue was snap frozen in liquid nitrogen and then stored at −80 °C until further use for experimental purposes. In a parallel study, wild type C57Bl/6J mice were randomly divided into 2 groups: a vehicle control and 100 µg/kg MC-LR exposed group, which were exposed as detailed above.

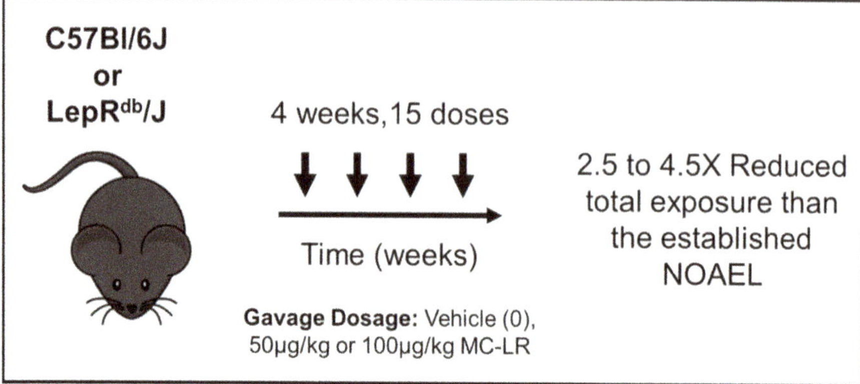

Figure 5. Schematic representation of the study design used to study the effect of low doses of MC-LR on healthy C57Bl/6J and NAFLD LepR^{db}/J mice. Ten-week-old mice were exposed with vehicle, 50 µg/kg or 100 µg/kg of MC-LR for a period of 4 weeks (15 total administrations) every 48 h. The mice in the vehicle control group were exposed with an equivalent volume of purified water.

4.3. Histological Studies

Liver sections were collected and immediately fixed in 4% formalin solution for 24 h, dehydrated in 70% ethanol, embedded in a paraffin block, and cut with a microtome to yield sections of 4-micron thickness. Hematoxylin & Eosin (H&E) and Periodic Acid-Schiff (PAS) staining were then performed on these sections. Whole slides were scored for hepatic injury by a pathologist, who was blinded to the study group assignments. Hepatic injury was assessed in both H&E and PAS was scored on a semi-quantitative scale of 0 (none or least hepatic injury) to 4 (maximum hepatic injury) and a composite injury score was taken as the average of the H&E and PAS injury score for each liver. Hepatic injury was assessed according to established methods including pale coloration of the tissue, infiltration of microvesicular lipid accumulation inside the hepatocytes, reduced glycogen content as well as ballooned hepatocytes with marked macro- and/or micro-vesicular lipid infiltration [45]. Hepatic damage was characterized by pale coloration of the tissue, infiltration of microvesicular lipid accumulation inside the hepatocytes, reduced glycogen content as well as ballooned hepatocytes with marked macro- and micro-vesicular lipid infiltration. A detailed description of the observations can be found in Table S5.

4.4. Blood Chemistry

Blood was collected via retro-orbital bleed in heparinized Micro-Hematocrit Capillary tubes (Cat no. 22-362-566, Fisher Scientific, Pittsburgh, PA, USA). Whole blood was then loaded onto Abaxis rotor with VetScan2 Chemistry Analyzer (Ref: 500-0038, Abaxis, Union City, CA, USA) with a comprehensive diagnostic profile for the quantitative analysis of Alanine aminotransferase (ALT), albumin (ALB), alkaline phosphatase (ALP), amylase (AMY), globulin (GLOB), glucose (GLU), blood urea nitrogen (BUN) and total protein (TP).

4.5. MC-LR Determination in Plasma and Urine

Both plasma and 24-h urine samples were collected post MC-LR exposure. Blood was collected by cardiac puncture in K3-EDTA microtubes (Sarstedt, Newton, NC, USA) and separated plasma was stored at −80 °C until further use. MC-LR in urine samples was quantified using both Ultra High Pressure Liquid Chromatography–triple Quadrupole–tandem Mass Spectrometry (UHPLC-QqQ-MS/MS, Columbia, MD, USA) and High Pressure Liquid Chromatography (Shimadzu Technologies, Addison, IL

USA)–Orbitrap Fusion Mass Spectrometry (Thermo Fisher Scientific, San Jose, CA, USA) system, while plasma samples were analyzed using HPLC-Orbitrap Fusion MS as we have previously described [43].

4.6. Genetic Analysis of Hepatotoxicity and Oxidative Stress

For genetic analysis, each exposure group shows data from 3 arrays (i.e., biological triplicates) and each array used a pooled cDNA sample from 4 mice/group. Hence, the data is representative of 12 mice from each exposure group. The data was analyzed using Qiagen analysis software (Qiagen, Germantown, MD, USA) and the criteria for fold change is based on any changes that are at least 2-fold above the normalized value for the vehicle exposed group. The p value for significance was considered as $p < 0.05$. RNA was extracted from liver tissues and isolated using the QIAzol/Chloroform extraction method. Approximately 500 ng of extracted RNA was used to synthesize cDNA (QIAGEN's RT2 First Strand Kit (Qiagen, Germantown, MD, USA). The cDNA was then used to run RT^2 Profiler^{TM} Mouse Hepatotoxicity Quantitative PCR Array (Cat No. PAMM-093Z, Qiagen, Germantown, MD, USA) as well as RT^2 Profiler^{TM} Mouse Oxidative Stress and Antioxidant Defense Quantitative PCR Array (Cat No. PAMM-065Z, Qiagen, Germantown, MD, USA) performed utilizing a QIAGEN Rotor-Gene Q thermo-cycler. Because previous reports have shown hepatotoxic and oxidative stress mediated injury induced by MC-LR, we selected targeted arrays to assess several measures of hepatotoxicity using the RT^2 Profiler^{TM} Mouse Hepatotoxicity Quantitative PCR Array (Cat No. PAMM-093Z which assesses genetic markers of Hepatotoxicity—Cxcl12, Cyp1a2, Casp3, Rb1; Nongenotoxic hepatocarcinogenicity—Ccng1; Necrosis—Fam214a, Mlxipl; Steatosis—CD36 and Cholestasis—Abcb1A, Abcb4) as well as RT^2 Profiler^{TM} Mouse Oxidative Stress and Antioxidant Defense Quantitative PCR Array (Cat No. PAMM-065Z which assesses genetic markers of peroxiredoxins; glutathione peroxidases; superoxide dismutases; oxygen transporters and oxidative stress response). Both arrays use Actin b, β2-microglobulin, GAPDH, Gusb and Hsp90ab1 as the housekeeping genes. All RT-qPCR was performed utilizing a QIAGEN Rotor-Gene Q thermo-cycler.

4.7. Proteomic Analyses of TiO_2 Enriched Phosphopeptides

Sample Preparation: Mouse liver tissue was homogenized in mass spectrometry grade water using a Potter Elvehjem Teflon on glass tissue homogenizer kept on ice. Homogenates were brought to 2% lithium dodecyl sulfate (LiDS) then heated to 95 °C for 5 minutes. Heat inactivated samples were exposed with 5 mM dithiothreitol (DTT) to reduce disulfide bonds, alkylated with 15 mM iodoacetamide (IAA) and then 5 mM additional DTT was added to quench the remaining IAA. Proteins were precipitated and LiDS removed by addition of 5 volumes of methanol. Methanol-washed pellets were resuspended in 0.5% deoxycholate (DOC) in 1X phosphate buffered saline plus 40 mM TEAB using sonication delivered by a QSonica cup-horn sonicator as necessary. Protein concentrations in the DOC solubilized samples were determined using a Bradford assay then 0.5 mg of protein was trypsinized using an overnight incubation with 2 µg Promega sequencing grade trypsin. Digested samples were centrifuged to remove particulates and the entire sample dried by speed-vac then resolubilized in 100 µL of 65% acetonitrile, 2% trifluoroacetic acid (TFA) and at 25% saturation with glutamic acid. Phosphopeptides were selected from the digests on an AssayMap Bravo (Agilent Technologies, Santa Clara, CA, USA) robot using TiO_2 cartridges. Phosphopeptides were analyzed at the Wayne State Proteomics Core using LC-MS/MS on an Orbitrap Fusion MS system (Thermo Fisher Scientific, San Jose, CA, USA). Each sample was analyzed independently using reversed-phase chromatography on an Acclaim PepMap RSLC (Thermo Fisher Scientific, San Jose, CA, USA), 75 µm × 25 cm column (Dionex, Sunnyvale, CA, USA). Peptides were eluted from the column in a 2-h gradient from 5% to 30% acetonitrile and analyzed directly by MS/MS using the Orbitrap Fusion.

4.8. Proteomic Data Analysis

MS spectra were searched against the Uniprot mouse complete database downloaded on 14 July 2017 (16,884 entries) using MaxQuant v1.6.2.10. (Max Planck Institute, Munich, Germany) [66]. Phosphorylation at Serine, Threonine and Tyrosine residues was set as a variable modification and the default penalty for modified peptide identification was reduced so that a minimum score of 20 and a minimum delta score of 3 were required for modified peptides. Match between runs was enabled. All other parameters were left at their default values. All analyses except the kinase enrichment analysis used phosphorylation sites without regard to localization confidence. Kinase enrichment analysis used all sites localized with > 80% confidence by Maxquant. Subsequent analysis used R v3.4.3 (http://www.R-project.org/) (The R Foundation, Vienna, Austria).

Phosphorylation site abundances were normalized so that each sample had the same median. Differentially abundant sites were identified using a moderated t-test [67]. A q-value was calculated for each site to account for multiple testing [68]. Gene Ontology (GO) biological processes that were affected by exposure were identified using Platform for Integrative Analysis of Omics Data (PIANO) [40]. T-statistics from the moderated t-test were submitted to PIANO and pathway enrichment was determined using the t-statistic mean. Phosphoproteins with multiple sites were not summarized and each site was submitted to PIANO individually. PIANO uses a permutation test to calculate an FDR corrected *p*-value. To reduce redundancy due to multiple GO categories with similar membership, affected pathways were clustered by phosphoprotein membership similarity and only the pathway with the lowest *p*-value in each cluster was reported. Pathways clustering was done using dynamic tree cut [69]. Pathway overlap was calculated as: (number of sites common to both pathways/number of sites in the smaller pathway). Kinase analysis was carried out using the regular expression-type kinase motifs distributed with Perseus software [70]. Identified phosphorylation sites that matched kinase criteria were converted to PIANO "kinase sets" and each kinase was tested for enrichment by PIANO analysis as above. Fuzzy c-means clustering was used to identify dose-response patterns in the data [71]. Sites were considered members of a cluster if they had greater than 0.5 membership. The number of clusters was set to 6 based on inspection of a plot of the number of sites clustered vs. the number of clusters. All sites were submitted to clustering without regard to their statistical significance. Gene Ontology (GO) biological processes were tested for enrichment in dose-response clusters using Fisher's exact test for an increased frequency of pathway components in cluster members versus cluster non-members. As for the samples from mice exposed with 100 µg/kg of MC-LR, Reactome pathways were tested for enrichment in dose-response clusters using Fisher's exact test for an increased frequency of pathway components in cluster members versus cluster non-members.

4.9. Statistical Analysis

Statistical analysis of all non-proteomic data was done using GraphPad PRISM 7 software (San Diego, CA, USA) and comparison within groups was done using Unpaired Student's t-test and Analysis of Variance (ANOVA) with Dunnette's multiple comparisons test. All data are presented as mean ± standard error of the mean (SEM) and a *p*-value of < 0.05 was considered to be statistically significant.

Supplementary Materials: The following are available online at http://www.mdpi.com/2072-6651/11/9/486/s1, Figure S1: Effect of MC-LR on survival and gross liver morphology, Figure S2: Effect of MC-LR exposure on liver injury enzymes in NAFLD mice, Figure S3: Phosphorylation sites affected by MC-LR exposure, Figure S4: Fuzzy c-means clusters of phosphorylation site abundance versus microcystin dose. 6 clusters were generated, Table S1: Effect of MC-LR exposure on tissue weights in $Lepr^{db}$/J mice, Table S2: Effect of MC-LR exposure on tissue weights in C57Bl/6J mice, Table S3: Effect of MC-LR exposure on blood chemistry in $Lepr^{db}$/J mice, Table S4: Effect of MC-LR exposure on blood chemistry in normal C57Bl/6J (WT) mice, Table S5: Hematoxylin & Eosin (H&E), Periodic Acid-Schiff (PAS), and Composite Liver Injury Scores in $Lepr^{db}$/J (db) mice, Table S6: Genetic analysis of hepatotoxicity in liver tissues, Table S7: Genetic analysis of oxidative stress response in liver tissues, Table S8: Identification of the clusters of pathways affected by 50 µg/kg MC-LR versus control using Reactome database, Table S9: GO Biological Process enrichment analysis, Table S10: REACTOME enrichment analysis, Table S11: Phosphorylation sites that were affected by microcystin exposure at 50 or 100 μg/kg.

Author Contributions: Conceptualization, A.L., D.J.K. and S.T.H.; data curation, A.L., R.C.S., J.D.B., P.M.S., N.J.C., N.K.S., F.K.K., S.Z., A.L.K., P.D., C.J.M., J.A.W., E.C., D.P., D.B.-R., D.I., B.L., N.M., A.F.G., D.M., S.T.H., D.J.K.; formal analysis, A.L., J.D.B., N.M., D.J.K. and S.T.H.; funding acquisition, S.T.H. and D.J.K.; investigation, A.L., P.M.S., N.J.C., N.K.S., D.P., D.B.R., D.I., B.L., N.M., D.J.K. and S.T.H.; methodology, A.L., D.J.K. and S.T.H.; project administration, D.M., S.T.H. and D.J.K.; resources, A.L., R.C.S., J.D.B., P.M.S., N.J.C., N.K.S., F.K.K., S.Z., A.L.K., P.D., C.J.M., J.A.W., E.C., D.P., D.B.R., D.I., B.L., N.M., A.F.G., D.M., S.T.H., D.J.K.; software, A.L., D.J.K. and S.T.H.; supervision, S.T.H. and D.J.K.; validation, P.M.S., N.J.C., J.A.W., D.I., B.L., N.M., A.F.G., D.M., D.J.K. and S.T.H.; visualization, A.L., J.D.B., P.M.S., N.J.C., D.P., D.B.R., D.I., B.L., N.M., S.T.H. and D.J.K.; writing—original draft preparation, A.L.; writing—review and editing, A.L., P.M.S., N.J.C., R.C.S., J.D.B., N.M., S.T.H., D.J.K.

Funding: This research was funded by Harmful Algal Bloom Research Initiative grants from the Ohio Department of Higher Education, David and Helen Boone Foundation Research Fund, University of Toledo Women and Philanthropy Genetic Analysis Instrumentation Center, The University of Toledo Medical Research Society, the Center for Urban Responses to Environmental Stressors (CURES) NIH Grant # P30 ES020957 and the Wayne State University Proteomics Core that is supported through NIH grants P30 ES020957, P30 CA 022453 and S10 OD010700.

Acknowledgments: Some of these data were presented in abstract form at the 2018 and 2019 Midwest Clinical and Translational Research Meeting in Chicago, IL. The authors are very grateful for the technical assistance provided by Pamela Brewster, Adam Spegele, Aaron Tipton and Dalal Mahmoud as well as instrumentation support provided by the Air Force Office of Scientific Research (DURIP 14RT0605) for the acquisition of the Orbitrap Fusion instrument.

Conflicts of Interest: The authors declare no conflict of interest.

References

1. Fawell, J.K.; Mitchell, R.E.; Everett, D.J.; Hill, R.E. The toxicity of cyanobacterial toxins in the mouse: I Microcystin-LR. *Hum. Exp. Toxicol.* **1999**, *18*, 162–167. [CrossRef] [PubMed]
2. Sedan, D.; Laguens, M.; Copparoni, G.; Aranda, J.O.; Giannuzzi, L.; Marra, C.A.; Andrinolo, D. Hepatic and intestine alterations in mice after prolonged exposure to low oral doses of Microcystin-LR. *Toxicon* **2015**, *104*, 26–33. [CrossRef] [PubMed]
3. Vasas, G.; Farkas, O.; Borics, G.; Felföldi, T.; Sramkó, G.; Batta, G.; Bácsi, I.; Gonda, S. Appearance of Planktothrix rubescens Bloom with [D-Asp3, Mdha7]MC–RR in Gravel Pit Pond of a Shallow Lake-Dominated Area. *Toxins* **2013**, *5*, 2434–2455. [CrossRef] [PubMed]
4. Mohamed, Z.A. First report of toxic *Cylindrospermopsis raciborskii* and *Raphidiopsis mediterranea* (Cyanoprokaryota) in Egyptian fresh waters. *FEMS Microbiol. Ecol.* **2007**, *59*, 749–761. [CrossRef] [PubMed]
5. Mantzouki, E.; Lürling, M.; Fastner, J.; Domis, L.D.S.; Wilk-Wo'zniak, E.; Koreiviene, J.; Seelen, L.; Teurlincx, S.; Verstijnen, Y.; Krztoń, W.; et al. Temperature Effects Explain Continental Scale Distribution of Cyanobacterial Toxins. *Toxins* **2018**, *10*, 156. [CrossRef] [PubMed]
6. He, J.; Li, G.; Chen, J.; Lin, J.; Zeng, C.; Chen, J.; Deng, J.; Xie, P. Prolonged exposure to low-dose microcystin induces nonalcoholic steatohepatitis in mice: A systems toxicology study. *Arch. Toxicol.* **2017**, *91*, 465–480. [CrossRef] [PubMed]
7. Ueno, Y.; Nagat, S.; Suttajit, M.; Mebs, D.; Vasconcelos, V. Immunochemical Survey of Microcystins in Environmental Water in Various Countries. In *Mycotoxins and Phycotoxins: Developments in Chemistry, Toxicology and Food Safety*; Alaken Inc.: Fort Collins, CO, USA, 1998; pp. 449–453.
8. Rinehart, K.L.; Harada, K.; Namikoshi, M.; Chen, C.; Harvis, C.A.; Munro, M.H.G.; Blunt, J.W.; Mulligan, P.E.; Beasley, V.R. Nodularin, microcystin, and the configuration of Adda. *J. Am. Chem. Soc.* **1988**, *110*, 8557–8558. [CrossRef]
9. Chen, L.; Xie, P. Mechanisms of microcystin-induced cytotoxicity and apoptosis. *Mini Rev. Med. Chem.* **2016**, *16*, 1018–1031. [CrossRef]
10. Yoshizawa, S.; Matsushima, R.; Watanabe, M.F.; Harada, K.I.; Ichihara, A.; Carmichael, W.W.; Fujiki, H. Inhibition of protein phosphatases by microcystis and nodularin associated with hepatotoxicity. *J. Cancer Res. Clin. Oncol.* **1990**, *116*, 609–614. [CrossRef]
11. Ding, W.X.; Shen, H.M.; Zhu, H.G.; Ong, C.N. Studies on Oxidative Damage Induced by Cyanobacteria Extract in Primary Cultured Rat Hepatocytes. *Environ. Res.* **1998**, *78*, 12–18. [CrossRef]

12. Solter, P.F.; Wollenberg, G.K.; Huang, X.; Chu, F.S.; Runnegar, M.T. Prolonged Sublethal Exposure to the Protein Phosphatase Inhibitor Microcystin-LR Results in Multiple Dose-Dependent Hepatotoxic Effects. *Toxicol. Sci.* **1998**, *44*, 87–96. [CrossRef] [PubMed]
13. Codd, G.A. Mechanisms of action and health effects associated with cyanobacterial toxins. *Toxicol. Lett.* **1996**, *88*, 21. [CrossRef]
14. Milutinović, A.; Živin, M.; Zorc-Pleskovič, R.; Sedmak, B.; Šuput, D. Nephrotoxic effects of chronic administration of microcystins-LR and-YR. *Toxicon* **2003**, *42*, 281–288. [CrossRef]
15. Pahan, K.; Gu, F.; Gruenberg, J.; Sheikh, F.G.; Namboodiri, A.M.S.; Singh, I. Inhibitors of Protein Phosphatase 1 and 2A Differentially Regulate the Expression of Inducible Nitric-oxide Synthase in Rat Astrocytes and Macrophages. *J. Biol. Chem.* **1998**, *273*, 12219–12226. [CrossRef] [PubMed]
16. Guzman, R.E.; Solter, P.F. Characterization of Sublethal Microcystin-LR Exposure in Mice. *Veter. Pathol.* **2002**, *39*, 17–26. [CrossRef] [PubMed]
17. Valério, E.; Vasconcelos, V.; Campos, A. New insights on the mode of action of microcystins in animal cells –A review. *Mini. Rev. Med. Chem.* **2016**, *16*, 1032–1041. [CrossRef] [PubMed]
18. Clarke, J.D.; Dzierlenga, A.; Arman, T.; Toth, E.; Li, H.; Lynch, K.D.; Tian, D.-D.; Goedken, M.; Paine, M.F.; Cherrington, N. Nonalcoholic fatty liver disease alters microcystin-LR toxicokinetics and acute toxicity. *Toxicon* **2019**, *162*, 1–8. [CrossRef] [PubMed]
19. Lone, Y.; Bhide, M.; Koiri, R.K. Microcystin-LR Induced Immunotoxicity in Mammals. *J. Toxicol.* **2016**, *2016*, 1–5. [CrossRef]
20. Campos, A.; Vasconcelos, V. Molecular Mechanisms of Microcystin Toxicity in Animal Cells. *Int. J. Mol. Sci.* **2010**, *11*, 268–287. [CrossRef]
21. Bell, S.G.; Codd, G.A. Cyanobacterial toxins and human health. *Rev. Med. Microbiol.* **1994**, *5*, 256–264. [CrossRef]
22. Takahashi, S.; Kaya, K. Quail spleen is enlarged by microcystin RR as a blue-green algal hepatotoxin. *Nat. Toxins* **1993**, *1*, 283–285. [CrossRef] [PubMed]
23. Chen, T.; Zhao, X.; Liu, Y.; Shi, Q.; Hua, Z.; Shen, P. Analysis of immunomodulating nitric oxide, iNOS and cytokines mRNA in mouse macrophages induced by microcystin-LR. *Toxicology* **2004**, *197*, 67–77. [CrossRef] [PubMed]
24. Soares, R.M.; Yuan, M.; Servaites, J.C.; Delgado, A.; Magalhães, V.F.; Hilborn, E.D.; Carmichael, W.W.; Azevedo, S.M.F.O. Sublethal exposure from microcystins to renal insufficiency patients in Rio de Janeiro, Brazil. *Environ. Toxicol.* **2006**, *21*, 95–103. [CrossRef] [PubMed]
25. Zhao, Y.; Xie, P.; Tang, R.; Zhang, X.; Li, L.; Li, D. In vivo studies on the toxic effects of microcystins on mitochondrial electron transport chain and ion regulation in liver and heart of rabbit. *Comp. Biochem. Physiol. Part C Toxicol. Pharmacol.* **2008**, *148*, 204–210. [CrossRef] [PubMed]
26. Lone, Y.; Koiri, R.K.; Bhide, M. An overview of the toxic effect of potential human carcinogen Microcystin-LR on testis. *Toxicol. Rep.* **2015**, *2*, 289–296. [CrossRef] [PubMed]
27. McLellan, N.L.; Manderville, R.A. Toxic mechanisms of microcystins in mammals. *Toxicol. Res.* **2017**, *6*, 391–405. [CrossRef] [PubMed]
28. Sarkar, S.; Kimono, D.; Albadrani, M.; Seth, R.K.; Busbee, P.; Alghetaa, H.; Porter, D.E.; Scott, G.I.; Brooks, B.; Nagarkatti, M.; et al. Environmental microcystin targets the microbiome and increases the risk of intestinal inflammatory pathology via NOX_2 in underlying murine model of Nonalcoholic Fatty Liver Disease. *Sci. Rep.* **2019**, *9*, 8742. [CrossRef] [PubMed]
29. Chen, J.; Xie, P.; Li, L.; Xu, J. First Identification of the Hepatotoxic Microcystins in the Serum of a Chronically Exposed Human Population Together with Indication of Hepatocellular Damage. *Toxicol. Sci.* **2009**, *108*, 81–89. [CrossRef] [PubMed]
30. Dietrich, D.R.; Ernst, B.; Day, B.W. Human Consumer Death and Algal Supplement Consumption: A Post Mortem Assessment of Potential Microcystin-Intoxication Via Microcystin Immunoistochemical (MCICH) Analyses. In Proceedings of the 7th International Conference on Toxic Cyanobacteria, Rio de Janeiro State, Brazil, 5–10 August 2007.
31. D'Anglada, L.V.; Joyce, M.D.; Jamie, S.; Belinda, H. *Health Effects Support Document for the Cyanobacterial Toxin Cylindrospermopsin*; US Environmental Protection Agency, Office of Water, Health and Ecological Central Division: Washington, DC, USA, 2015.

32. Angulo, P.; Kleiner, D.E.; Dam-Larsen, S.; Adams, L.A.; Bjornsson, E.S.; Charatcharoenwitthaya, P.; Mills, P.R.; Keach, J.C.; Lafferty, H.D.; Stahler, A.; et al. Liver fibrosis, but no other histologic features, is associated with long-term outcomes of patients with nonalcoholic fatty liver disease. *Gastroenterology* **2015**, *149*, 389–397. [CrossRef] [PubMed]
33. Le, M.H.; Devaki, P.; Ha, N.B.; Jun, D.W.; Te, H.S.; Cheung, R.C.; Nguyen, M.H. Prevalence of non-alcoholic fatty liver disease and risk factors for advanced fibrosis and mortality in the United States. *PLoS ONE* **2017**, *12*, e0173499. [CrossRef]
34. Brown, G.T.; Kleiner, D.E. Histopathology of nonalcoholic fatty liver disease and nonalcoholic steatohepatitis. *Metabolism* **2016**, *65*, 1080. [CrossRef] [PubMed]
35. Fazel, Y.; Koenig, A.B.; Sayiner, M.; Goodman, Z.D.; Younossi, Z.M. Epidemiology and natural history of non-alcoholic fatty liver disease. *Metabolism* **2016**, *65*, 1017–1025. [CrossRef] [PubMed]
36. Bedogni, G.; Miglioli, L.; Masutti, F.; Tiribelli, C.; Marchesini, G.; Bellentani, S. Prevalence of and risk factors for nonalcoholic fatty liver disease: The Dionysos nutrition and liver study. *Hepatology* **2005**, *42*, 44–52. [CrossRef] [PubMed]
37. Takahashi, Y.; Soejima, Y.; Fukusato, T. Animal models of nonalcoholic fatty liver disease/nonalcoholic steatohepatitis. *World J. Gastroenterol.* **2012**, *18*, 2300–2308. [CrossRef] [PubMed]
38. Albadrani, M.; Alhasson, F.; Dattaroy, D.; Chandrashekaran, V.; Seth, R.; Nagarkatti, M.; Nagarkatti, P.; Chatterjee, S. Microcystin exposure Exacerbates Non-alcoholic Fatty Liver Disease (NAFLD) via NOX2 Dependent Activation of miR21-induced Inflammatory Pathways. *Free. Radic. Biol. Med.* **2017**, *112*, 61. [CrossRef]
39. Park, S.; Rich, J.; Hanses, F.; Lee, J.C. Defects in innate immunity predispose C57BL/6J-Leprdb/Leprdb mice to infection by Staphylococcus aureus. *Infect. Immun.* **2009**, *77*, 1008–1014. [CrossRef]
40. Väremo, L.; Nielsen, J.; Nookaew, I. Enriching the gene set analysis of genome-wide data by incorporating directionality of gene expression and combining statistical hypotheses and methods. *Nucleic Acids Res.* **2013**, *41*, 4378–4391. [CrossRef]
41. Fabregat, A.; Jupe, S.; Matthews, L.; Sidiropoulos, K.; Gillespie, M.; Garapati, P.; Haw, R.; Jassal, B.; Korninger, F.; May, B.; et al. The Reactome Pathway Knowledgebase. *Nucleic Acids Res.* **2017**, *46*, D649–D655. [CrossRef]
42. Lau, J.K.; Zhang, X.; Yu, J. Animal models of non-alcoholic fatty liver disease: Current perspectives and recent advances. *J. Pathol.* **2017**, *241*, 36–44. [CrossRef]
43. Palagama, D.S.; Baliu-Rodriguez, D.; Lad, A.; Levison, B.S.; Kennedy, D.J.; Haller, S.T.; Westrick, J.; Hensley, K.; Isailovic, D. Development and applications of solid-phase extraction and liquid chromatography-mass spectrometry methods for quantification of microcystins in urine, plasma, and serum. *J. Chromatogr. A* **2018**, *1573*, 66–77. [CrossRef]
44. Giannini, E.G.; Testa, R.; Savarino, V. Liver enzyme alteration: A guide for clinicians. *Can. Med Assoc. J.* **2005**, *172*, 367–379. [CrossRef] [PubMed]
45. Kumar, V.; Abbas, A.K.; Aster, J.C. *Robbins and Cotran Pathologic Basis of Disease*, 9th ed.; Elsevier Saunders: Philadelphia, PA, USA.
46. Toivola, D.M.; Eriksson, J.E.; Brautigan, D.L. Identification of protein phosphatase 2A as the primary target for microcystin-LR in rat liver homogenates. *FEBS Lett.* **1994**, *344*, 175–180. [CrossRef]
47. Andrinolo, D.; Sedan, D.; Telese, L.; Aura, C.; Masera, S.; Giannuzzi, L.; Marra, C.A.; De Alaniz, M.J. Hepatic recovery after damage produced by sub-chronic intoxication with the cyanotoxin microcystin LR. *Toxicon* **2008**, *51*, 457–467. [CrossRef] [PubMed]
48. Zhang, J.; Chen, J.; Xia, Z. Microcystin-LR exhibits immunomodulatory role in mouse primary hepatocytes through activation of the NF-kappaB and MAPK signaling pathways. *Toxicol. Sci.* **2013**, *136*, 86–96. [CrossRef] [PubMed]
49. Li, X.; Liu, Y.; Song, L.; Liu, J. Responses of antioxidant systems in the hepatocytes of common carp (Cyprinus carpio L.) to the toxicity of microcystin-LR. *Toxicon* **2003**, *42*, 85–89. [CrossRef]
50. Svirčev, Z.; Baltić, V.; Gantar, M.; Juković, M.; Stojanović, D.; Baltic, M. Molecular Aspects of Microcystin-induced Hepatotoxicity and Hepatocarcinogenesis. *J. Environ. Sci. Health Part C* **2010**, *28*, 39–59. [CrossRef] [PubMed]
51. Puerto, M.; Pichardo, S.; Jos, A.; Prieto, A.I.; Sevilla, E.; Frías, J.E.; Cameán, A.M. Differential oxidative stress responses to pure Microcystin-LR and Microcystin-containing and non-containing cyanobacterial crude extracts on Caco-2 cells. *Toxicon* **2010**, *55*, 514–522. [CrossRef]

52. Li, Y.; Han, X. Microcystin–LR causes cytotoxicity effects in rat testicular Sertoli cells. *Environ. Toxicol. Pharmacol.* **2012**, *33*, 318–326. [CrossRef]
53. Ma, J.; Li, Y.; Duan, H.; Sivakumar, R.; Li, X. Chronic exposure of nanomolar MC-LR caused oxidative stress and inflammatory responses in HepG2 cells. *Chemosphere* **2018**, *192*, 305–317. [CrossRef]
54. Sedan, D.; Giannuzzi, L.; Rosso, L.; Marra, C.A.; Andrinolo, D. Biomarkers of prolonged exposure to microcystin-LR in mice. *Toxicon* **2013**, *68*, 9–17. [CrossRef]
55. Li, H.; Cai, Y.; Xie, P.; Li, G.; Hao, L.; Xiong, Q. Identification and Expression Profiles of IL-8 in Bighead Carp (*Aristichthys nobilis*) in Response to Microcystin-LR. *Arch. Environ. Contam. Toxicol.* **2013**, *65*, 537–545. [CrossRef] [PubMed]
56. Chen, Y.; Wang, J.; Zhang, Q.; Xiang, Z.; Li, D.; Han, X. Microcystin-leucine arginine exhibits immunomodulatory roles in testicular cells resulting in orchitis. *Environ. Pollut.* **2017**, *229*, 964–975. [CrossRef] [PubMed]
57. Zhang, Z.; Zhang, X.X.; Wu, B.; Yin, J.; Yu, Y.; Yang, L. Comprehensive insights into microcystin-LR effects on hepatic lipid metabolism using cross-omics technologies. *J. Hazard Mater.* **2016**, *315*, 126–134. [CrossRef] [PubMed]
58. Hillebrandt, S.; Goos, C.; Matern, S.; Lammert, F. Genome-wide analysis of hepatic fibrosis in inbred mice identifies the susceptibility locus Hfib1 on chromosome 15. *Gastroenterology* **2002**, *123*, 2041–2051. [CrossRef] [PubMed]
59. Shi, Z.; Wakil, A.E.; Rockey, D.C. Strain-specific differences in mouse hepatic wound healing are mediated by divergent T helper cytokine responses. *Proc. Natl. Acad. Sci. USA* **1997**, *94*, 10663–10668. [CrossRef]
60. Sediment and Algae Color the Great Lakes. Available online: https://coastalscience.noaa.gov/news/sediment-and-algae-color-the-great-lakes-image-of-the-day/ (accessed on 6 August 2019).
61. Bullerjahn, G.S.; McKay, R.M.; Davis, T.W.; Baker, D.B.; Boyer, G.L.; D'Anglada, L.V.; Doucette, G.J.; Ho, J.C.; Irwin, E.G.; Kling, C.L.; et al. Global solutions to regional problems: Collecting global expertise to address the problem of harmful cyanobacterial blooms. A Lake Erie case study. *Harmful Algae* **2016**, *54*, 223–238. [CrossRef] [PubMed]
62. Pham, T.L.; Utsumi, M. An overview of the accumulation of microcystins in aquatic ecosystems. *J. Environ. Manag.* **2018**, *213*, 520–529. [CrossRef]
63. Svirčev, Z.; Drobac, D.; Tokodi, N.; Mijović, B.; Codd, G.A.; Meriluoto, J. Toxicology of microcystins with reference to cases of human intoxications and epidemiological investigations of exposures to cyanobacteria and cyanotoxins. *Arch. Toxicol.* **2017**, *91*, 621–650. [CrossRef]
64. Stewart, I.; Webb, P.M.; Schluter, P.J.; Fleming, L.E.; Burns, J.W.; Gantar, M.; Backer, L.C.; Shaw, G.R. Epidemiology of recreational exposure to freshwater cyanobacteria—An international prospective cohort study. *BMC Public Health* **2006**, *6*, 93. [CrossRef]
65. Su, R.C.; Blomquist, T.M.; Kleinhenz, A.L.; Khalaf, F.K.; Dube, P.; Lad, A.; Breidenbach, J.D.; Mohammed, C.J.; Zhang, S.; Baum, C.E.; et al. Exposure to the Harmful Algal Bloom (HAB) Toxin Microcystin-LR (MC-LR) Prolongs and Increases Severity of Dextran Sulfate Sodium (DSS)-Induced Colitis. *Toxins* **2019**, *11*, 371. [CrossRef]
66. Cox, J.; Mann, M. MaxQuant enables high peptide identification rates, individualized ppb-range mass accuracies and proteome-wide protein quantification. *Nat. Biotechnol.* **2008**, *26*, 1367–1372. [CrossRef] [PubMed]
67. Smyth, G.K. Linear models and empirical bayes methods for assessing differential expression in microarray experiments. *Stat. Appl. Genet. Mol. Biol.* **2004**, *3*, 1–25. [CrossRef] [PubMed]
68. Storey, J.D.; Tibshirani, R. Statistical significance for genomewide studies. *Proc. Natl. Acad. Sci. USA* **2003**, *100*, 9440–9445. [CrossRef] [PubMed]
69. Langfelder, P.; Zhang, B.; Horvath, S. Defining clusters from a hierarchical cluster tree: The Dynamic Tree Cut package for R. *Bioinformatics* **2007**, *24*, 719–720. [CrossRef] [PubMed]
70. Tyanova, S.; Temu, T.; Sinitcyn, P.; Carlson, A.; Hein, M.Y.; Geiger, T.; Mann, M.; Cox, J. The Perseus computational platform for comprehensive analysis of (prote) omics data. *Nat. Methods* **2016**, *13*, 731–740. [CrossRef] [PubMed]
71. Carruthers, N.J.; Rosenspire, A.J.; Caruso, J.A.; Stemmer, P.M. Low level Hg^{2+} exposure modulates the B-cell cytoskeletal phosphoproteome. *J. Proteom.* **2018**, *173*, 107–114. [CrossRef] [PubMed]

Article

Exposure to the Harmful Algal Bloom (HAB) Toxin Microcystin-LR (MC-LR) Prolongs and Increases Severity of Dextran Sulfate Sodium (DSS)-Induced Colitis

Robin C. Su [1], Thomas M. Blomquist [2], Andrew L. Kleinhenz [1], Fatimah K. Khalaf [1], Prabhatchandra Dube [1], Apurva Lad [3], Joshua D. Breidenbach [3], Chrysan J. Mohammed [1], Shungang Zhang [1], Caitlin E. Baum [2], Deepak Malhotra [1], David J. Kennedy [1,3,*] and Steven T. Haller [1,3,*]

1 Department of Medicine, The University of Toledo College of Medicine and Life Sciences, Toledo, OH 43614, USA; Robin.Su@rockets.utoledo.edu (R.C.S.); Andrew.Kleinhenz@utoledo.edu (A.L.K.); Kareem.Khalaf@rockets.utoledo.edu (F.K.K.); Prabhatchandra.Dube@utoledo.edu (P.D.); Chrysan.Mohammed@rockets.utoledo.edu (C.M.); Shungang.Zhang@rockets.utoledo.edu (S.Z.); Deepak.Malhotra@utoledo.edu (D.M.)

2 Department of Pathology, The University of Toledo College of Medicine and Life Sciences, Toledo, OH 43614, USA; Thomas.Blomquist@utoledo.edu (T.M.B.); Caitlin.Baum@utoledo.edu (C.E.B.)

3 Department of Medical Microbiology and Immunology, The University of Toledo College of Medicine and Life Sciences, Toledo, OH 43614, USA; Apurva.Lad@rockets.utoledo.edu (A.L.); Joshua.Breidenbach@rockets.utoledo.edu (J.D.B.)

* Correspondence: David.Kennedy@utoledo.edu (D.J.K.); Steven.Haller@utoledo.edu (S.T.H.); Tel.: 419-383-6822 (D.J.K. & S.T.H.)

Received: 31 May 2019; Accepted: 22 June 2019; Published: 25 June 2019

Abstract: Inflammatory Bowel Disease (IBD) represents a collection of gastrointestinal disorders resulting from genetic and environmental factors. Microcystin-leucine arginine (MC-LR) is a toxin produced by cyanobacteria during algal blooms and demonstrates bioaccumulation in the intestinal tract following ingestion. Little is known about the impact of MC-LR ingestion in individuals with IBD. In this study, we sought to investigate MC-LR's effects in a dextran sulfate sodium (DSS)-induced colitis model. Mice were separated into four groups: (a) water only (control), (b) DSS followed by water (DSS), (c) water followed by MC-LR (MC-LR), and (d) DSS followed by MC-LR (DSS + MC-LR). DSS resulted in weight loss, splenomegaly, and severe colitis marked by transmural acute inflammation, ulceration, shortened colon length, and bloody stools. DSS + MC-LR mice experienced prolonged weight loss and bloody stools, increased ulceration of colonic mucosa, and shorter colon length as compared with DSS mice. DSS + MC-LR also resulted in greater increases in pro-inflammatory transcripts within colonic tissue (TNF-α, IL-1β, CD40, MCP-1) and the pro-fibrotic marker, PAI-1, as compared to DSS-only ingestion. These findings demonstrate that MC-LR exposure not only prolongs, but also worsens the severity of pre-existing colitis, strengthening evidence of MC-LR as an under-recognized environmental toxin in vulnerable populations, such as those with IBD.

Keywords: inflammatory bowel disease; dextran sulfate sodium; colitis; microcystin; colon

Key Contribution: This study is the first investigation assessing the effects of the Harmful Algal Bloom toxin, microcystin-LR in a model of pre-existing inflammatory bowel disease.

1. Introduction

Inflammatory bowel disease (IBD) is a collection of disorders characterized by both acute and chronic inflammation of the gastrointestinal (GI) tract [1]. IBD has become a global health burden, with an estimated 1 million individuals in the USA and 2.5 million individuals in Europe being affected by IBD [2]. Total indirect and direct costs of IBD in the US were estimated to have been between $14.6 and $31.6 billion in 2014 [3] and a recent study found total costs of IBD patients in 2015 to be three times higher than non IBD patients [4]. In addition, IBD is rapidly growing in prevalence within newly industrialized countries around the world [2].

Two of the most common forms of IBD are Crohn's disease (CD) and ulcerative colitis (UC), which share certain characteristics but also exhibit key differences. CD can affect any region of the GI tract and is characterized by discontinuous "skip lesions" while UC most commonly manifests within the distal GI tract, starting in the rectum and progressing proximally in a continuous manner along the distal colon [5]. Hallmarks of CD are transmural acute and chronic inflammation, non-caseating granulomas, strictures, and fistulas, while UC is characterized by mucosal and submucosal acute inflammation, crypt abscesses, ulcerations, depletion of goblet cells and mucin, bloody diarrhea, and weight loss [1]. These two conditions are frequently accompanied by other comorbidities and complications, making overall clinical presentations complex, difficult to manage, financially burdensome, and symptomatically debilitating for affected individuals.

IBD is a complex disease with a multifactorial etiology. There are genetic components that predispose individuals to develop IBD, as well as a dysregulation within the host immune system and GI microbial environment [6]. However, there is also an important role that the environment plays in IBD's pathogenesis and disease severity. This is highlighted by its dominant prevalence in industrialized countries and its growing prevalence in newly developed countries [6]. Many of these environmental factors have been explored, including environmental sanitation and hygiene; behavioral factors, such as smoking, diet, stress management, and breastfeeding; and use of medications, such as antibiotics, non-steroidal anti-inflammatory drugs, and oral contraceptives [6]. Additional environmental factors include microorganism infections, such as those caused by *Helicobacter pylori*, *Mycobacterium avium*, *Escherichia coli*, *Yersinia enterocolitica*, *Listeria monocytogenes*, and *Candida albicans* [6]. The identification of these different triggers of IBD disease progression have allowed for the establishment of appropriate preventative and therapeutic measures, however, there is still an urgent need to continue investigating other potential offenders.

One growing global environmental concern that has not been studied for its effects in IBD populations is microcystin. Microcystins (MCs) are a collection of potent toxins produced by cyanobacteria, also known as blue-green algae [7]. Of these toxins, microcystin-LR (MC-LR) is one of the most commonly produced forms and is also one of the most toxic variants [8]. Harmful algal blooms (HABs) contaminate freshwater environments and have affected every region of the USA. Globally, more than 40% of lakes and reservoirs in Europe, Asia, and America have favorable conditions for HABs, with 25–75% of blooms being considered toxic [9]. In addition, these HABs are increasing exponentially in frequency and severity worldwide [10]. The acute and chronic effects of microcystin exposure in humans have been recently reviewed [11]. Notable events of MC-LR exposure and toxicity in humans have been documented around the world. One of the most notable events occurred in 1996, where 116 of 130 patients at a dialysis center experienced acute liver failure and death within one week of exposure to water sources contaminated with microcystin [12,13]. Previous studies have identified the gastrointestinal tract to be a potentially important target of MC-LR toxicity and have even shown the intestines to be the site of greatest MC-LR bioaccumulation [10,14–16]. While MC-LR has been shown to cause severe liver damage [15], found to be a potential human carcinogen [8], and documented to be fatal in humans in some complicated cases [12], there is a critical need to investigate MC-LR's effects within the intestines, especially in more vulnerable settings, such as IBD.

In this current study, we aimed to address these pressing gaps in knowledge by utilizing the well-established dextran sulfate sodium (DSS) model of colitis in C57BL/6 mice [17]. This model has been developed to mimic characteristics of both CD and UC, and has been extensively validated by the use of several therapeutic agents used to treat human IBD [18]. The DSS model induces acute colitis and has been shown to sustain chronic levels of inflammation [19]. Because of MC-LR's known bioaccumulation in the intestines, we hypothesized that MC-LR would prolong and/or worsen the severity of DSS-induced colitis.

2. Results

2.1. Body Weight and Survival

Mice in the MC-LR group showed no significant differences in weight throughout the 14-day study as compared with control mice. Mice in the DSS group and in the DSS + MC-LR group showed decreases in body weight starting at day 6 (Figure 1). Starting at day 7, the body weights of DSS and DSS + MC-LR mice were significantly lower than the control and MC-LR mice ($p < 0.05$). No differences were observed between DSS and DSS + MC-LR mice. Body weights of the DSS and DSS + MC-LR mice continued to decrease until day 10 ($p < 0.001$). DSS mice progressively regained their weight starting at day 11. At day 13, DSS mice had significantly greater body weights than DSS + MC-LR mice. DSS mice continued to regain their weight until day 14, at which point they showed no significant differences in body weight versus the control and MC-LR mice, but were significantly greater in body weight versus DSS + MC-LR mice. While DSS mice regained their weight, mice in the DSS + MC-LR group continued to show significantly lower body weights until day 14 as compared with the control and MC-LR mice ($p < 0.05$). One mouse in the DSS + MC-LR group was found dead on day 14. Analysis on this mouse was not possible as organs were not attainable through standard procedures.

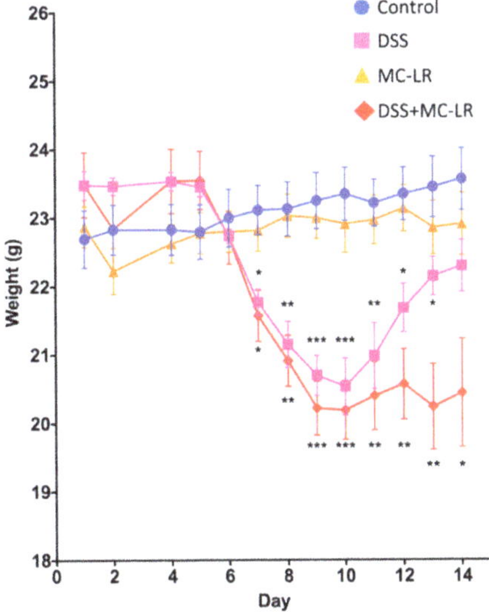

Figure 1. Mouse body weights taken daily throughout the 14-day study. Data presented indicate the mean ± SEM (n = 6–10 mice per group). * $p < 0.05$, ** $p < 0.01$, and *** $p < 0.001$ vs. the control group.

2.2. Stool Grading

Mice in the control and MC-LR groups showed no occult or gross blood in their stool throughout the 14-day study (Figure 2). Mice in the DSS group and in the DSS + MC-LR group began showing signs of occult blood in their stool on day 4 and gross blood starting on day 5. No significant differences in the occurrence of occult blood was observed between DSS and DSS + MC-LR groups. All mice in both DSS and DSS + MC-LR groups exhibited gross blood in their stool from days –8. Mice in DSS and DSS + MC-LR groups showed a decrease in gross blood and in increase in occult blood in their stool starting on day 9. DSS mice showed complete resolution of gross and occult blood by day 10. By day 11, there was a significantly lower occurrence of occult blood in the DSS group as compared with the DSS + MC-LR group (Figure 2). DSS + MC-LR mice showed no resolution of blood in their stool, with persistent occult blood still being detectable at day 14.

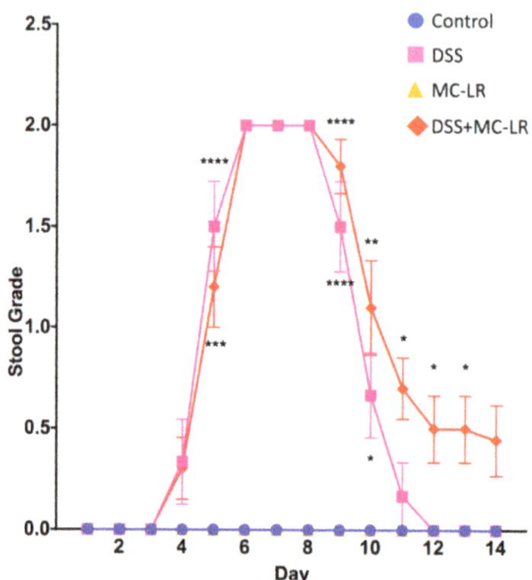

Figure 2. Daily stool grading throughout the 14-day study. 0 = no occult or gross blood, 1 = occult blood present, and 2 = gross blood present. Data presented indicate the mean ± SEM (n = 6–10 mice per group). * $p < 0.05$, ** $p < 0.01$, *** $p < 0.001$, and **** $p < 0.0001$ vs. control group.

2.3. Colon Length and Spleen Weight

Colons were measured from the distal end to the colon-cecum junction as exemplified by the representative images in Figure 3A. Colons of mice in the DSS, MC-LR, and DSS + MC-LR groups showed significant decreases in length as compared with the control colons ($p < 0.0001$) (Figure 3B). It was also noted that the colon lengths of the DSS + MC-LR mice were significantly shorter than those in the DSS ($p < 0.01$) and MC-LR groups ($p < 0.001$).

Spleen weights were significantly greater in the DSS group ($p < 0.001$) and the DSS + MC-LR group ($p < 0.05$) as compared with the control group (Figure 4). No differences were observed in spleen weight between the DSS and DSS + MC-LR groups. Spleen weights in the MC-LR group were not increased as compared with the control group.

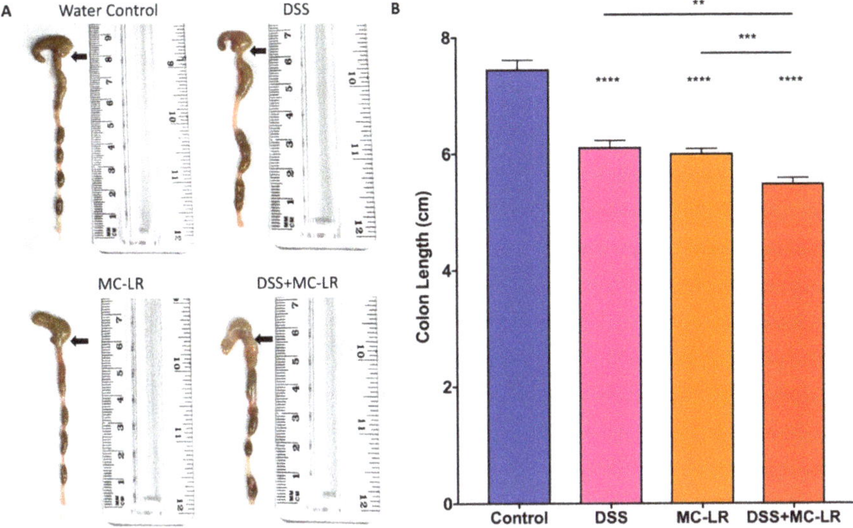

Figure 3. Effect of DSS and MC-LR on colon length. (**A**) Representative gross images and measurements of mouse colons with cecums still attached. Black arrows indicate the colon-cecum junction as a landmark for colon length measurement. (**B**) Diagram of colon lengths. Data presented indicate the mean ± SEM (n = 6–10 mice per group). ** $p < 0.01$, *** $p < 0.001$, and **** $p < 0.0001$.

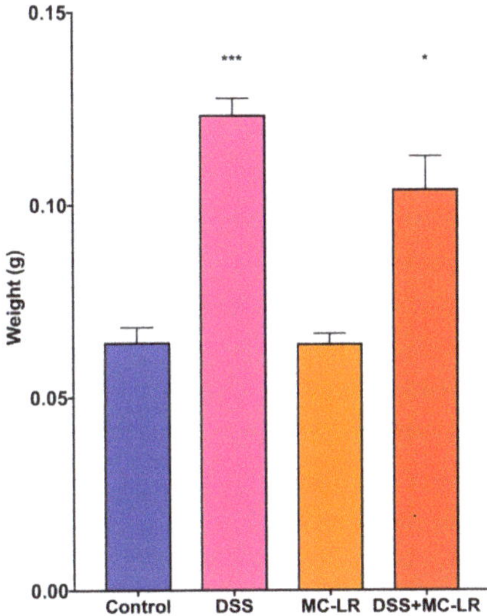

Figure 4. Spleen weights measured at the time of organ harvesting. Data presented indicate the mean ± SEM (n = 6–10 mice per group). * $p < 0.05$, and *** $p < 0.001$ vs. control group.

2.4. Histopathology

Histopathological analysis of hematoxylin and eosin (H & E) stained colon sections revealed that DSS exposure led to segmental regions of ulceration, crypt abscesses, marked acute inflammatory cell infiltration, and early architectural distortion with gland branching and budding (i.e., early chronic changes), as compared with the normal colonic tissue of the control group (Figure 5A). Analysis of

Periodic acid–Schiff (PAS) stained colon sections highlighted loss of goblet cells with mucin depletion, especially within and flanking the ulcer beds (Figure 5B).

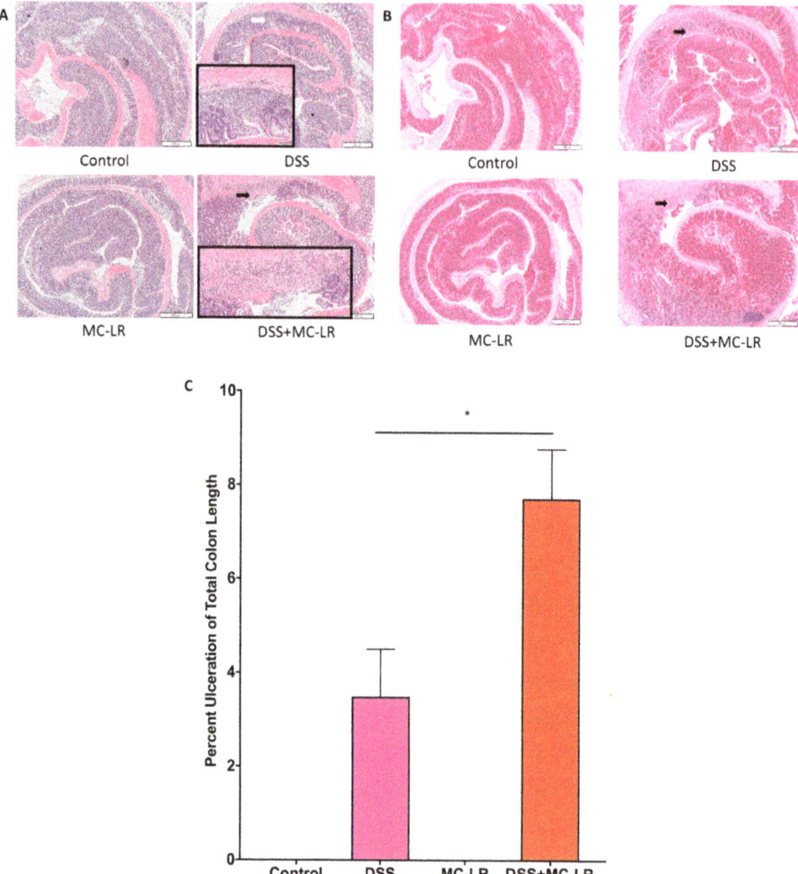

Figure 5. Effect of DSS and MC-LR on histopathological changes. (**A**) Representative images of H&E stained colon sections. Colon tissue from control and MC-LR mice did not show any specific pathologic changes. DSS exposure led to disruption of the epithelium, segmental regions of marked acute inflammatory cell infiltration, ulceration and branching and budding of glands (early chronic changes). A white arrow demarcates a representative area of tissue ulceration that has begun early re-epithelialization (magnified in the black framed area) and shows obliteration of crypts and infiltration by inflammatory cells. DSS + MC-LR combined exposure demonstrates increased severity of the same histopathological changes seen in the DSS group. A black arrow demarcates a representative area of tissue ulceration (magnified in the black framed area), which is significantly larger in length than that found in the DSS group. Scale Bar: 100 µm. (**B**) Representative PAS stained colon sections (same samples as the representative H and E sections). Colon tissue from control and MC-LR mice show continuous staining of goblet cells and mucin throughout the length of the colon. DSS exposure led to the loss of goblet cells and mucin, especially flanking the areas of ulceration (black arrow). The decrease in goblet cells and mucin depletion is more exaggerated in the DSS + MC-LR group (black arrow). Scale Bar: 100 µm (**C**) Quantification of tissue ulceration. Total ulcer length throughout the colon was normalized to total colon length. Data presented indicate the mean ± SEM (n = 6–10 mice per group). * $p < 0.05$ vs. the DSS group.

H and E analysis of the colons in the DSS + MC-LR group revealed greater severity of the same histopathological findings observed in the DSS group. Measurement of the length of ulcerated mucosa to total colon length demonstrates a greater percentage of ulceration within the DSS + MC-LR group as compared to the DSS group (Figure 5C). The MC-LR and control groups did not show evidence of pathological changes.

2.5. Gene Expression in the Colon

As seen in Figure 6, qPCR analysis demonstrated that the mRNA levels of the pro-inflammatory cytokines TNF-α and IL-1β were significantly upregulated in the DSS group as compared with the control group ($p < 0.01$ and $p < 0.0001$, respectively). The upregulation of TNF-α and IL-1β was further increased in the DSS + MC-LR group ($p < 0.0001$ and $p < 0.05$, respectively) and were significantly higher than the levels observed in the DSS group ($p < 0.01$ and $p < 0.05$, respectively).

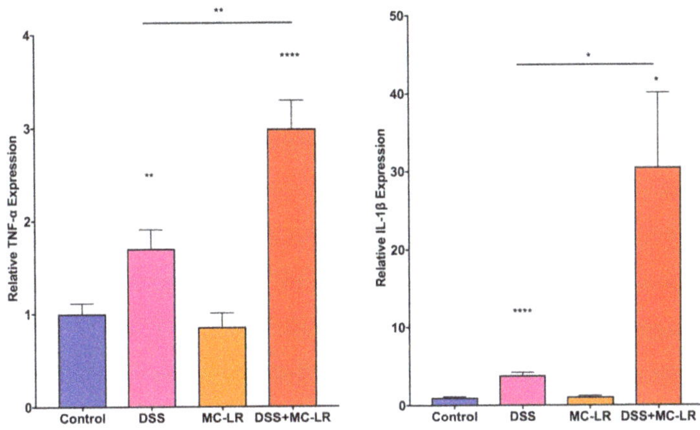

Figure 6. RT-qPCR analysis of proinflammatory cytokines TNF-α and IL-1β mRNA. Data presented indicate the mean ± SEM (n = 6–10 mice per group). * $p < 0.05$, ** $p < 0.01$, and **** $p < 0.0001$ vs. the control group.

CD40 mRNA levels (Figure 7) were increased in the DSS group and significantly elevated in the DSS + MC-LR group compared to control ($p < 0.001$). CD40 expression was not elevated in the MC-LR group.

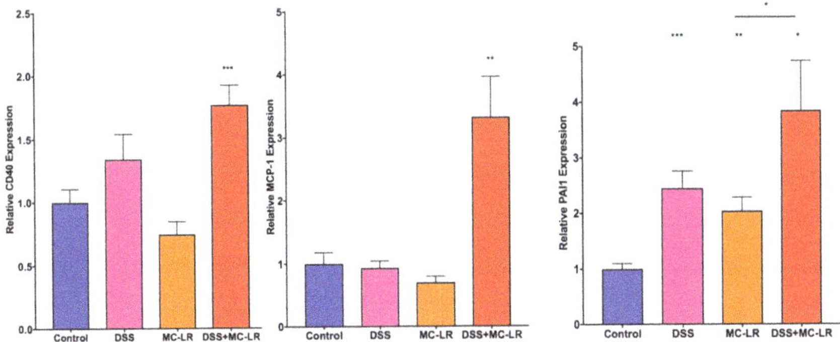

Figure 7. RT-qPCR analysis of CD40 and its downstream products MCP-1 and PAI-1. Data presented indicate the mean ± SEM (n = 6–10 mice per group). * $p < 0.05$, ** $p < 0.01$, and *** $p < 0.001$ vs. the control group.

Expression of MCP-1 (Figure 7) was significantly upregulated in the DSS + MC-LR group ($p < 0.01$). PAI-1 was upregulated within the DSS ($p < 0.001$), MC-LR ($p < 0.01$), and DSS + MC-LR ($p < 0.05$) groups as compared with the control group. PAI-1 levels were significantly higher in the DSS + MC-LR group as compared with the MC-LR group ($p < 0.05$).

3. Discussion

This study is the first to describe MC-LR's effects within the gastrointestinal tract of mice with pre-existing colitis. Here, we demonstrate that MC-LR can prolong and potentiate the severity of colitis within a DSS colitis mouse model. By utilizing a well-established colitis protocol, we found DSS to induce weight loss, splenomegaly, and severe colitis marked by transmural acute inflammation, ulceration, shortened colon length, and bloody stools. These gross effects were accompanied by the upregulation of key pro-inflammatory transcripts within colonic tissue, including TNF-α and IL-1β. While MC-LR alone only resulted in modest colonic shortening and increases in PAI-1 expression, MC-LR in the setting of DSS-induced colitis resulted in the prolongation and exacerbation of disease state as compared with DSS alone. Within the DSS + MC-LR group, we observed prolonged weight loss and bloody stools, increased ulceration of colonic mucosa, shorter colon length, and a greater increase in the pro-inflammatory transcripts of TNF-α and IL-1β as compared with DSS alone. It is also important to note that the heightened disease state in the DSS + MC-LR group resulted in the death of one of the mice before the conclusion of the study, a finding not found in any of the other groups or reported in any other previous study.

It has been well established that MC-LR enters cells through organic anion transporting polypeptides (OATP) [20]. Once in cells, MC-LR inhibits serine/threonine protein phosphatases (PPs), especially PP1 and PP2A [20]. Such inhibition of PPs lead to hyperphosphorylation of various enzymes and cytoskeletal elements, leading to a disruption of cellular processes [20]. In Sertoli cells, it has been found that one of the consequences of MC-LR + PP complex formation is the inhibition of miR-98-5p and miR-758, leading to the enhanced expression of MAPK11 [21]. Enhanced expression and activation of MAPK11 leads to the phosphorylation of transcription factor ATF-2, which binds to the promotor of TNF-α and leads to TNF-α expression [21].

Another potential mechanism by which MC-LR enhances inflammatory cytokine production has recently been investigated by Adegoke et al. [22]. Again in Sertoli cells, it was found that MC-LR induces the upregulation and activation of toll-like receptor 4 (TLR4) and its downstream effector, nuclear factor-kappaB (NF-kB), in a dose-dependent manner [22]. It has been proposed that the activation of NF-kB by TLR4 is mediated by either myeloid differentiation primary response 88 (MyD88) or TIR-domain-containing adapter-inducing interferon-β (TRIF) [22]. NF-kB activation subsequently leads to the upregulation of pro-inflammatory cytokines, including TNF-α and IL-1β [22]. A follow up study by Adegoke et al. utilized TLR4-IN-C34 (C34) to inhibit TLR4 in order to demonstrate an attenuation of MC-LR toxicity by the TLR4/NF-kB pathway [23]. Inhibition with C34 was found to attenuate damage caused by MC-LR and attenuate the production of inflammatory cytokines, including TNF-α and IL-1β [23].

In addition, MC-LR activates the innate immune system by the recruitment of lymphocytes, neutrophils, and macrophages into affected tissues [24,25]. Such activation leads to downstream cytokine production, including TNF-α and IL-1β by macrophages. The upregulation of TNF-α and IL-1β due to MC-LR exposure has been previously confirmed [26]. Given that disruption of the intestinal epithelial barrier serves as a major driver of IBD pathogenesis [27], our results may highlight the importance of intestinal barrier integrity in MC-LR toxicity. Pre-existing colitis, coupled with barrier dysfunction, could lead to greater tissue uptake of MC-LR, greater recruitment of innate immune cells, and greater production of inflammatory cytokines, a phenomenon that is not activated in the setting of intact intestinal barriers in healthy wild type mice. This concept is illustrated in Figure 8.

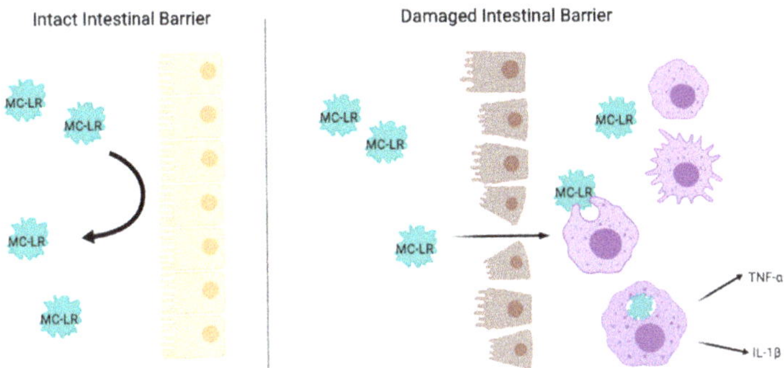

Figure 8. Schematic diagram displaying a potential mechanism by which MC-LR entry is facilitated by intestinal epithelial damage. An intact intestinal barrier helps to prevent the tissue uptake of foreign matter, including MC-LR. A damaged intestinal barrier may facilitate the tissue uptake of MC-LR. MC-LR, like other foreign material is scavenged by macrophages, which release inflammatory cytokines, such as TNF-α and IL-1β, and further recruits additional innate immune cells driving inflammation.

As CD40 signaling has been recently implicated in active regions of IBD in human patients, we evaluated CD40 as a potential mechanism for this disease prolongation and exacerbation [28]. CD40 is part of the TNF superfamily, one of three families that can be exploited in T cell co-stimulation [29]. CD40 has been evaluated in the setting of IBD (both UC and CD) and has been found to be overexpressed in IBD. Interestingly, previous studies in IBD patients have shown that CD40 overexpression is directly proportional to the extent of disease severity and is only found in actively inflamed regions of intestinal tissue [30]. In addition to their overexpression on T cells, CD40 has also been shown to be expressed on intestinal fibroblasts and on intestinal epithelial cells in actively inflamed colonic tissue of IBD patients, which was further confirmed in a cell model [28,31]. In the present study, we observed a significant upregulation of CD40 in the DSS + MC-LR colons, signifying its potential role in disease exacerbation. Of note, we also observed the upregulation of MCP-1 and PAI-1, two downstream products of CD40 activation. These help to confirm the initiation of this pathway as a result of MC-LR exposure in the setting of DSS-induced colitis.

In this study, it is important to also note the relevance of the dosage of MC-LR used. The World Health Organization (WHO) has set a limit on permissible levels of microcystin in finished drinking water at one part per billion (ppb), which is commonly understood as 1 μg/kg [32]. The WHO has also established that when conducting research in an animal model, an uncertainty factor of 1000 can be applied in order to account for intra- and interspecies variation, allowing for 1 ppb to be extrapolated to 1000 μg/kg in an animal model [33], which is the concentration we utilized in the present study. In addition, this dose of 1000 μg/kg has previously been studied by Fawell et al. [34]. While our current study exposed mice to 1000 μg MC-LR/ kg body weight daily for one week, Fawell et al. exposed mice to 40, 200, and 1000 μg MC-LR/ kg body weight daily for 13 weeks, a prolonged, chronic exposure timeframe. While no significant increases in pathology was noted in the 40 μg/kg male mouse exposure group as compared with controls after 13 weeks, increased trends in liver pathology (including chronic inflammation, hepatocyte degeneration, and hemosiderin deposition), and blood chemistry (including alanine aminotransferase and aspartate aminotransferase) were noted in the 200 and 1000 μg/kg male mouse exposure groups at the conclusion of 13 weeks of daily MC-LR exposure [34]. While these hepatic findings may be significant with chronic exposure (13 weeks) in healthy mouse models, we aimed to investigate the gastrointestinal effects of acute exposure (one week) to MC-LR. Interestingly, while we saw limited effects in WT mice, MC-LR had significant effects in mice with pre-existing colitis. Our findings suggest that more stringent regulations for MC-LR in finished drinking water should be considered in order to protect populations that may be more susceptible to MC-LR toxicity, such as those with pre-existing gastrointestinal diseases.

In summary, while MC-LR exposure alone was not found to induce significant inflammation, histopathology, changes in stool, or changes in body or spleen weight, it was found to have a profound effect in the setting of pre-existing colitis by prolonging and exacerbating disease conditions. The use of relevant MC-LR exposure levels further highlights the clinical relevance and urgent need for stricter guidelines in order to protect vulnerable populations. Future studies will help to further elucidate the role of CD40 in DSS-induced colitis, its role in disease exacerbation in the presence of MC-LR exposure, and its potential as a therapeutic target for disease prevention and reversal.

4. Materials and Methods

4.1. Mice

All animal experimentation was conducted in accordance with the National Institutes of Health (NIH) Guide for the Care and Use of Laboratory Animals under protocols approved by The University of Toledo Institutional Animal Care and Use Committee (IACUC protocol #108663, approval date 9 February 2016). All mice were housed in a specific pathogen free facility, maintained at standard conditions of 23 ± 1 °C under a 12-h light cycle and were allowed to eat a normal chow diet ad libitum. Male C57BL/6 mice were purchased from The Jackson Laboratory at five weeks of age. The mice were immediately assigned randomly to one of four groups: (a) water only (control), (b) DSS followed by water (DSS), (c) water followed by MC-LR (MC-LR), and (d) DSS followed by MC-LR (DSS + MC-LR). The control water group consisted of 6 mice, the DSS group consisted of six mice, the MC-LR group consisted of 10 mice, and the DSS + MC-LR group consisted of 10 mice. All 32 mice were allowed to acclimatize to their new environment until eight weeks of age.

4.2. Colitis Induction and MC-LR Exposure Protocol

At eight weeks of age, mice were initiated into the study as outlined in Figure 9. The control mice were allowed to drink water ad libitum for 14 days. Between days 8–14, control mice were orally gavaged daily with water as a sham procedure. To induce colitis, mice in the DSS group were given 3% DSS (MP Biomedical, Solon, OH, USA, Item No. 0216011080) in drinking water ad libitum for seven days according to established protocols [17]. Following this, the mice were allowed to drink untreated water ad libitum for 7 days, while also being given water by daily sham oral gavage. The MC-LR group of mice were allowed to drink water ad libitum for 14 days. During days 8–14, these mice were orally gavaged 1000 μg/kg MC-LR (Cayman Chemical, Ann Arbor, MI, USA, item no. 10007188) daily. The DSS + MC-LR group of mice were allowed to drink 3% DSS water ad libitum for seven days and then were allowed to drink untreated water ad libitum for seven days. During days 8–14, these mice were orally gavaged 1000 μg/kg MC-LR daily. Weight of each mouse was taken daily. Stool was evaluated daily for the presence of occult and gross blood. Occult blood was measured using the Beckman Coulter Hemoccult Single Slides kit (Med Plus Physician Supplies, Edison, New Jersey, USA, Catalog #BC-60151). Stool was graded by the following: 0 = no occult or gross blood, 1 = occult blood present, and 2 = gross blood present. All mice were euthanized on day 15 and organs were harvested and weighed immediately following euthanasia. Colons (with cecum still intact) were measured adjacent to a standard ruler and photographs were taken. Following removal of the cecum and thorough washing of the colon with PBS, sections of distal and proximal colon were taken from each sample, flash frozen in liquid nitrogen, and subsequently stored at −80 °C for future qPCR analysis.

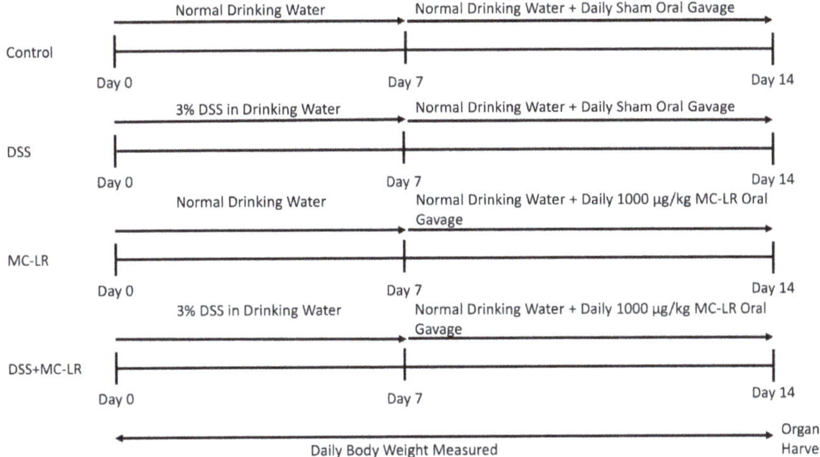

Figure 9. Experimental design evaluating the effects of DSS exposure, MC-LR exposure, and combined DSS + MC-LR exposure within colons of C57BL/6 mice. The study was conducted on eight-week old mice. DSS mice were given 3% DSS in drinking water for seven days followed by seven days of normal drinking water and daily sham oral gavage. MC-LR mice were given normal water for seven days followed by seven days of continued normal water and daily 1000 µg/kg MC-LR oral gavage. DSS + MC-LR mice were given 3% DSS in drinking water for seven days followed by seven days of normal drinking water and daily 1000 µg/kg MC-LR oral gavage. Body weights were measured daily and organ harvesting was conducted immediately after euthanasia.

4.3. Histology

Remaining colonic tissue from these mice were cut longitudinally, wrapped around a rigid holder, placed in cassettes, and fixed in 10% neutral buffered formalin for 24 h. The cassettes were then transferred to 70% ethanol. This formalin fixed tissue was then processed and embedded in paraffin (FFPE). Five (5) micron tissue sections were placed on glass slides and stained with hematoxylin and eosin (H & E) and Periodic acid–Schiff (PAS). Images of histology slides were taken using an Olympus CKX53 microscope and Olympus CellSens software (Standard 1.15) (Center Valley, PA, USA).

Severity of colon ulceration was further quantified using the Olympus CellSens software by measuring total length of ulcerated colon and normalizing to the total length of colon to give percent of colonic mucosa with ulceration.

4.4. RNA Extraction and RT-qPCR Method

RNA extraction, cDNA preparation, and RT-qPCR were all performed utilizing the QIAGEN (Germantown, MD, USA) automated liquid handling workflow system (QIAcube HT and QIAgility). RNA from distal colonic tissue was isolated utilizing the QIAzol/chloroform extraction methodology. RNA was purified using the lithium chloride method as previously published [35]. Approximately 500 ng of extracted RNA was used to synthesize cDNA (QIAGEN's RT2 First Strand Kit). RT-qPCR was performed utilizing QIAGEN's Rotor-Gene Q thermo-cycler. Calculation of gene expression was conducted by comparing the relative change in cycle threshold value (ΔCt). Fold change in expression was calculated using the 2-ΔΔCt equation as previously described [36]. The following Taqman primers were used and obtained from Thermo Fisher Scientific: TNF-α (Mm00443258_m1), IL-1β (Mm00434228_m1), PAI-1 (Mm00435858_m1), MCP-1 (Mm00441242_m1), and CD40 (Mm00441891_m1). 18s rRNA from Thermo Fisher Scientific was used as a housekeeping gene for normalization of transcript expression (catalog no. 4319413E).

4.5. Statistical Analysis

All data is presented as mean ± SEM. Statistical analysis was conducted with GraphPad Prism 7.0d software (San Diego, CA, USA) using the unpaired two-tailed Student's t-test. Significance was determined if p values were < 0.05.

Author Contributions: Conceptualization: R.C.S., D.J.K., and S.T.H.; data curation: R.C.S., T.M.B., A.L.K., F.K.K., P.D., A.L., J.D.B., C.J.M., S.Z., D.J.K., and S.T.H.; formal analysis: R.C.S., T.M.B., D.J.K., and S.T.H.; funding acquisition: D.J.K. and S.T.H.; investigation: R.C.S., T.M.B., D.J.K., and S.T.H.; methodology: R.C.S., D.J.K., and S.T.H.; project administration: D.M., D.J.K., and S.T.H.; resources: R.C.S., T.M.B., A.L.K., F.K.K., P.D., A.L., J.D.B., C.J.M., S.Z., D.J.K., and S.T.H.; software: R.C.S., D.J.K., and S.T.H.; supervision: D.J.K. and S.T.H.; validation: T.M.B., D.J.K., and S.T.H.; visualization: R.C.S., T.M.B., D.J.K., and S.T.H.; writing—original draft: R.C.S.; writing—review and editing, R.C.S., T.M.B., C.E.B., D.J.K., and S.T.H.

Funding: This research was funded by a Harmful Algal Bloom Research Initiative grant from the Ohio Department of Higher Education, the David and Helen Boone Foundation Research Fund, the University of Toledo Women and Philanthropy Genetic Analysis Instrumentation Center, and the University of Toledo Medical Research Society.

Conflicts of Interest: The authors declare no conflict of interest.

References

1. Xavier, R.J.; Podolsky, D.K. Unravelling the pathogenesis of inflammatory bowel disease. *Nature* **2007**, *448*, 427–434. [CrossRef] [PubMed]
2. Kaplan, G.G. The global burden of IBD: from 2015 to 2025. *Nat. Rev. Gastroenterol. Hepatol.* **2015**, *12*, 720. [CrossRef] [PubMed]
3. Mehta, F. Report: Economic implications of inflammatory bowel disease and its management. *Am. J. Manag. Care* **2016**, *22*, s51–s60. [PubMed]
4. Park, K.T.; Ehrlich, O.G.; Allen, J.I.; Meadows, P.; Szigethy, E.M.; Henrichsen, K.; Kim, S.C.; Lawton, R.C.; Murphy, S.M.; Regueiro, M.; et al. The Cost of Inflammatory Bowel Disease: An Initiative From the Crohn's & Colitis Foundation. *Inflamm. Bowel. Dis.* **2019**. [CrossRef]
5. Jeengar, M.K.; Thummuri, D.; Magnusson, M.; Naidu, V.G.M.; Uppugunduri, S. Uridine Ameliorates Dextran Sulfate Sodium (DSS)-Induced Colitis in Mice. *Sci. Rep.* **2017**, *7*, 3924. [CrossRef] [PubMed]
6. Molodecky, N.A.; Kaplan, G.G. Environmental risk factors for inflammatory bowel disease. *Gastroenterol. Hepatol.* **2010**, *6*, 339–346.
7. Campos, A.; Vasconcelos, V. Molecular Mechanisms of Microcystin Toxicity in Animal Cells. *Int. J. Mol. Sci.* **2010**, *11*, 268–287. [CrossRef]
8. Lone, Y.; Koiri, R.K.; Bhide, M. An overview of the toxic effect of potential human carcinogen Microcystin-LR on testis. *Toxicol. Rep.* **2015**, *2*, 289–296. [CrossRef]
9. Chorus, I. Introduction: Cyanotoxins—Research for Environmental Safety and Human Health. In *Cyanotoxins: Occurrence, Causes, Consequences*; Chorus, I., Ed.; Springer Berlin Heidelberg: Berlin/Heidelberg, Germany, 2001; pp. 1–4. [CrossRef]
10. Greer, B.; Meneely, J.P.; Elliott, C.T. Uptake and accumulation of Microcystin-LR based on exposure through drinking water: An animal model assessing the human health risk. *Sci. Rep.* **2018**, *8*, 4913. [CrossRef]
11. Svircev, Z.; Drobac, D.; Tokodi, N.; Mijovic, B.; Codd, G.A.; Meriluoto, J. Toxicology of microcystins with reference to cases of human intoxications and epidemiological investigations of exposures to cyanobacteria and cyanotoxins. *Arch. Toxicol.* **2017**, *91*, 621–650. [CrossRef]
12. Jochimsen, E.M.; Carmichael, W.W.; An, J.S.; Cardo, D.M.; Cookson, S.T.; Holmes, C.E.; Antunes, M.B.; de Melo Filho, D.A.; Lyra, T.M.; Barreto, V.S.; et al. Liver failure and death after exposure to microcystins at a hemodialysis center in Brazil. *N. Engl. J. Med.* **1998**, *338*, 873–878. [CrossRef]
13. Carmichael, W.W.; Azevedo, S.M.; An, J.S.; Molica, R.J.; Jochimsen, E.M.; Lau, S.; Rinehart, K.L.; Shaw, G.R.; Eaglesham, G.K. Human fatalities from cyanobacteria: chemical and biological evidence for cyanotoxins. *Environ. Health Perspect.* **2001**, *109*, 663–668. [CrossRef] [PubMed]
14. Gaudin, J.; Huet, S.; Jarry, G.; Fessard, V. In vivo DNA damage induced by the cyanotoxin microcystin-LR: comparison of intra-peritoneal and oral administrations by use of the comet assay. *Mutat. Res.* **2008**, *652*, 65–71. [CrossRef] [PubMed]

15. Sedan, D.; Laguens, M.; Copparoni, G.; Aranda, J.O.; Giannuzzi, L.; Marra, C.A.; Andrinolo, D. Hepatic and intestine alterations in mice after prolonged exposure to low oral doses of Microcystin-LR. *Toxicon* **2015**, *104*, 26–33. [CrossRef]
16. Zhang, H.-J.; Zhang, J.-Y.; Hong, Y.; Chen, Y.-X. Evaluation of organ distribution of microcystins in the freshwater phytoplanktivorous fish Hypophthalmichthys molitrix. *J. Zhejiang Univ. Sci. B* **2007**, *8*, 116–120. [CrossRef] [PubMed]
17. Chassaing, B.; Aitken, J.D.; Malleshappa, M.; Vijay-Kumar, M. Dextran sulfate sodium (DSS)-induced colitis in mice. *Curr. Protoc. Immunol.* **2014**, *104*, 15–25. [CrossRef]
18. Melgar, S.; Karlsson, L.; Rehnström, E.; Karlsson, A.; Utkovic, H.; Jansson, L.; Michaëlsson, E. Validation of murine dextran sulfate sodium-induced colitis using four therapeutic agents for human inflammatory bowel disease. *Int. Immunopharmacol.* **2008**, *8*, 836–844. [CrossRef] [PubMed]
19. Melgar, S.; Karlsson, A.; Michaëlsson, E. Acute colitis induced by dextran sulfate sodium progresses to chronicity in C57BL/6 but not in BALB/c mice: correlation between symptoms and inflammation. *Am. J. Physiol. Gastrointest. Liver Physiol.* **2005**, *288*, G1328–G1338. [CrossRef]
20. Fischer, A.; Hoeger, S.J.; Stemmer, K.; Feurstein, D.J.; Knobeloch, D.; Nussler, A.; Dietrich, D.R. The role of organic anion transporting polypeptides (OATPs/SLCOs) in the toxicity of different microcystin congeners in vitro: a comparison of primary human hepatocytes and OATP-transfected HEK293 cells. *Toxicol. Appl. Pharmacol.* **2010**, *245*, 9–20. [CrossRef]
21. Chen, Y.; Zhou, Y.; Wang, J.; Wang, L.; Xiang, Z.; Li, D.; Han, X. Microcystin-Leucine Arginine Causes Cytotoxic Effects in Sertoli Cells Resulting in Reproductive Dysfunction in Male Mice. *Sci. Rep.* **2016**, *6*, 39238. [CrossRef]
22. Adegoke, E.O.; Wang, C.; Machebe, N.S.; Wang, X.; Wang, H.; Adeniran, S.O.; Zhang, H.; Zheng, P.; Zhang, G. Microcystin-leucine arginine (MC-LR) induced inflammatory response in bovine sertoli cell via TLR4/NF-kB signaling pathway. *Environ. Toxicol. Pharmacol.* **2018**, *63*, 115–126. [CrossRef] [PubMed]
23. Adegoke, E.O.; Adeniran, S.O.; Zeng, Y.; Wang, X.; Wang, H.; Wang, C.; Zhang, H.; Zheng, P.; Zhang, G. Pharmacological inhibition of TLR4/NF-kappaB with TLR4-IN-C34 attenuated microcystin-leucine arginine toxicity in bovine Sertoli cells. *J. Appl. Toxicol.* **2019**, *39*, 832–843. [CrossRef] [PubMed]
24. Lin, W.; Hou, J.; Guo, H.; Qiu, Y.; Li, L.; Li, D.; Tang, R. Dualistic immunomodulation of sub-chronic microcystin-LR exposure on the innate-immune defense system in male zebrafish. *Chemosphere* **2017**, *183*, 315–322. [CrossRef] [PubMed]
25. Guzman, R.E.; Solter, P.F. Hepatic oxidative stress following prolonged sublethal microcystin LR exposure. *Toxicol. Pathol.* **1999**, *27*, 582–588. [CrossRef] [PubMed]
26. Ma, J.; Li, Y.; Duan, H.; Sivakumar, R.; Li, X. Chronic exposure of nanomolar MC-LR caused oxidative stress and inflammatory responses in HepG2 cells. *Chemosphere* **2018**, *192*, 305–317. [CrossRef] [PubMed]
27. Vindigni, S.M.; Zisman, T.L.; Suskind, D.L.; Damman, C.J. The intestinal microbiome, barrier function, and immune system in inflammatory bowel disease: a tripartite pathophysiological circuit with implications for new therapeutic directions. *Ther. Adv. Gastroenterol.* **2016**, *9*, 606–625. [CrossRef] [PubMed]
28. Borcherding, F.; Nitschke, M.; Hundorfean, G.; Rupp, J.; von Smolinski, D.; Bieber, K.; van Kooten, C.; Lehnert, H.; Fellermann, K.; Buning, J. The CD40-CD40L pathway contributes to the proinflammatory function of intestinal epithelial cells in inflammatory bowel disease. *Am. J. Pathol.* **2010**, *176*, 1816–1827. [CrossRef] [PubMed]
29. Danese, S.; Sans, M.; Fiocchi, C. The CD40/CD40L costimulatory pathway in inflammatory bowel disease. *Gut* **2004**, *53*, 1035–1043. [CrossRef] [PubMed]
30. Polese, L.; Angriman, I.; Cecchetto, A.; Norberto, L.; Scarpa, M.; Ruffolo, C.; Barollo, M.; Sommariva, A.; D'Amico, D.F. The role of CD40 in ulcerative colitis: histochemical analysis and clinical correlation. *Eur. J. Gastroenterol. Hepatol.* **2002**, *14*, 237–241. [CrossRef]
31. Gelbmann, C.M.; Leeb, S.N.; Vogl, D.; Maendel, M.; Herfarth, H.; Schölmerich, J.; Falk, W.; Rogler, G. Inducible CD40 expression mediates NFkappaB activation and cytokine secretion in human colonic fibroblasts. *Gut* **2003**, *52*, 1448–1456. [CrossRef]
32. Miller, M.A.; Kudela, R.M.; Mekebri, A.; Crane, D.; Oates, S.C.; Tinker, M.T.; Staedler, M.; Miller, W.A.; Toy-Choutka, S.; Dominik, C.; et al. Evidence for a novel marine harmful algal bloom: cyanotoxin (microcystin) transfer from land to sea otters. *PLoS ONE* **2010**, *5*, e12576. [CrossRef] [PubMed]

33. Organization, W.H. Cyanobacterial toxins: Microcystin-LR in Drinking Water. *Guidel. Drink. Water Qual.* **1998**, *2*, 83–127.
34. Fawell, J.K.; Mitchell, R.E.; Everett, D.J.; Hill, R.E. The toxicity of cyanobacterial toxins in the mouse: I microcystin-LR. *Hum. Exp. Toxicol.* **1999**, *18*, 162–167. [CrossRef] [PubMed]
35. Viennois, E.; Tahsin, A.; Merlin, D. Purification of Total RNA from DSS-treated Murine Tissue via Lithium Chloride Precipitation. *Bio Protoc.* **2018**, *8*, e2829. [CrossRef] [PubMed]
36. Kennedy, D.J.; Khalaf, F.K.; Sheehy, B.; Weber, M.E.; Agatisa-Boyle, B.; Conic, J.; Hauser, K.; Medert, C.M.; Westfall, K.; Bucur, P.; et al. Telocinobufagin, a Novel Cardiotonic Steroid, Promotes Renal Fibrosis via Na(+)/K(+)-ATPase Profibrotic Signaling Pathways. *Int. J. Mol. Sci.* **2018**, *19*, 2566. [CrossRef] [PubMed]

MDPI
St. Alban-Anlage 66
4052 Basel
Switzerland
Tel. +41 61 683 77 34
Fax +41 61 302 89 18
www.mdpi.com

Toxins Editorial Office
E-mail: toxins@mdpi.com
www.mdpi.com/journal/toxins

www.ingramcontent.com/pod-product-compliance
Lightning Source LLC
LaVergne TN
LVHW071014170726
843515LV00006B/1134
* 9 7 8 3 0 3 9 4 3 6 7 9 8 *